T0192372

Vortex Methods

The goal of this book is to present and analyze vortex methods as a tool for the direct numerical simulation of incompressible viscous flows.

Vortex methods have matured in recent years, offering an interesting alternative to finite-difference and spectral methods for high-resolution numerical solutions of the Navier–Stokes equations. In the past two decades research in the numerical analysis aspects of vortex methods has provided a solid mathematical background for understanding the convergence features of the method and several new tools have been developed to generalize its application. At the same time vortex methods retain their appealing physical character that was the motivation for their introduction.

Scientists working in the areas of numerical analysis and fluid mechanics will benefit from this book, which may serve both communities as both a reference monograph and a textbook for computational fluid dynamics courses.

Georges-Henri Cottet received his Ph.D. and Thèse d'Etat in Applied Mathematics from Université Pierre et Marie Curie in Paris. He is currently professor of mathematics at the Université Joseph Fourier in Grenoble, France.

Petros Koumoutsakos received his Ph.D. in aeronautics and applied mathematics from the California Institute of Technology. He is currently a professor at ETH-Zürich and a senior research fellow at the Center for Turbulence Research at NASA Ames/Stanford University.

Vortex Methods:
Theory and Practice

GEORGES-HENRI COTTET

Université Joseph Fourier in Grenoble

PETROS D. KOUMOUTSAKOS

ETH-Zürich
and
CTR, NASA Ames/Stanford University

CAMBRIDGE
UNIVERSITY PRESS

CAMBRIDGE UNIVERSITY PRESS
Cambridge, New York, Melbourne, Madrid, Cape Town, Singapore, São Paulo

Cambridge University Press
The Edinburgh Building, Cambridge CB2 8RU, UK

Published in the United States of America by Cambridge University Press, New York

www.cambridge.org
Information on this title: www.cambridge.org/9780521621861

First published 2000
This digitally printed version 2008

A catalogue record for this publication is available from the British Library

Library of Congress Cataloguing in Publication data
Cottet, G.-H. (Georges-Henri). 1956–
Vortex methods: theory and practice / Georges-Henri Cottet.
Petros D. Koumoutsakos.
p.   cm.
ISBN 0-521-62186-0
1. Navier–Stokes equations – Numerical solutions.   2. Vortex-motion.
I. Koumoutsakos, Petros D.   II. Title.
QA925.C68   1999
532´.0533´01515353–dc21   99-12277
                                                CIP

ISBN 978-0-521-62186-1 hardback
ISBN 978-0-521-06170-4 paperback

Contents

Preface

The goal of this book is to present and analyze vortex methods as a tool for the direct numerical simulation of incompressible viscous flows. Its intended audience is scientists working in the areas of numerical analysis and fluid mechanics. Our hope is that this book may serve both communities as a reference monograph and as a textbook in a course of computational fluid dynamics in the schools of applied mathematics and engineering.

Vortex methods are based on the discretization of the vorticity field and the Lagrangian description of the governing equations that, when solved, determine the evolution of the computational elements. Classical vortex methods enjoy advantages such as the use of computational elements only in cases in which the vorticity field is nonzero, the automatic adaptivity of the computational elements, and the rigorous treatment of boundary conditions at infinity. Until recently, disadvantages such as the computational cost and the inability to treat accurately viscous effects had limited their application to modeling the evolution of the vorticity field of unsteady high Reynolds number flows with a few tens to a few thousands computational elements. These difficulties have been overcome with the advent of fast summation algorithms that have optimized the computational cost and recent developments in numerical analysis that allow for the accurate treatment of viscous effects. Vortex methods have reached today a level of maturity, offering an interesting alternative to finite-difference and spectral methods for high-resolution numerical solutions of the Navier–Stokes equations. In the past two decades research in numerical analysis aspects of vortex methods has provided a solid mathematical background for understanding the accuracy and the stability of the method. At the same time vortex methods retain their appealing physical character that, we believe, was the motivation for their introduction.

Historically, simulations with vortex methods date back to the 1930s, with Rosenhead's calculations by hand of the Kelvin–Helmholtz instabilities. For several decades the grid-free character and the physical attributes of vortex methods were exploited in the simulation of unsteady separated flows. Simultaneously, the close relative of vortex methods, the surface singularity (panel) methods were developed and still remain as a powerful engineering tool for the prediction of loads in aerodynamic configurations. The modern developments of vortex methods originate in the works of Chorin in the 1970s (in particular for the design of random-walk methods), and in the three-dimensional calculations of Leonard (in the USA) and Rehbach (in France). These numerical works soon motivated the interest of applied mathematicians for understanding the convergence properties of these methods in the early 1980s. The very first complete convergence analysis was done in the USA by Hald, followed by Beale and Majda. In Europe, at about the same time, the group of Raviart undertook this research in parallel with the analysis of particle methods for plasma physics.

The past two decades have seen significant developments in the design of fast multiple methods for the efficient evaluation of the velocity field by Greengard and Rohklin; design and numerical analysis of new accurate methods for the treatment of viscous effects (in the group of Raviart); a number of benchmark applications demonstrating the capabilities of vortex methods for Direct Numerical Simulations of unsteady separated flows in Leonard's group at Caltech; and finally a deeper understanding of convergence properties, with convergence proofs of random-walk methods by Long and Goodman, and convergence proof for point vortex methods by Hou and co-workers. In this book we discuss these recent developments by mixing as much as possible the points of view of numerical analysis and fluid mechanics. We indeed believe that a remarkable feature of vortex methods is that, unlike other numerical methods, such as finite differences and finite elements, they are fundamentally linked to the physics they aim to reproduce.

Concerning the numerical analysis in the inviscid case, several approaches are now available since the pioneering work of Hald. In this book we focus on a convergence proof based on the tools developed approximately 10 years ago around the concept of weak solutions to advection equations in distribution spaces. There are three reasons that motivated this choice: the notion of weak measure solution is the central mathematical concept in particle methods; second, within this framework the convergence analysis is inherently linked to the structure of the equations, and it applies in many apparently different situations: two- and three-dimensional grid-free methods, including vortex filament methods, and vortex-in-cell methods. Convergence properties of contour dynamics methods, which are not explicitly covered in this book, are also easily understood with these tools. Finally, we believe that the present convergence proof gives optimal results, in particular with respect to the smoothness of the flow, leading to error estimates similar to those of more traditional numerical methods. In Chapter 2 we present this convergence theory for two-dimensional inviscid flows.

Throughout the book, we have tried to maintain a balance between plain numerical analysis and a more qualitative description of the methods. We have given particular attention to conservation properties that are essential in the design of vortex methods. For instance, the energy conservation in two-dimensional schemes, which follows from the Hamiltonian character of the particle motion, is a feature that distinguishes vortex methods from Eulerian schemes. For three-dimensional schemes, covered in Chapter 3, the conservation of circulation has long been an argument in favor of vortex filament methods against the vortex particle methods. We discuss several ways now available to enforce conservation in this second class of methods. We also address practical issues related to the constraint of divergence-free vorticity fields when using vortex particles in three dimensions.

In Chapter 4 we discuss boundary conditions for inviscid flow vortex simulations. We present this in a formal way by considering the Poincaré identity that basically provides the kinematic boundary conditions (no-through-flow) for the Biot–Savart law for flows around solid boundaries. We discuss the method of surface singularities and panel methods as special types of vortex methods. The incorporation of these techniques along with vortex shedding models and the Kutta condition in engineering calculations is discussed.

For viscous flows, besides the popular random-walk method, we have emphasized in Chapter 5 the so-called deterministic vortex methods. These schemes, started at Ecole Polytechnique in 1983, have now given rise to several variants. Applications for two- and three-dimensional flows demonstrate the practicality of viscous schemes and demonstrate the ability of vortex methods to simulate viscous effects accurately, while maintaining the Lagrangian character of the method. Concerning the numerical analysis, we postulate that convergence for the Navier–Stokes equations can be understood in the light of convergence for the Euler equations and for linear convection–diffusion equations. This approach is somewhat biased, as the technical difficulties in the numerical analysis for the full Navier–Stokes equations are much more than the sum of difficulties for the Euler and linear equations, but it makes the presentation more cohesive. We thus focused our attention on linear convection–diffusion equations.

In Chapter 6 we discuss viscous vortex methods for flows evolving in a domain containing solid boundaries. Here the proof of convergence is a far less easy task. It is possible to carry out a numerical analysis, but at the cost of doing constructions (like extending the vorticity support outside the domain) that are not possible for practical applications. We have thus preferred to stress here the difficulties and indicate various attempts to overcome them. To our knowledge there is no completely satisfactory solution for general geometries, in particular because of the need to regularize vortices near the boundary. This problem, already present for inviscid flows, is even more crucial for viscous flows since one has to design, and to implement in the context of a viscous scheme, vorticity boundary conditions. In that context we discuss the no-slip boundary condition and its equivalence with the vorticity boundary condition. We emphasize the case of Neumann-type conditions and investigate their links with classical vorticity generation algorithms. Integral techniques for the implementation of these boundary conditions are then presented and illustrated by applications of vortex methods in the direct numerical simulation of bluff body flows.

In Chapter 7 we discuss the issue of particle distortion inherent in all Lagrangian methods. We argue the necessity of maintaining a somewhat regular Lagrangian grid, from the point of view of numerical analysis, as well as of its practical ramifications. We present methodologies to achieve this goal either by manipulating the particle locations or by processing the circulation of the particles.

In Chapters 4 and 6 we stress the difficulties of vortex methods in dealing with bounded flows. We are indeed convinced that one can get most of the power of vortex methods by combining them, in what we would call hybrid schemes, with Eulerian methods that precisely may avoid difficulties inherent in particle methods near boundaries. A broad class of hybrid schemes, including domain decomposition techniques, is described and illustrated in Chapter 8.

Chapter 1 is an introduction to the notation and the main properties of incompressible fluids. Finally, in the appendices we have included some key concepts of numerical analysis that would help make this book self contained. We have also included a description of what we consider the muscle of vortex methods, the fast summation technique.

As a final thought, we stress that in this book we do not attempt a thorough review of the progress in vortex methods and their applications in the past decades. In particular, applications to the important field of reacting and compressible flows or free surface flows are not explicitly covered. For this we refer to the proceedings [9, 10, 11, 22, 39, 87] of the workshops that have been devoted since 1987 to vortex methods and, more generally, vortex dynamics. We hope, however, that the book demonstrates some important recent advances in these methods and helps make them recognized as a valuable tool in computational fluid dynamics.

Grenoble,
Zürich,
December 1998.

Acknowledgment

We are indebted to many of our colleagues for their path breaking and continuing research efforts in the fields of numerical analysis and flow simulation using particle and vortex methods. We have been fortunate to be exposed to their works through conferences and publications and we hope in return to contribute to their efforts with this book.

We wish to express our gratitude to the people who have contributed directly to this project, by correcting earlier versions of the manuscript, by providing us with stimulating suggestions, and by graciously making available illustrations from their work: Nikolaus Adams, Jonathan Freund, Ahmed Ghoniem, Ron Henderson, Robert Krasny, Tony Leonard, Tom Lundgren, Sylvie Mas-Gallic, Eckart Meiburg, Iraj Mortazavi, Monika Nitsche, Mohamed Lemine Ould-Salihi, Bartosz Protas, Karim Shariff, John Strain, Jens Walther, and Gregoire Winckelmans.

Furthermore, we have benefited tremendously and wish to deeply acknowledge the resources and the spirited environment created by our colleagues at the Laboratoire de Modelisation et Calcul at the University Joseph Fourier in Grenoble, the Institute for Fluid Dynamics at ETH Zurich, and the Center for Turbulence Research at NASA Ames and Stanford University.

We wish to thank Alan Harvey of Cambridge University Press for his indispensable enthusiasm and care throughout this project and TechBooks for their patience and professionalism during production of the book.

This book is dedicated to Catherine, Giulia, Helene, Martin, Ann-Louise and Sophie for guiding us through this and many other adventures with their love.

1

Definitions and Governing Equations

Vorticity plays an important role in fluid dynamics analysis, and in many cases it is advantageous to describe dynamic events in a flow in terms of the evolution of the vorticity field.

The vorticity field (ω) is related to the velocity field (**u**) of a flow as

$$\omega = \nabla \times \mathbf{u}. \tag{1.0.1}$$

It follows from this definition that vorticity is a solenoidal field:

$$\nabla \cdot \omega = 0. \tag{1.0.2}$$

In a Cartesian coordinate system (x, y, z) this relation yields the following relationships between the velocity componenets (u_x, u_y, u_z) and the vorticity components $(\omega_x, \omega_y, \omega_z)$:

$$\omega_x = \frac{\partial u_z}{\partial y} - \frac{\partial u_y}{\partial z}, \quad \omega_y = \frac{\partial u_x}{\partial z} - \frac{\partial u_z}{\partial x}, \quad \omega_z = \frac{\partial u_y}{\partial x} - \frac{\partial u_x}{\partial y}. \tag{1.0.3}$$

In two dimensions the vorticity field has only one nonzero component (ω_z) orthogonal to the (x, y) plane, thus automatically satisfying solenoidal condition (1.0.2).

The circulation Γ of the vorticity field around a closed curve L, surrounding a surface S with unit normal **n** is defined by

$$\Gamma = \int_L \mathbf{u} \cdot d\mathbf{r} = \int_S \omega \cdot \mathbf{n} \, dS, \tag{1.0.4}$$

where $d\mathbf{r}$ denotes an element of the curve.

There are several physical interpretations of the definition of vorticity. We will adopt the point of view that vorticity is a solid-body-like rotation that can

1

be imparted to the elements because of a stress distribution in the fluid. Hence when we consider a vorticity-carrying fluid element, the increment of angular velocity $(d\Omega)$ across an infinitesimal distance $(d\mathbf{r})$ over the element is given by

$$d\Omega = \frac{1}{2}\omega \times d\mathbf{r}. \qquad (1.0.5)$$

When we can track the translation and deformation of vorticity-carrying fluid elements, because of the kinematics and dynamics of the flow field we are able to obtain a complete description of the flow field. Considering the vorticity-carrying fluid elements as computational elements is the basis of the vortex methods that we analyze in this book. The close link of numerics and physics is the essense of vortex methods, and it is a point of view that will be emphasized throughout this book.

In this introductory chapter we present fundamental definitions and equations relating to the kinematics and the dynamics of the vorticity field. In Section 1.1 we introduce the description of flow phenomena in terms of Eulerian and Lagrangian points of view. Using these two descriptions, we present in Section 1.2 the dynamic laws governing the evolution of the vorticity field in a viscous, incompressible flow field. In Section 1.3 we present Helmholtz's and Kelvin's laws governing the motion of the vorticity field.

1.1. Kinematics of Vorticity

There are two different ways of expressing the behavior of the fluid that may be classified as the Lagrangian and the Eulerian point of view. Their difference lies in the choice of coordinates we wish to use to describe flow phenomena.

1.1.1. Lagrangian Description

When the fluid is viewed as a collection of fluid elements that are freely translating, rotating, and deforming, then we may identify the dependent quantities of the flow field (such as the velocity, temperature, etc.) with these individual fluid elements. In that sense the Lagrangian viewpoint is a natural extension of particle mechanics. To obtain a full description of the flow we need to identify the initial location of the fluid elements and the initial value of the dependent variable. The independent variables are then the initial location of a point $(\mathbf{x}_p^0)$ and time (T). By following the trajectories of the collection of fluid elements, we are able to sample at every location in space and instant in time the quantity of interest.

The primary flow quantity in this description is the velocity of the individual fluid elements. The velocity of a fluid element that is residing in an inertial frame of reference at $\mathbf{X}_p$ is expressed as

$$\mathbf{u}_p = \frac{\partial \mathbf{X}_p}{\partial T}. \tag{1.1.1}$$

The acceleration of a fluid particle in a Lagrangian frame is expressed as

$$\mathbf{a}_p = \frac{\partial \mathbf{u}_p}{\partial T}. \tag{1.1.2}$$

The Lagrangian description is ideally suited to describing phenomena in terms of the vorticity of the flow field.

1.1.2. Eulerian Description

In this description of the flow, our observation point is fixed at a certain location $\mathbf{x}$ of the flow field. The flow quantities as they are changing with time t are considered as functions of $\mathbf{x}$. Unlike in Lagrangian methods the location of our observation point remains unchanged by time, and it is the change of the values of the dependent variables at the observation point that describes the flow field.

The Eulerian and the Lagrangian quantities of the flow are related as

$$\mathbf{x} = \mathbf{X}(\mathbf{x}^0, T), \tag{1.1.3}$$

$$t = T. \tag{1.1.4}$$

The Eulerian description of the flow is the most commonly used method to describe flow phenomena in the fluid mechanics literature. In this description, individual fluid elements and their history are not tracked explicitly, but rather it is the global picture of the field that is changing with time that provides us with the description of the flow.

1.1.3. The Material Derivative

The material derivative allows us to relate the Eulerian and the Lagrangian time derivatives of a dependent variable. Let Q be a quantity of the flow expressed in a Lagrangian frame as $Q(\mathbf{x}^0, T)$ and let q be the same quantity expressed in an Eulerian frame, that is, $q(\mathbf{x}, t)$. Then we would have that

$$Q(\mathbf{x}^0, T) = q[\mathbf{x} = \mathbf{X}(\mathbf{x}^0, T), t]. \tag{1.1.5}$$

So the rate of change of Q with time T may be related to the rate change of q with time t with the chain rule for differentiation as

$$\frac{\partial Q}{\partial T} = \frac{\partial q}{\partial \mathbf{x}} \cdot \frac{\partial \mathbf{x}}{\partial T} + \frac{\partial q}{\partial t} \frac{\partial t}{\partial T}, \tag{1.1.6}$$

and since we have for the velocity of a fluid particle that $\mathbf{u} = \partial \mathbf{x}/\partial T$ then

$$\frac{\partial Q}{\partial T} = \frac{\partial q}{\partial t} + \mathbf{u} \cdot \frac{\partial q}{\partial \mathbf{x}}. \tag{1.1.7}$$

The first term is the local rate of change of a variable, and the second term is the convective change of the dependent variable. The substantial derivative (i.e., the rate of change of quantity in a Lagrangian frame) is a convenient way of understanding several phenomena in fluid mechanics, and Stokes has given it a special symbol:

$$\frac{D()}{Dt} = \frac{\partial()}{\partial t} + (\mathbf{u} \cdot \nabla)(). \tag{1.1.8}$$

From the definition of the substantial derivative we may easily see then that

$$\frac{D\mathbf{x}}{Dt} = \mathbf{u}. \tag{1.1.9}$$

We may also determine the rate of change of a material line element ($d\mathbf{r}$) by using the definition of the substantial derivative as

$$\frac{D(d\mathbf{r})}{Dt} = d\mathbf{u} = \partial_j \mathbf{u}\, dr_j = d\mathbf{r} \cdot \nabla \mathbf{u}. \tag{1.1.10}$$

1.1.4. Reynold's Transport Theorem

As an illustrative example of the Lagrangian and the Eulerian descriptions of the flow, we may consider the rate of change of the volume integral of the quantity Q in a material volume $[V(t)]$ with surface $[S(t)]$ having normal $\mathbf{n}$ and velocity $\mathbf{u}$, i.e.,

$$\frac{d}{dt} \int_{V(t)} Q \, dV. \tag{1.1.11}$$

Contributions for this rate of change are given by the local rate of change of Q, $\int_{V(t)} \partial Q/\partial t \, dV$, as well as from the motion of the boundary $\int_{S(t)} Q(\mathbf{u} \cdot \mathbf{n}) \, dS$

[note that for small times dt we may write $dV = dS(\mathbf{u} \cdot \mathbf{n})\, dt$] so that we have

$$\frac{d}{dt} \int_{V(t)} Q\, dV = \int_{V(t)} \frac{\partial Q}{\partial t}\, dV + \int_{S(t)} Q(\mathbf{u} \cdot \mathbf{n})\, dS, \qquad (1.1.12)$$

By using vector calculus we may write

$$\frac{d}{dt} \int_{V(t)} Q\, dV = \int_{V(t)} \frac{\partial Q}{\partial t}\, dV + \int_{V(t)} \nabla \cdot (Q\mathbf{u})\, dV, \qquad (1.1.13)$$

or by using the expression for the substantial derivative we may write that

$$\frac{d}{dt} \int_{V(t)} Q\, dV = \int_{V(t)} \frac{DQ}{Dt}\, dV + \int_{V(t)} Q\nabla \cdot \mathbf{u}\, dV. \qquad (1.1.14)$$

which is known as Reynold's transport theorem for the quantity Q.

1.2. Dynamics of Vorticity

The motion of an incompressible Newtonian fluid is governed by the following equations that express the conservation of mass and momentum of fluid in Eulerian and Lagrangian frames [160]. In the Eulerian description we consider the development of the flow field as it is observed at a fixed point P of the domain, while in the Lagrangian description we consider the equations from the point of view of a material fluid element that moves with the local velocity of the flow.

The conservation of mass can be expressed as
Eulerian Description:

$$\frac{\partial \rho}{\partial t} \qquad + \qquad \nabla \cdot (\rho \mathbf{u}) \quad = 0. \qquad (1.2.1)$$

Rate of accumulation of mass per unit volume at P	Net flow rate of mass out of P per unit volume

Lagrangian Description:

$$\frac{D\rho}{Dt} \quad = \quad -\rho \qquad \nabla \cdot \mathbf{u}. \qquad (1.2.2)$$

Rate of change of the density of a fluid element	Mass per unit volume	Particle-volume expansion rate

The conservation of momentum can be expressed in terms of the velocity (**u**) and the pressure P of the flow field as

Eulerian Description:

$$\rho\frac{\partial \mathbf{u}}{\partial t} \quad + \quad \rho\mathbf{u}\cdot\nabla\mathbf{u} \quad = \quad -\nabla P \quad + \quad \mu\Delta\mathbf{u}, \tag{1.2.3}$$

| Rate of increase of momentum at P | Net flow rate of momentum carried in P by $\rho\mathbf{u}$ | Net pressure force | Net viscous force |

where μ denotes the dynamic viscosity of the fluid, and $\nu = \mu/\rho$ denotes the kinematic viscosity of the fluid with density ρ.

Lagrangian Description:

$$\rho\frac{D\mathbf{u}}{Dt} \quad = \quad -\nabla P \quad + \quad \mu\Delta\mathbf{u}. \tag{1.2.4}$$

| Acceleration of a fluid particle | Net pressure force | Net viscous force |

With definition of vorticity (1.0.1) the momentum equations for an incompressible, Newtonian fluid of uniform density can be expressed in Lagrangian and Eulerian forms as

Eulerian Description:

$$\rho\frac{\partial\omega}{\partial t} \quad + \quad \rho\mathbf{u}\cdot\nabla\omega \quad = \quad \rho\omega\cdot\nabla\mathbf{u} \quad + \quad \mu\Delta\omega. \tag{1.2.5}$$

| Rate of increase of vorticity | Net flow rate of vorticity | Vortex stretching | Viscous diffusion |

Lagrangian Description:

$$\rho\frac{D\omega}{Dt} \quad = \quad \rho\omega\cdot\nabla\mathbf{u} \quad + \quad \mu\Delta\cdot\omega. \tag{1.2.6}$$

| Rate of change of particle vorticity | Rate of deforming vortex lines | Net rate of viscous diffusion |

Note that in the velocity–vorticity formulation the pressure of the flow can be recovered from the equation

$$\frac{1}{\rho}\Delta P \quad = \quad -\nabla\cdot\left(\frac{1}{2}|\mathbf{u}|^2 - \mathbf{u}\times\omega\right). \tag{1.2.7}$$

In the case of a viscous, Newtonian flow of a fluid with nonuniform density, rotation can be imparted to the fluid elements because of the baroclinic generation of vorticity. In this case the equation for the vorticity field is

$$\frac{D(\omega/\rho)}{Dt} = \left(\frac{1}{\rho}\omega \cdot \nabla\right)\mathbf{u} + \nu\Delta\omega + \frac{1}{\rho}\nabla P \times \nabla\frac{1}{\rho}. \qquad (1.2.8)$$

1.3. Helmholtz's and Kelvin's Laws for Vorticity Dynamics

In order to characterize the kinematic evolution of the vorticity field it is useful to introduce some geometrical concepts. We consider the vector of the vorticity field and we identify the lines that are tangential to this vector as vortex lines. In turn, a collection of these lines can form vortex surfaces or vector tubes. The motions of fluid elements carrying vorticity obey certain laws that were first outlined by Helmholtz for the inviscid evolution of the vorticity and further extended by Kelvin to include the effects of viscosity.

From the solenoidal condition for the vorticity field, integrating over a volume of fluid with nonzero vorticity, and using the Gauss theorem, we obtain that

$$\int_V \nabla \cdot \omega \, dV = \int_S \omega \cdot \mathbf{n} \, dS = 0, \qquad (1.3.1)$$

where V denotes the volume of the fluid encompassed by the surface S. When we consider a vortex tube, Eq. (1.3.1) dictates that the strength of the vortex tube is the same at all cross sections. This is Helmholtz's first theorem. When Eq. (1.3.1) is applied to a vorticity tube with cross sections A_1 and A_2 with respective uniform normal vorticity components $\omega_1 = \omega \cdot \mathbf{n}_1$ and $\omega_2 = \omega \cdot \mathbf{n}_2$ (Fig. 1.1) we obtain that

$$|\omega_1|A_1 = |\omega_2|A_2 = |\Gamma| \qquad (1.3.2)$$

independently of the behavior of the vorticity field between the two cross-sections of the vortex tube. Equation (1.3.2) defines the circulation (Γ) of the vortex tube.

When we consider the Lagrangian description of the inviscid evolution of the vorticity field in an incompressible flow (with $\rho = 1$), Eq. (1.2.6) can be expressed as

$$\frac{D\omega}{Dt} = \omega \cdot \nabla\mathbf{u}. \qquad (1.3.3)$$

Comparing Eqs. (1.3.3) and (1.1.10) for the evolution of material lines,

$$\frac{D d\mathbf{r}}{Dt} = d\mathbf{r} \cdot \nabla\mathbf{u}, \qquad (1.3.4)$$

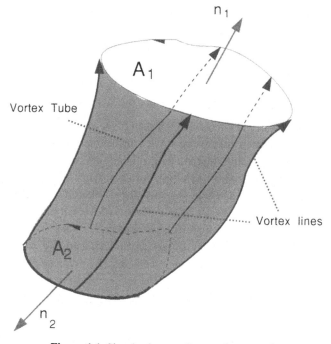

Figure 1.1. Sketch of vortex lines and vortex tube.

we observe that in a circulation-preserving motion the vortex lines are material lines. This is Helmholtz's second theorem for the motion of vorticity elements. As a result of this law, fluid elements that at any time belong to one vortex line, however they may be translated, remain on the vortex line. A result of the first and the second laws is the property of vortex lines and tubes: that no matter how they evolve, they must always form closed curves or they must have their ends in the bounding surface of the fluid.

Kelvin extended the laws of Helmholtz in order to account for the effects of viscosity and at the same time provide a different physical interpretation for the motion of vorticity-carrying fluid elements in terms of the circulation around a closed curve. From the definition of circulation for a line around a cross section of a vortex tube we obtain that

$$\Gamma = \int_L \mathbf{u} \cdot d\mathbf{r}. \tag{1.3.5}$$

Now by using the Lagrangian form of the velocity–pressure formulation for the acceleration of the material particles we obtain

$$\frac{D\Gamma}{Dt} = \frac{D}{Dt} \int_L \mathbf{u} \cdot d\mathbf{r} \tag{1.3.6}$$

$$= \int_L \frac{D\mathbf{u}}{Dt} \cdot d\mathbf{r} + \int_L \frac{Dd\mathbf{r}}{Dt} \cdot d\mathbf{u}. \tag{1.3.7}$$

As we are tracking material lines we obtain that

$$\int_L \frac{Dd\mathbf{r}}{Dt} \cdot d\mathbf{u} = \int_L \mathbf{u} \cdot d\mathbf{u} = 0. \tag{1.3.8}$$

Using Eq. (1.3.8) and momentum equation (Eq. 1.2.4), we can express Eq. (1.3.7) as

$$\frac{D\Gamma}{Dt} = \int_L \frac{D\mathbf{u}}{Dt} \cdot d\mathbf{r} \tag{1.3.9}$$

$$= -\int_L \nabla P \cdot d\mathbf{r} + v \int_L \Delta \mathbf{u} \cdot d\mathbf{r}. \tag{1.3.10}$$

Noting that the pressure term integrates to zero, we obtain that

$$\frac{D\Gamma}{Dt} = v \int_L (\Delta \mathbf{u}) \cdot d\mathbf{r}. \tag{1.3.11}$$

In the case of an inviscid flow, the right-hand side of Eq. (1.3.11) is zero and the circulation of material elements is conserved. This is Kelvin's theorem for the modification of circulation of fluid elements.

In the case of baroclinic flow the circulation around a material line can be modified because of the baroclinic generation of vorticity, and Kelvin's theorem is modified as

$$\frac{D\Gamma}{Dt} = v \int_L (\Delta \mathbf{u}) \cdot d\mathbf{r} + \int \frac{1}{\rho^2} \nabla \rho \times \nabla P \cdot \mathbf{n} \, dS. \tag{1.3.12}$$

Note that the second term on the right-hand side is an integral over the area encompassed by the material curve. Equation (1.3.12) is known as Bjerken's theorem.

2

Vortex Methods for Two-Dimensional Flows

The simulation of phenomena governed by the two-dimensional Euler equations are the first and simplest example in which vortex methods have been successfully used. The reason can be found in Kelvin's theorem, which states that the circulation in material – or Lagrangian – elements is conserved. Mathematically, this comes from the conservative form of the vorticity equation. Following markers – or particles – where the local circulation is concentrated is thus rather natural. At the same time, the nonlinear coupling in the equations resulting from the velocity evaluation immediately poses the problem of the mollification of the particles into blobs and of the overlapping of the blobs, which soon was realized to be a central issue in vortex methods.

The two-dimensional case thus encompasses some of the most important features of vortex methods. We first introduce in Section 2.1 the properties of vortex methods by considering the classical problem of the evolution of a vortex sheet. We present in particular the results obtained by Krasny in 1986 [129, 130] that demonstrated the capabilities of vortex methods and played an important role in the modern developments of the method. We then give in Sections 2.2 to 2.4 a more conventional exposition of vortex methods and of the ingredients needed for their implementation: choice of cutoff functions, initialization procedures, and treatment of periodic boundary conditions. Section 2.6 is devoted to the convergence analysis of the method and to a review of its conservation properties.

2.1. An Introduction to Two-Dimensional Vortex Methods: Vortex Sheet Computations

The origin of vortex methods may be traced back to the 1920s and 1930s in the works of Prager (1928) [163] and Rosenhead (1931) [172, 173]. They utilized vortex methods in two seemingly unrelated contexts that reveal the multifaceted character of the method. In order to solve the problem of potential flow around bodies, Prager [163] considered the boundary of the body as a surface of discontinuity, i.e., a vortex sheet. In order to satisfy the appropriate boundary equations, the problem is formulated as a boundary integral equation for the strength of the vortex sheet. The determination of the vortex sheet strength provides a complete description of the potential flow field. The surface-singularity method for the solution of the potential flows has served as a predictive tool for the calculation of loads around aerodynamic configurations. The vortex sheet on the surface is consistent with the limit of viscous flow at infinite Reynolds number, and it may be viewed as the limit of a boundary layer of infinitesimal thickness. Note that for a nondeformable body the shape of the vortex sheet remains the same and it is only its strength that is determined so that the appropriate boundary conditions are satisfied (this is discussed in more detail when we come to the boundary-value problem in Chapters 4 and 6). One may wish to consider, however, how the vorticity that exists on the surface of the body may enter the fluid. It is evident that this is possible through the action of viscous or other nonconservative forces. However, for the purposes of this chapter we are not interested in invoking viscosity. The Kafeelöffel experiment of Klein (1910) provides a mechanism by which the vorticity may enter the fluid in an inviscid flow. As discussed by Saffman (1992) "the Kelvin and Helmholtz theorems preclude the generation of piecewise continuous vorticity, but do not prevent the formation of vortex sheets or the generation of circulation." According to Klein's experiment, a two-dimensional plate in an incompressible ideal fluid is set in steady motion with velocity U normal to the plate. In order to enforce the no-through-flow boundary conditions, one may use then the method of surface singularities, as mentioned above, and replace the surface of the plate by a vortex sheet, the strength of which can be determined analytically.

Klein's experiment consists in removing the plate, either by pulling it abruptly out of the fluid or by dissolving it instantaneously in the fluid. Hence the bound vortex sheet enters the fluid because of this topological change. Because of the absence of the confining body surface (and the respective boundary conditions), the vortex sheet is then allowed to evolve, following an integral equation for its motion. Rosenhead, in 1931 [172], was the first to consider the evolution of a vortex sheet and to compute its evolution by discretizing it into elemental vortices by using their locations as quadrature points. Performing his calculations by hand, he was not able to conduct simulations for extended times. At the same time the use of a limited number of discretization elements prevented him from facing further complications with the evolution of the vortex sheet, namely, that the evolution of these discrete vortices does not necessarily represent the evolution of a continuous vortex sheet.

This issue was addressed in a rigorous way by Krasny [129, 130], who elucidated at the same time some important mathematical and numerical analysis points of the method. Krasny's calculations contain most of the features that all rigorous vortex methods calculations wish to retain. As they are concerned also with a fundamental problem in fluid mechanics, we wish to give a more detailed presentation of these calculations.

Let us consider an initial vorticity field that is confined on a curve Γ_0, parametrized by $\xi \in [0, 1] \to \gamma_0(\xi)$ with a normal $\mathbf{n}(\xi, t)$. We consider then the (scalar) vorticity field defined as $\omega_0(\xi) = \alpha(\xi)\delta[\mathbf{n}(\xi, t)]$. With this notation we mean that, whenever one has to integrate ω_0 against a test function ϕ, one considers this function on Γ_0 and then integrates $\alpha \phi$ along this curve, or, in short,

$$\langle \omega_0, \phi \rangle = \int_0^1 \alpha(\xi)\phi[\gamma_0(\xi)]\,d\xi.$$

It is readily seen, by use of, for example, the formalism of weak solutions to advection equations as developed in Appendix A, that the solution to the Euler equations with this initial vorticity is a time-dependent vortex sheet supported by a curve $\Gamma(t)$ with density α. The curve $\Gamma(t)$ is a material line that carries constant circulation. In other words the vorticity satisfies for all positive time

$$\langle \omega(\cdot, t), \phi \rangle = \int_0^1 \alpha(\xi)\phi[\gamma(\xi, t)]\,d\xi, \tag{2.1.1}$$

where $\gamma(\xi, t)$ is a parameterization of $\Gamma(t)$, thus satisfying

$$\frac{\partial \gamma}{\partial t}(\xi, t) = \mathbf{u}[\gamma(\xi, t), t]. \tag{2.1.2}$$

The velocity $\mathbf{u}$ is coupled at all times in a self-consistent way with ω through the relations $(\nabla \cdot \mathbf{u}) = 0$ and $\nabla \times \mathbf{u} = \omega$ or, equivalently, by the Poisson equation:

$$\Delta \mathbf{u} = -\nabla \times \omega.$$

The definition of the vorticity field $\omega = \alpha\delta[\mathbf{n}(\cdot, t)]$ and the relation $\nabla \times \mathbf{u} = \omega$ clearly indicate that the velocity field induced by a vortex sheet has a component parallel to the sheet that is discontinuous across the sheet. This confirms that the vortex sheet can be viewed as the limit case of a thin layer with a rapid transition between two velocity profiles.

Equations (2.1.1) and (2.1.2) are valid as long as their solution remains a smooth curve. It is believed that a singularity in the curvature of the sheet appears after a finite time, and if a solution persists afterwards, it is no longer a vortex sheet solution in the sense of Eqs. (2.1.1) and (2.1.2).

One goal of numerical simulations is to try first to confirm this singularity formation and then to describe what kind of solution might persist afterward. To be more specific, let us assume that we are dealing with a problem that is periodic in one direction, with unit period. In addition, we are looking for velocities that have the following behavior at infinity:

$$\lim_{x_2 \to +\infty} \mathbf{u}(x_1, x_2) + \mathbf{u}(x_1, -x_2) = 0,$$

where x_1 is the direction of periodicity. The velocity of the sheet is determined with the Green's function solution of the Poisson equation with periodic boundary conditions:

$$\mathbf{u} = \mathbf{K}_p \star \omega \tag{2.1.3}$$

where $\star$ denotes convolution and the periodic vector-valued kernel $\mathbf{K}_p$ has the form

$$\mathbf{K}_p(x_1, x_2) = -\frac{1}{2} \left(\frac{\sinh 2\pi x_2}{\cosh 2\pi x_2 - \cos 2\pi x_1}, \frac{\sin 2\pi x_1}{\cosh 2\pi x_2 - \cos 2\pi x_1} \right).$$

To evaluate the integrals involved in the above convolution we refer to the meaning of integrating the vorticity of the vortex sheet as discussed above. So, with $\gamma(\xi, t) = [x(\xi, t), y(\xi, t)]$, combining Eqs. (2.1.2) and (2.1.3) together with definition (2.1.1) of the sheet yields the so-called Birkoff–Rott equations:

$$\frac{\partial x}{\partial t} = -\frac{1}{2} \int_0^1 \alpha(\xi) \frac{\sinh 2\pi [y - y(\xi, t)]}{\cosh 2\pi [y - y(\xi, t)] - \cos 2\pi [x - x(\xi, t)]} d\xi,$$
$$\tag{2.1.4}$$

$$\frac{\partial y}{\partial t} = \frac{1}{2} \int_0^1 \alpha(\xi) \frac{\sin 2\pi [x - x(\xi, t)]}{\cosh 2\pi [y - y(\xi, t)] - \cos 2\pi [x - x(\xi, t)]} d\xi,$$
$$\tag{2.1.5}$$

where the integrals have to be understood in the sense of principal values. These equations completely determine the solution as long as it remains a vortex sheet. As we already mentioned, these solutions can develop singularities. This is a result of the integrand singularity on the right-hand side of Eqs. (2.1.4) and (2.1.5). The linear stability analysis of this system around the steady-state trivial solution of a flat sheet with $\alpha \equiv 1$ shows that small perturbations of the form

$\exp(2i\pi k\xi)$ are amplified with an exponential rate $\exp(2\pi f)$ with $f \equiv k/2$. This mechanism is known as the Kelvin–Helmholtz instability. Mathematically speaking, this means that a linear combination of such modes will develop a bounded solution, up to a finite time T, only if the coefficients of the modes decrease exponentially, or, in other words, if the initial condition is analytic. The width of the band of analyticity is linked to the exponential decay of the Fourier modes, which is in turn related to the finite time of existence (we refer to Ref. 129 and the references therein for more detailed mathematical discussions).

In order to examine numerically this singularity formation and to investigate whether a solution persists (in a weak sense) beyond this singularity, a natural procedure is to

1. regularize the Birkhoff–Rott equations (2.1.4) and (2.1.5) by getting rid of the singularity of the integrand,
2. replace the continuous sheet by a discrete set of points.

The effect of step 1 is to provide a smooth velocity field, leading to a well-posed problem for all times, the solution of which will be close to the analytical solution as long as one exists.

The interpretation of step 2 is that continuous integrals are replaced by numerical quadratures. The accuracy of this procedure is conditioned by the smoothness of the integrand. This smoothness obviously deteriorates when the regularization parameter ε tends to 0, therefore making it necessary to increase the number N of computational points.

The convergence procedure of Krasny [129, 130], inspired by an earlier work of Anderson [4], is precisely to fix the regularization parameter and increase N until convergence is obtained, then to decrease ε and repeat the convergence study with respect to N.

The specific type of regularization used in these calculations is not of central importance. Krasny removed the singularity of the kernel by the simple addition of a positive constant in the denominator (we will discuss later in Section 2.3 some general criteria to design efficient regularizations):

$$\mathbf{K}_p^\varepsilon(x_1, x_2)$$
$$= -\frac{1}{2}\left(\frac{\sinh 2\pi x_2}{\cosh 2\pi x_2 - \cos 2\pi x_1 + \varepsilon^2}, \frac{\sin 2\pi x_1}{\cosh 2\pi x_2 - \cos 2\pi x_1 + \varepsilon^2}\right).$$

As for the quadrature rule, the most efficient one is the midpoint rule, which is spectrally accurate if the integrand is analytic (see Ref. 109 for a discussion of this accuracy issue in the absence of regularization). For a given value of ε and N, one now has to solve the following system of ordinary differential equations

(ODEs):

$$\frac{dx_i}{dt} = -\frac{1}{2N} \sum_{j=1}^{N} \frac{\sinh 2\pi(y_i - y_j)}{\cosh 2\pi(y_i - y_j) - \cos 2\pi(x_i - x_j) + \varepsilon^2}, \qquad (2.1.6)$$

$$\frac{dy_i}{dt} = \frac{1}{2N} \sum_{j=1}^{N} \frac{\sin 2\pi(x_i - x_j)}{\cosh 2\pi(y_i - y_j) - \cos 2\pi(x_i - x_j) + \varepsilon^2}. \qquad (2.1.7)$$

We give in Figure 2.1 the converged (with respect to N) results obtained in Ref. 129 for successive times. In these simulations, the initial condition was a

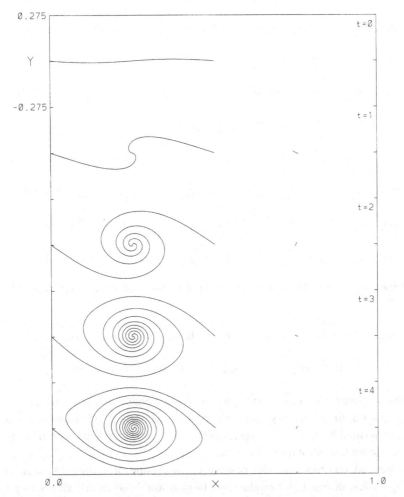

Figure 2.1. Sheet evolutions for $\varepsilon = 0.1$ (the dotted spirals on the right-hand side indicate the particle locations). (Courtesy of R. Krasny.)

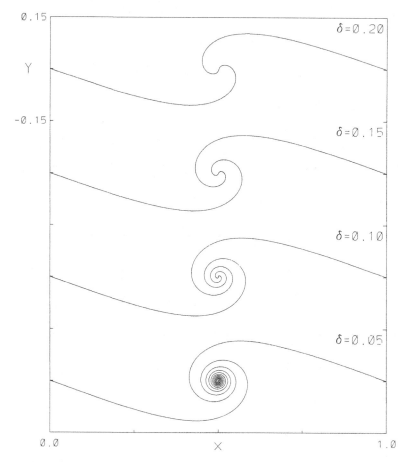

Figure 2.2. Vortex sheet at a given time for decreasing values of ε. (Courtesy of R. Krasny.)

sinusoidal perturbation of a flat sheet of the kind mentioned above, namely,

$$x(\xi, 0) = \xi + 0.01 \, \sin 2\pi\xi, \quad y(\xi, 0) = -0.01 \sin 2\pi\xi.$$

The maximum time of analyticity for this initial condition can be estimated on the basis of linear stability analysis to be approximately 0.375. This value was later verified by numerical experiments with $\varepsilon = 0$ combined with filtering techniques to control round-off errors [130].

Beyond this time one can observe the formation of a spiral; as seen in Figure 2.2, decreasing the value of ε (δ in Krasny's notations) allows to get sharper details inside the core of this spiral. This is confirmed in Figure 2.3

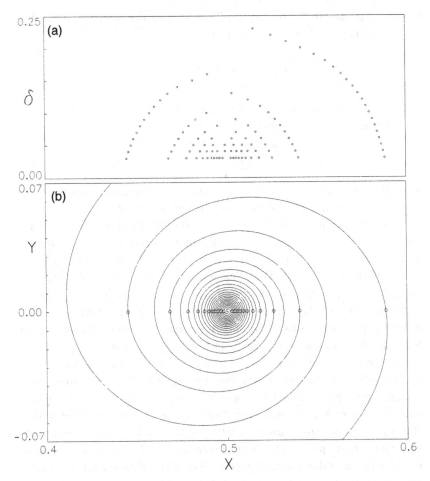

Figure 2.3. Convergence study for the spiral intersections with the x_1 axis. (Courtesy of R. Krasny.)

where are plotted, for decreasing values of ε, the x_1 coordinates of the successive intersection of the sheet with the first axis. One notes that reducing the value of ε only introduces further roll-up inside the spirals without changing significantly the position of the existing turns (some recent calculations by Krasny with even smaller values of ε seem, however, to indicate that the limiting process might be less straightforward).

The first important conclusion of these calculations is that there is no dissipative time scale associated with the parameter ε. This parameter gives only the scale under which the solution is not described. We can already state that the reason for this absence of numerical dissipation is that the discrete dynamical system of equations (2.1.6) and (2.1.7) underlying the numerical method is Hamiltonian. If we denote by E the quantity

$$E[(x_i, y_i)_i] = \frac{1}{N} \sum_{i,j} \log \left[\cosh 2\pi (y_i - y_j) - \cos 2\pi (x_i - x_j) + \varepsilon^2 \right],$$

then it is readily seen that the system of equations (2.1.6) and (2.1.7) can be rewritten under the Hamiltonian form

$$\frac{dx_i}{dt} = \frac{\partial E}{\partial y_i}; \quad \frac{dy_i}{dt} = -\frac{\partial E}{\partial x_i}; \quad i \in [1, N].$$

We will emphasize this fact in Section 2.6 and show that this Hamiltonian form yields the conservation of kinetic energy by vortex methods.

Another important feature of the numerical method just described is that it takes full advantage of the fact that the vorticity remains on a small support, here a one-dimensional curve; moreover, the far field is explicitly taken into account in velocity formula (2.1.3). This explains the early attempts of Rosenhead to simulate the vortex sheet evolution with a vortex method. Moreover, refinement techniques can easily take advantage of the geometry of the problem by inserting points in the curve in between neighboring particles whenever they go too far apart (as a matter of fact, as it is done in Krasny's calculations).

These two features – lack of numerical dissipation and localization of the method – should be compared with what would happen if a more conventional, say finite-difference, method was used. In this case it would not be possible to deal anymore with a one-dimensional problem and a more subtle refinement strategy would have to be used to capture accurately the vortex sheet without wasting too many points. The far-field boundary conditions would have to be modeled by an artificial boundary condition; finally it would not be straightforward to avoid creating numerical diffusion and preserve stability (e.g., for finite differences, upwinding is often introduced to stabilize the methods, with the potential pitfall of numerical dissipation).

The spirit of the vortex methods discussed in this book will always try to retain as much as possible the basic features of Krasny's vortex sheet calculations, while introducing additional phenomena such as diffusion effects, the vortex stretching in three-dimensional flows, and boundary conditions.

2.2. General Definition

We now turn to a more clear-cut description of vortex methods for inviscid two-dimensional unbounded flows (Section 2.5 and, more generally, Chapter 4 will be devoted to boundary conditions).

Let us first recall the vorticity–velocity formulation of the Euler equations on which these methods are based:

$$\frac{\partial \omega}{\partial t} + \text{div}\,(\mathbf{u}\omega) = 0, \tag{2.2.1}$$

$$\omega(\cdot, 0) = \omega_0, \tag{2.2.2}$$

$$\text{div}\,\mathbf{u} = 0, \tag{2.2.3}$$

$$\text{curl}\,\mathbf{u} = \omega, \tag{2.2.4}$$

$$|\mathbf{u}| \stackrel{\infty}{\to} \mathbf{u}_\infty. \tag{2.2.5}$$

One interesting feature of the velocity–vorticity formulation is the fact that the pressure and the associated divergence-free constraint on the velocity have disappeared.

A classical way to relate the velocity and the vorticity [Eqs. (2.2.3)–(2.2.5)] is through an integral representation. If we denote by G the Green's function for the Laplacian operator in two dimensions and by $\mathbf{K}$ its rotational counterpart, then, if $\mathbf{x} = (x_1, x_2)$, we have:

$$G(\mathbf{x}) = -\frac{1}{2\pi}\log(|\mathbf{x}|); \quad \mathbf{K}(\mathbf{x}) = (2\pi|\mathbf{x}|^2)^{-1}(-x_2, x_1),$$

and the Biot–Savart law reads

$$\mathbf{u} = \mathbf{u}_\infty + \mathbf{K} \star \omega. \tag{2.2.6}$$

As already noted for the vortex sheet calculations, this formula accounts explicitly for the far-field boundary condition.

From now on we are going to focus on the forms of Eqs. (2.2.1), (2.2.2), and (2.2.6) of the Euler equations. Vortex methods are based on the Lagrangian formulation of these equations, in particular on Kelvin's theorem (see Section 1.3), which asserts the conservation of circulation along material elements moving with the fluid:

$$\frac{d}{dt}\int_{V(t)} \omega(\cdot, t)\,d\mathbf{x} = 0. \tag{2.2.7}$$

The basic idea in vortex methods is then to sample the computational domain into cells in which the initial circulation is concentrated on a single point – or particle. The resulting approximation can be written as

$$\omega_0 \simeq \omega_0^h = \sum_p \alpha_p \delta(\mathbf{x} - \mathbf{x}_p),$$

where the value of α_p is an estimate of the initial circulation around the particle $\mathbf{x}_p$.

Assume next that a smooth approximation $\mathbf{u}^h$ of $\mathbf{u}$ is given, and denote by $\mathbf{x}_p^h$ the trajectories of the particles along the flow $\mathbf{u}^h$. In view of Eq. (2.2.7), it is natural to require that α_p remains for all time the local circulation around $\mathbf{x}_p^h$. Thus an approximation of the vorticity at time t is given by

$$\omega^h(\mathbf{x}, t) = \sum_p \alpha_p \delta \left[\mathbf{x} - \mathbf{x}_p^h(t) \right]. \qquad (2.2.8)$$

The precise mathematical justification of this approximation is given in Appendix A, based on the principle of weak solutions to advection equations.

The numerical approximation is therefore completely defined once the coupling between ω^h and $\mathbf{u}^h$ is restored through some approximation of the Biot–Savart law. Because of the singularity of the kernel $\mathbf{K}$, implementing directly Eq. (2.2.6) with the vorticity field ω^h can lead to very large values when two particles approach each other.

Although it is possible to show that, under some assumptions, in particular on the regularity of the underlying flow, this cannot happen (see Section 2.6 below), the usual strategy to overcome this difficulty is to remove the singularity of $\mathbf{K}$. The simplest way to do it is to add some small positive constant to prevent the denominator of $\mathbf{K}$ from vanishing – as was done in Krasny's calculations. A more general approach, which allows a precise control of the accuracy of the procedure, is to replace $\mathbf{K}$ by a mollification $\mathbf{K}_\varepsilon$ obtained in the following way: first a smooth cutoff function ζ is chosen that satisfies $\int \zeta(\mathbf{x}) d\mathbf{x} = 1$. If ε is a small parameter, one denotes by ζ_ε the function defined by

$$\zeta_\varepsilon(\mathbf{x}) = \varepsilon^{-2} \zeta \left(\frac{\mathbf{x}}{\varepsilon} \right).$$

Finally one sets

$$\mathbf{K}_\varepsilon = \mathbf{K} \star \zeta_\varepsilon,$$

and the numerical particles $\mathbf{x}_p^h(t)$ are computed by numerically integrating the system of ODEs,

$$\frac{d\mathbf{x}_p^h}{dt} = \mathbf{u}^h \left(\mathbf{x}_p^h, t \right); \quad \mathbf{x}_p^h(0) = \mathbf{x}_p \qquad (2.2.9)$$

with

$$\mathbf{u}^h = \mathbf{K}_\varepsilon \star \omega^h. \qquad (2.2.10)$$

The complete numerical method is then defined by Eqs. (2.2.9) and (2.2.10). In practice it amounts to the resolution of a (often large) set of coupled differential equations.

Note that one can view the mollified velocity field of Eq. (2.2.10) as the exact velocity associated with a vorticity ω_ε^h consisting of mollified particles:

$$\omega_\varepsilon^h(\mathbf{x}) = \sum_p \alpha_p \zeta_\varepsilon \left(\mathbf{x} - \mathbf{x}_p^h \right). \tag{2.2.11}$$

These mollified particles are called vortex blobs, with core size ε. So far, we have implicitly assumed that this core size has a constant value. It may be natural to allow this value to vary in space and time, depending on the scales sought to be resolved in different zones of the flow. A variable-blob method would consist of defining a function $\varepsilon(\mathbf{x}) \ll 1$ and modifying Eq. (2.2.11) into

$$\omega_\varepsilon^h(\mathbf{x}) = \sum_p \alpha_p \zeta_{\varepsilon(\mathbf{x}_p)} \left(\mathbf{x} - \mathbf{x}_p^h \right), \tag{2.2.12}$$

which gives the velocity formula

$$\mathbf{u}^h(\mathbf{x}) = \sum_p \alpha_p \mathbf{K}_{\varepsilon(\mathbf{x}_p)} \left(\mathbf{x} - \mathbf{x}_p^h \right). \tag{2.2.13}$$

The efficiency of vortex methods is conditioned in particular by

- the choice of the cutoff function ζ,
- the way locations and circulations of particles are initially set.

These two issues will be respectively addressed in Sections 2.3 and 2.4.

The numerical time-advancing scheme required for solving Eq. (2.2.9) is an additional important factor. The numerical solution of systems of ODEs is a well-covered subject and we will not address it further in this book. In practice it is important to use schemes that are at least second order (Adams–Bashforth or Runge–Kutta schemes are commonly used).

A striking difference between vortex methods and grid-based methods such as spectral, finite difference, or finite element lies in the treatment of the vorticity transport equation. The philosophy of grid-based methods is more or less to project this equation on a finite dimensional functional space – typically functions that are locally (in the case of finite element) or globally (for spectral methods) polynomials of a given degree. In vortex methods the transport equation is dealt with exactly, and the approximation amounts to only replacing the initial vorticity by a set of particles and smoothing the velocity field that carries these particles. The fact that the smoothing does not appear directly in the vorticity transport equation is the reason that vortex methods do not face the usual dilemma of grid-based methods between accuracy and stability. On the other hand, the fact that the vorticity field is sampled on a moving grid makes vortex methods sensitive to the smoothness of the velocity field. This important issue will be addressed in Chapter 7.

In closing this section let us mention that, besides the initial vorticity field, vortex methods also allow us to handle source terms – besides linear zero-order or second-order terms which will be discussed in subsequent chapters. Assume one has to solve the system of equations (2.2.1)–(2.2.5) with a source term S on the right-hand side. The vortex approximation of this system will consist of updating the circulation of particles $\mathbf{x}_p(t)$ by the amount of local circulation produced by S. This circulation can be obtained by multiplying the value of S at the particle locations $\mathbf{x}_p$ with the volume of the Lagrangian computational cell around $\mathbf{x}_p$. In the case of an incompressible flow this volume remains constant in time, and its value v_p depends on how particles were initialized. The circulation is finally updated by

$$\frac{d\alpha_p}{dt} = v_p S(\mathbf{x}_p, t).$$

2.3. Cutoff Examples and Construction of Mollified Kernels

The desirable features sought for cutoff functions are smoothness and accuracy. Accuracy in particular means that one wishes to avoid, in the Biot–Savart law, the production of too much smearing in the particle trajectories by the mollified kernel.

As we will see in Section 2.6 the moment properties of the cutoff are a convenient way to measure its accuracy. We say that a cutoff is of order r if the following assumptions hold:

$$\int \zeta(\mathbf{x}) \, d\mathbf{x} = 1,$$

$$\int \mathbf{x}^i \zeta(\mathbf{x}) \, d\mathbf{x} = 0 \quad \text{if } |\mathbf{i}| \leq r - 1, \qquad (2.3.1)$$

$$\int |\mathbf{x}|^r |\zeta(\mathbf{x})| \, d\mathbf{x} < \infty.$$

The meaning of the above conditions is that the cutoff ζ has, up to the power $r-1$, the same moment properties as the Dirac measure. In other words, the vortex blobs in formula (2.2.11) and the vorticity particles share the same momentum, starting with total circulation, linear impulse, and so on, up to order $r - 1$

[if $\mathbf{i} = (i_1, i_2)$ is a multi-index, $|\mathbf{i}| = i_1 + i_2$ and $\mathbf{x}^{\mathbf{i}} = x_1^{i_1} x_2^{i_2}$]. Notice that, for symmetry reasons, an even positive cutoff function can always be scaled in order to be at least of second order. Conversely, higher order requires giving up the positivity of the cutoff, which means that positive circulations may locally produce negative values in the mollified vorticity field (as, most often in numerical methods, positivity and high-order accuracy have difficulty in coexisting).

The simplest examples of cutoff functions are provided by the one-dimensional characteristic top-hat function χ in the interval $[-1/2, 1/2]$. In dimension d one can choose either

$$\zeta = \chi \otimes \chi \cdots \otimes \chi$$

or

$$\zeta(\mathbf{x}) = \frac{2^d}{S_d} \chi(|\mathbf{x}|),$$

where S_d denotes the volume of the unit sphere in $\mathbf{R}^d$. If one wishes smoother functions, the simplest is to take successive convolutions of χ by itself, which has also the effect of increasing the size of the support of ζ and thus of the blobs. For example $\chi \star \chi$ is the classical triangle-hat function with support in $[-1, 1]$, while $\chi \star \chi \star \chi$ is piecewise quadratic with support in $[-3/2, 3/2]$.

Of course all these functions are positive and therefore cannot be of order more than 2. For higher-order cutoff functions, one in general favors functions without a compact support.

Let us first give an example, borrowed from Ref. 25, showing that it is possible to construct C^∞ cutoff functions at any order. These cutoff functions are obtained from the so-called generalized Gaussian functions, defined by

$$\Gamma_l(\mathbf{x}) = c_l \mathcal{F}^{-1}[\exp(-|\boldsymbol{\xi}|^{2l})],$$

where $l \geq 1$, $\mathcal{F}^{-1}$ denotes the inverse Fourier transform, and c_l are normalization coefficients that ensure that the integrals of these functions are unity. If $l = 1$, one recovers the usual Gaussian function. It is a simple matter to check that Γ_l is infinitely differentiable. It is of order $r = 2l$. This is because, if α is a multi-index, $\mathbf{x}^\alpha \Gamma_l$ is, up to a multiplication by a constant, the inverse Fourier transform of $\partial^\alpha \exp(-|\boldsymbol{\xi}|^{2l})$. But these derivatives can be written as the product of $\exp(-|\boldsymbol{\xi}|^{2l})$ by a polynomial in $\boldsymbol{\xi}$ containing no power below $2l - |\alpha|$. Thus

$$\partial^\alpha \exp(-|\boldsymbol{\xi}|^{2l}) = \sum_{2l-|\alpha| \leq |\beta|} \lambda_\beta \boldsymbol{\xi}^\beta \exp(-|\boldsymbol{\xi}|^{2l}),$$

so that

$$\mathcal{F}^{-1}[\partial^\alpha \exp(-|\xi|^{2l})] = \sum_{2l-|\alpha|\leq|\beta|} \mu_\beta \partial^\beta \{\mathcal{F}^{-1}[\exp(-|\xi|^{2l})]\},$$

and its integral over $\mathbf{R}^d$ vanishes.

Although interesting from a theoretical point of view, these cutoffs are of little practical interest, unless one wishes to perform computations of velocities in the Fourier space. A more efficient way to obtain cutoff functions of order greater than 2 is to choose a positive cutoff function ζ with spherical symmetry (that is, one that depends on only the radius), decaying fast enough at infinity and to combine properly different scales in the same cutoff.

More precisely, let ζ be any radially symmetric function with $\int |\mathbf{x}|^4 \zeta(\mathbf{x}) < \infty$ and $a \neq 0, 1$; to obtain a fourth-order cutoff, it suffices to find two coefficients λ and μ such that, if we set $\tilde{\zeta}(\mathbf{x}) = \lambda\zeta(\mathbf{x}) + \mu\zeta(\mathbf{x}/a)$,

$$\int \tilde{\zeta}(\mathbf{x})\,d\mathbf{x} = 1, \quad \int |x_1|^2 \tilde{\zeta}(\mathbf{x})\,d\mathbf{x} = 0. \tag{2.3.2}$$

Indeed for symmetry reasons we have for all coefficients λ and μ

$$\int x_1 x_2 \tilde{\zeta}(\mathbf{x})\,d\mathbf{x} = 0, \quad \int |x_2|^2 \tilde{\zeta}(\mathbf{x})\,d\mathbf{x} = \int |x_1|^2 \tilde{\zeta}(\mathbf{x})\,d\mathbf{x},$$

so that conditions (2.3.2) are enough to provide a fourth-order cutoff function. If we denote by S_0 and S_2, respectively, the integrals of ζ and $|\mathbf{x}|^2\zeta$, straightforward calculations show that these conditions are satisfied if

$$(\lambda + a^2\mu)S_0 = 1, \quad (\lambda + a^4\mu)S_2 = 0,$$

that is,

$$\mu = \left(\frac{1}{S_2} - \frac{1}{S_0}\right)\frac{1}{a^4 - a^2}, \quad \lambda = \left(\frac{a^2}{S_0} - \frac{1}{S_2}\right)\frac{1}{a^2 - 1}.$$

Another strategy to construct fourth-order cutoff functions is to combine ζ and its radial derivative (provided this latter decays fast enough at infinity) with the proper coefficient, which can be computed through calculations similar to those given above.

If even higher accuracy is desired, one can combine three different scales to obtain cutoff of order 6, and so on. This method is used in Ref. 26 to obtain high-order cutoff functions starting from the Gaussian. It is also possible to combine Gaussian with polynomials of degree $r - 2$ to obtain a cutoff function of order r.

Besides smoothness and order of accuracy, computational cost is also a criterion for choosing a cutoff. Gaussian-based cutoffs are often favored because of their smoothness and fast decay, but they are expensive to compute, and it is advisable to tabulate their values in actual computations. Algebraic cutoff functions avoid this additional task.

Once the choice of a cutoff is made, it remains to compute the associated kernel $\mathbf{K}_\varepsilon$. To avoid the explicit computations of the convolution it is customary to use a radially symmetric cutoff together with the definition of Δ in spherical coordinates. In two dimensions, given a cutoff such that $\zeta(\mathbf{x}) = \bar{\zeta}(|\mathbf{x}|)$, the calculations are as follows. The kernel G_ε $(G_\varepsilon = G \star J_\varepsilon)$ satisfies $G_\varepsilon(\mathbf{x}) = \bar{G}_\varepsilon(|\mathbf{x}|)$ where,

$$-\frac{1}{r}\frac{\partial}{\partial r}\left(r\frac{\partial \bar{G}_\varepsilon}{\partial r}\right) = \bar{\zeta}_\varepsilon.$$

This yields

$$\frac{\partial \bar{G}_\varepsilon}{\partial r} = \frac{1}{r}\int_0^r -s\bar{\zeta}_\varepsilon(s)\,ds.$$

To get $\mathbf{K}_\varepsilon$ it suffices to write

$$\mathbf{K}_\varepsilon(\mathbf{x}) = \frac{1}{r}(x_2, -x_1)\frac{\partial \bar{G}_\varepsilon}{\partial r} = \frac{1}{r^2}(x_2, -x_1)\int_0^r -s\bar{\zeta}_\varepsilon(s)\,ds.$$

In many cases (as for Gaussian-based or algebraic cutoffs) this leads to explicit algebraic forms of the kernel. For example, in the case of a Gaussian function $\bar{\zeta}(r) = \pi^{-1}\exp(-r^2)$ one obtains

$$\mathbf{K}_\varepsilon(\mathbf{x}) = \frac{1}{2\pi r^2}(-x_2, x_1)[1 - \exp(-r^2\varepsilon^{-2})]. \tag{2.3.3}$$

Starting from a fourth-order Gaussian-based kernel constructed as indicated above, with $a = 2$, one obtains the formula [26]

$$\mathbf{K}_\varepsilon^{(4)}(\mathbf{x}) = \frac{(-x_2, x_1)}{2\pi r^2}[1 - \exp(-r^2/2\varepsilon^2)][1 + 2\exp(-r^2/2\varepsilon^2)]. \tag{2.3.4}$$

These examples clearly show that $\mathbf{K}_\varepsilon$ is a mollification of $\mathbf{K}$ that acts in only an ε neighborhood of the singularity. This is actually a general property: if $\bar{\zeta}_\varepsilon$ decays like $|\mathbf{x}|^{-M}$ at infinity, we can write

$$\frac{\partial \bar{G}_\varepsilon}{\partial r} = \frac{1}{r}\left[\int_0^\infty -s\bar{\zeta}_\varepsilon(s)\,ds + \int_r^\infty -s\bar{\zeta}_\varepsilon(s)\,ds\right]$$

$$= -\frac{1}{2\pi r}\left[1 + 0\left[\left(\frac{r}{\varepsilon}\right)^{2-M}\right]\right],$$

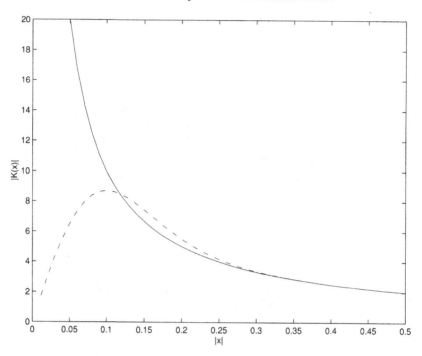

Figure 2.4. Modulus versus r for the exact kernel **K** (solid curve) and its mollifications of Eqs. (2.3.3) (dotted curve) and (2.3.4) (dotted–dashed curve) for $\varepsilon = 0.1$.

which shows that $\mathbf{K}_\varepsilon = \mathbf{K}g_\varepsilon$ with $|1 - g_\varepsilon(\mathbf{x})| \leq C(\mathbf{x}/\varepsilon)^{2-M}$. In the particular case of a cutoff function with compact support of size R, it is also readily seen that $\mathbf{K}_\varepsilon(\mathbf{x}) = \mathbf{K}(\mathbf{x})$ if $|\mathbf{x}| \geq R\varepsilon$. Figure 2.4 shows the profiles of the mollified kernels given by formulas (2.3.3) and (2.3.4) for $\varepsilon = 0.1$. One can see that the choice of high-order cutoff essentially translates into sharper short-range particle interactions. Table 2.1 indicates some additional cutoff shapes with their order and the formulas of the associated kernels. Note that the first cutoff, introduced by Chorin [49], is not bounded, but does give a continuous kernel. Note also that the algebraic fourth-order cutoff given in the Table 2.1 is not strictly speaking fourth-order, as the fourth-order moment is not finite.

2.4. Particle Initializations

We describe here several ways to initialize particles in order to approximate the exact initial vorticity field as accurately as possible.

A natural choice is to draw cells of uniform size h inside the support of ω_0 and to initialize a particle at the center of each of these cells. The circulation assigned

Table 2.1. *Examples of cutoff functions and mollified kernels*

ζ	$\mathbf{K}_\varepsilon$	Order
$\begin{cases} \frac{1}{2\pi r} & r \le 1 \\ 0 & r > 1 \end{cases}$	$\begin{cases} \frac{(-y,x)}{2\pi\varepsilon} & r \le \varepsilon \\ \frac{(-y,x)}{2\pi r^2} & r > \varepsilon \end{cases}$	2
$\frac{2}{\pi} \frac{2-r^2}{(1+r^2)^4}$	$\frac{(-y,x)}{2\pi} \frac{r^4+3(\varepsilon r)^2+4\varepsilon^4}{(\varepsilon^2+r^2)^3}$	4
$\frac{1}{2\pi}(4 - r^2)e^{-r^2}$	$\frac{1}{2\pi} \frac{(-y,x)}{r^2} \left[1 + (r^2/\varepsilon^2 - 1)\exp\left(\frac{-r^2}{\varepsilon^2}\right)\right]$	4
$\frac{1}{\pi}(6 - 6r^2 + r^4)e^{-r^2}$	$\frac{(-y,x)}{2\pi r^2}\left[1 - \left(1 - 2\frac{r^2}{\varepsilon^2} + \frac{r^4}{2\varepsilon^4}\right)e^{-r^2/\varepsilon^2}\right]$	6

to this point is the local value of the vorticity multiplied by the volume h^d of the cell, where $d = 2, 3$ is the dimension. If this quantity is available, one can also directly assign the integral of the vorticity on the cell. A third possibility consists of choosing the particle circulations such that the mollified vorticity, as given by formula (2.2.11), takes the prescribed values at the particle locations; this variant will be discussed in more detail in the context of circulation processing schemes in Chapter 7.

The accuracy of these initializations relies very much on the order of quadrature formulas in which particle locations are used as quadrature points (see Section A.1). The first choice is related to the midpoint rule. For periodic or unbounded geometries, it is thus infinite-order accurate in the sense that the distance in some appropriate distribution space between the exact vorticity and its particle approximation can be bounded by Ch^m for all m, provided the vorticity has derivatives of an order up to m bounded. For rectangular domains without periodicity assumptions, the midpoint rule is only second order, but higher-order initializations can be obtained if initial particle locations coincide with quadrature points associated with Gauss-type quadrature formulas. Obviously what we just said applies as well to all geometries that can be smoothly mapped into rectangles; in this case the actual particles will of course have to be images by the inverse mapping of points on a uniform (or Gauss-type) grid.

A second class of initializations is based on random choices. The first method consists of initializing particles randomly in the support of the vorticity and assigning them the local value of the vorticity multiplied by the average volume around the particle. We evaluate this volume by dividing the size of the support of the vorticity by the number of particles. The convergence of this initialization relies on the laws of large-numbers (see Lemma A.1.3 in Appendix A). The rate of convergence, defined in a statistical sense, is governed by $1/\sqrt{N}$, where N is the number of particles. Since in two dimensions N is of the order of h^{-2}, where h is the average spacing of the particles, this convergence rate compares poorly against the deterministic choice. The random choice, however, can prove to be useful in case there is no guarantee of any kind concerning the smoothness of the vorticity.

It is finally possible to combine deterministic and random choices. One first splits the vorticity support in cells of size h, then chooses randomly one particle inside each cell. In two dimensions, it is possible to prove (see Lemma A.1.4) that, still in a statistical sense, one gets second order just as for the midpoint formula in rectangular domains. However, this applies to C^1 functions, instead of C^2 for the midpoint formula. For vortex methods, it turns out that somehow this kind of smoothness can be considered as free, because of the regularization effect of the velocity computation (this argument is more developed in the convergence proof of Section 2.6 below). The quadrature estimate related to this latter kind of initialization, which we will call quasi-random, can thus be seen as optimal.

To evaluate the relative performance of deterministic and random choices let us now give two simple numerical experiments. In the first one we have chosen an initial vorticity field with radial symmetry, leading therefore to a stationary solution of the Euler equations in the plane. It is given by

$$\omega_0(\mathbf{x}) = (1 - |\mathbf{x}|^2)^3 \quad \text{if } |\mathbf{x}| \leq 1 , \qquad \omega_0(\mathbf{x}) = 0 \quad \text{if } |\mathbf{x}| > 1.$$

Examples of this type have become classical tests of the accuracy of vortex methods. Particle trajectories would be circles if the motion equations were solved exactly. However, because of the gradient of the vorticity, the various circles move with different velocities, yielding after some time strong distortions in the particle distribution. One can observe in Figure 2.5 that if particles are initialized in a completely deterministic way, the computation of the velocity at the early stages is rather accurate, because this choice takes advantage of all the regularity of the initial vorticity. However, for longer times, as strong shears develop in the particle distribution, the accuracy deteriorates, then remains more or less steady. If one uses the mixed random–deterministic method, it appears clearly that the error is bigger initially but remains rather stable, so that its accuracy turns out to be comparable with the completely deterministic case. As for the completely random choice, the accuracy is also stable, but lower than in the two other choices. All calculations are based on a mesh size $h = 0.1$, resulting in ~ 300 particles in support of the vorticity, a fourth-order cutoff function with $\varepsilon = 2h$, and a fourth-order Runge–Kutta time-stepping scheme with a time step $\Delta t = 1$. An explanation of these results is that, although the vorticity is smooth, the flow map giving the particle locations develops large derivatives that deteriorate the quadrature estimates as time goes on. This will be further

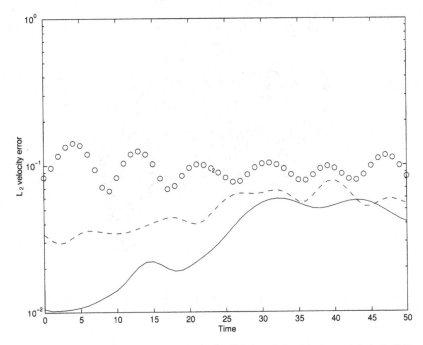

Figure 2.5. Error curves for the evolution of a circular patch with deterministic (solid), quasi-random (dashed), and random (circles) initializations.

illustrated by our numerical analysis in Section 2.6, and accuracy issues related to particle distortions will be discussed in more detail in Chapter 7. The saturation in the deterioration of the accuracy observed for the deterministic initialization can be explained by a transition between a high-order quadrature formula requiring smoothness and a low-order one that is valid for nonsmooth data.

A visual explanation of the good performance of random choices can be given by consideration of the evolution of a non-uniform elliptical vortex. This type of initial condition produces even stronger shears than the radial vortex just considered, resulting in the ejection of thin filaments that are difficult to capture. Besides its own interest, this test is enlightening as it produces features that are generic to many flows of interest (shear layers, boundary layers, etc.). Figure 2.6(a) shows that a uniform initialization results in a layered distribution of the particles, with large holes that indicate that the flow is not correctly resolved everywhere. On the contrary, the quasi-random and random choices [Figures 2.6(b) and 2.6(c), respectively] maintain a homogeneous distribution of points for a longer time. We will also come back to this test in Chapter 7.

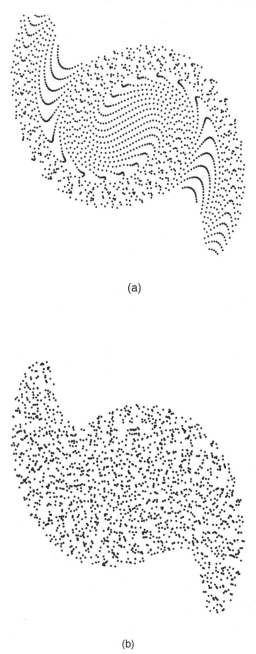

(a)

(b)

Figure 2.6. Locations of particles initialized on an elliptical vorticity profile with (a) a deterministic, (b) a quasi-random, or (c) completely random formula.

(c)

Figure 2.6. (*Continued*)

2.5. The Case of No-Through-Flow or Periodic Boundary Conditions

We have so far considered the case of flows evolving in free space. For other boundary conditions, one must modify the Biot–Savart law to take into account the desired boundary condition. The case of computational domains with solid walls will be the topic of Chapter 4. Our goal here is to outline only the modifications that must be done in the definitions of vortex methods in two simple cases, which will be analyzed in Section 2.6: the case of a bounded flow in a simply connected domain and the case of flows in periodic boxes.

In the first case, we assume that the normal component of the velocity is zero, the so-called no-through-flow boundary condition. In terms of the stream function ψ such that $\mathbf{u} = \nabla \times \psi$, this means that the tangential derivative of ψ vanishes at the boundary. Since the boundary has a single connected component, ψ is a constant, which can be set to zero as ψ is obviously determined up to an additive constant, at the boundary. The relationship $\omega = \nabla \times \mathbf{u}$ is equivalent to $-\Delta \psi = \omega$ and we are finally led to the system

$$-\Delta \psi = \omega \quad \text{in } \Omega, \qquad (2.5.1)$$

$$\psi = 0 \quad \text{on } \partial \Omega. \qquad (2.5.2)$$

This system can be given an integral solution that, incorporates, in addition to the vorticity, a boundary term. The procedures to determine this boundary term properly will be described in Chapter 4.

Replacing ω in the above system with the vortex blobs means that the blobs at a distance less than ε from the boundary will be chopped there. One may of course wonder about the effect of this truncation on the overall accuracy of the method. It turns out that, if the initial vorticity vanishes sufficiently fast at the boundary, this effect is negligible. This is clearly seen if one considers the particular case in which ω_0 has a compact support inside the domain. In this case the support of the vorticity remains well inside the computational domain for all finite time, and it suffices to choose ε small enough to make sure that blobs that might intersect the boundary actually carry no vorticity at all, so that the chopping step does not affect the computation of the vorticity. In the general case $\omega_0 \in W_0^{m,\infty}(\Omega)$ (that is, when the vorticity and its derivatives vanish at the boundary) it indeed results from the analysis in Section 2.6 below.

Let us now turn to the case of periodic boundary conditions. If one has to solve the Euler equation in a square Ω of size L with periodic boundary conditions, the definitions given in Section 2.2 apply with the minor modification of replacing the kernel $\mathbf{K}$ with its periodic version $\mathbf{K}_L$:

$$\mathbf{K}_L(\mathbf{x}) = \sum_{\mathbf{k} \in Z^2} \mathbf{K}(x + \mathbf{k}L).$$

Actually the infinite sum

$$\mathbf{u}(\mathbf{x}) = \int_\Omega \mathbf{K}_L(\mathbf{x} - \mathbf{y})\omega(\mathbf{y}) \, d\mathbf{y}$$

is finite as soon as ω has zero mean. This condition is clearly necessarily satisfied by a periodic vorticity field, as shown by a straightforward integration by parts of $\omega = \text{curl } \mathbf{u}$ in Ω. To check that it is sufficient to guarantee the finiteness of the sum, we first observe that we can rewrite $\mathbf{u}$ as

$$\mathbf{u}(\mathbf{x}) = \int_\Omega [\mathbf{K}_L(\mathbf{x} - \mathbf{y}) - \mathbf{K}(\mathbf{x})]\omega(\mathbf{y}) \, d\mathbf{y}.$$

If we now consider the first component of the velocity, we note that combining the terms corresponding to values of $\mathbf{k}$ of the form (k_1, k_2) and $(k_1, -k_2)$ leads us to consider the series

$$\sum_{\mathbf{k}, k_2 > 0} \frac{x_2 - y_2}{|\mathbf{x} - \mathbf{y} - \mathbf{k}L|^2} - \frac{x_2}{|\mathbf{x} - \mathbf{y} - \mathbf{k}L|^2}.$$

This series is clearly bounded by

$$C \sum_{\mathbf{k}} \frac{|\mathbf{y}|}{|\mathbf{x} - \mathbf{k}L|^3}, \qquad (2.5.3)$$

which converges uniformly for $\mathbf{y} \in \Omega$.

The definition of vortex methods for periodic boundary conditions differs from the free-space methods only by the computation of the velocity, which is based on periodic kernels that are mollified versions of $\mathbf{K}_L$.

In practice it means that one has to initialize particles $\mathbf{x}_p$, $|p| \leq N$, in Ω and, whenever the computation of the velocity is needed, to consider image particles $\{\mathbf{x}_p^h + \mathbf{k}L, \mathbf{k} \in \mathbf{Z}^2\}$. The velocity at a particle $\mathbf{x}_p^h$ is then given by

$$\sum_{\mathbf{k} \in \mathbf{Z}^2} \sum_{|q| < N} \alpha_q \mathbf{K}_\varepsilon \left(\mathbf{x}_p^h - \mathbf{x}_q^h + \mathbf{k}L \right). \qquad (2.5.4)$$

If a particle happens to leave the computational box, it has to be replaced with its image that enters the box on the opposite side. Of course, the zero-mean condition on the vorticity needed for the velocity to be finite in the continuous case must be translated here into the the the discrete condition $\sum_p \alpha_p = 0$. Also, the symmetry property of $\mathbf{K}$ needed to prove the finiteness of $\mathbf{K}_L$ must be satisfied by the mollification $\mathbf{K}_\varepsilon$. This requires that the cutoff function ζ be even, which in practice is always the case.

Finally, one must note that the sum on the right-hand side of expression (2.5.4) can be truncated. Given an error tolerance of δ, since the remainder of series (2.5.3) decays like $1/|\mathbf{k}|L$, one can drop the image particles in boxes of indices $\mathbf{k}$ such that $|\mathbf{k}L| > \delta^{-1}$.

The case that was considered at the beginning of this chapter, that of a domain extending to infinity in one direction and periodic in the other direction, can be recovered as well. It is indeed simpler, because in this case the periodic kernel can be given in a closed form. In complex notations $z = x_1 + i x_2$ one can write

$$\mathbf{K}(x_1, x_2) = \frac{1}{2i\pi z}.$$

To derive a kernel that is periodic in the x_1 direction, we add the contributions of image boxes in this direction, and we obtain

$$\mathbf{K}_p(x_1, x_2) = \frac{1}{z} + 2\sum_{k=1}^{\infty} \frac{1}{z-k} = \frac{1}{z} + 2\sum_{k=1}^{\infty} \frac{z+k}{z^2-k^2}$$

$$= \frac{1}{z} + 2\sum_{k=1}^{\infty} \frac{z}{z^2-k^2} = \frac{1}{2} \cot(\pi z).$$

A little algebra to develop $\cot(\pi z)$ back into real coordinates allows us to end up precisely with the formulas used in Section 2.1.

2.6. Convergence and Conservation Properties

It is on purpose that we have put together the analysis of conservation and convergence features of vortex methods. While convergence analysis is the classical way to assess the validity of a numerical method, it does not tell everything about it, in particular because convergence results are true in only the asymptotic sense, that is, in the limit of small numerical parameters, while simulations often cannot afford high resolution. From the fluid mechanics point of view, a high-order numerical method that, for example, would not conserve the total circulation might be useless at low resolution.

The most striking feature of vortex methods, which has triggered interest among fluid dynamicists, is its built in conservation of many inviscid flow invariants. In some sense the conservation properties of vortex methods are a guarantee that, even if used in underresolved simulations, they give a correct qualitative answer. On the other hand, conventional numerical analysis, together with the extension of vortex methods to more general than simple inviscid two-dimensional flows, has given them the status of plain numerical methods, rather than merely fluid models, and allowed us to understand and quantify the role of the numerical parameters that govern the method.

This section is organized as follows: We first give a general convergence result based on the mathematical framework described in Appendix A. We then review other convergence results that correspond to limit cases of the general theory. Note that here we are concerned essentially with the theoretical aspects of the convergence of vortex methods. Numerical illustrations will be given in Chapter 7. We finally derive the conservation properties of vortex methods.

2.6.1. A General Convergence Result

The first complete convergence analysis of vortex method was given by Hald in 1979. This analysis [98] in particular showed for the first time the relative roles of the numerical parameters involved in the method, namely the particle spacing and the cutoff width. This convergence result was restricted, however, to rather severe overlapping conditions and cutoff assumptions. It was generalized in 1982 independently by Beale and Majda [25] and Cottet [58]. Beale and Majda included in their analysis [25] a three-dimensional scheme that will be discussed in Chapter 3, while Cottet [58] focused on a minimal smoothness requirement for the cutoff function and on a simplified consistency analysis relying on quadrature estimates. Anderson and Greengard included in 1985 the time discretization in their analysis [8] and described a totally grid-free three-dimensional scheme, which was later analyzed by Beale [19].

In all these references the analyses were based on estimations of how close numerical particles could remain from exact tracers of the flow, making heavy use of the exact and approximated flow maps. We follow here an alternative approach suggested in Ref. 60 and further developed in Refs. 44, 45. This analysis is based on the notion of weak solutions to advection equations, developed in Appendix A and does not explicitly involve the particle trajectories. The reason why we choose this point of view is that it easily extends to three-dimensional schemes, including filament methods, as well as vortex-in-cell schemes in which the particle velocity is based on a grid solver rather than on the Biot–Savart law. Another nice feature is that, while traditional analysis was always based on the assumption of a very smooth underlying flow, this approach allows us to account optimally for a given smoothness of the flow. More precisely, it shows that vortex methods enjoy the best rate of convergence one might expect: if the derivatives of order up to $r - 1$ of the vorticity are bounded and if one uses a cutoff of order r, the discrepancy between computed and exact velocity decays like ε^r, very much as for a grid-based method of the same order.

For simplicity our convergence result will be stated in the case of periodic boundary conditions. However, it will be clear from its proof that it applies as well to bounded domains with no-through-flow boundary conditions, provided the initial vorticity vanishes at the boundary and the calculation of the velocity from the vorticity incorporates the constraint of zero normal velocity at the boundary. We will also indicate how its proof has to be modified to address the free-space problem.

Finally we will restrict ourselves to the case of C^∞ cutoff functions. We have actually shown in Section 2.3 that in practice this kind of cutoff is natural and easy to implement, so there is no reason to focus on less regular cases.

The numerical error in vortex methods comes basically from two sources: the regularization needed in the computation of the velocity and the particle approximation. It is easy to control the first one in terms of the cutoff order and of the smoothness of the flow. To show that the second one is negligible compared with the regularization error, we will need to construct the particle approximation from a regularized initial vorticity field instead of the original one. This will introduce an additional regularization error, but of the same order as the one involved in the reconstruction of the velocity and will allow us to use high-order quadrature estimates to control the particle approximation. Moreover, the analysis will show that there is no added price, in terms of the error estimate, to pay for this regularization. We will comment further at the end of this section on some practical aspects of this initialization technique.

Let us now come to the precise definition of the method we plan to analyze. From now on, we make constant use of the notations and results given in Appendix A. Let $\Omega =]0, 1[^2$, $r \geq 1$, and let ω_0 be a periodic function in Ω. Let ζ be a C^∞ cutoff function of order r. We define the periodic convolution of a periodic function (or distribution) f defined on Ω with a function g defined in $\mathbf{R}^2$ by $g \star_p f = g \star \bar{f}$, where $\bar{f}$ denotes the periodic extension of f in $\mathbf{R}^2$. Then, if ζ is a cutoff function, we set

$$\zeta_\varepsilon(\mathbf{x}) = \varepsilon^{-2}\zeta(\mathbf{x}/\varepsilon), \quad \omega_0^\varepsilon = \zeta_\varepsilon \star_p \omega_0, \quad \mathbf{K}_\varepsilon = \mathbf{K} \star \zeta_\varepsilon.$$

We now initialize particles $\mathbf{x}_p$ on a uniform mesh of size h in the box Ω and we set

$$\alpha_p = h^2\omega_0^\varepsilon(\mathbf{x}_p), \quad \omega_0^h(\mathbf{x}) = \sum \alpha_p \delta(\mathbf{x} - \mathbf{x}_p)$$

(this is the deterministic initialization described in Section 2.4; for random initializations, as far as we know, the convergence of the genuine vortex methods is an open problem, even in the smooth case; however, see Subsection 2.6.2 below for some indications of the convergence of a modified vortex scheme with random choice).

The numerical method is then defined, as explained in Section 2.2 through Eq. (2.2.9), for the motion of the particles, with a velocity field defined by $\mathbf{u}^h = \mathbf{K}_\varepsilon \star_p \omega^h$.

Theorem 2.6.1. *Let $r \geq 2$ and $T > 0$. Assume that $\omega_0 \in W^{r-1,\infty}(\Omega)$ and that ζ is a cutoff function of the order of r, in the sense of Eqs. (2.3.1). Assume further that there exists $s > 0$ such that $h \leq \varepsilon^{1+s}$; then for h small enough the following estimate holds for $t \in [0, T]$.*

$$\|(\mathbf{u} - \mathbf{u}^h)(\cdot, t)\|_{L^p(\Omega)} \leq C(T)\varepsilon^r.$$

The condition linking the parameters h and ε means that the blobs around the particles must overlap, with, at least in principle, an increase in overlapping as ε tends to 0. We will see below that actually this condition can be relaxed to some extent and that the vortex method without smoothing can be proved to be convergent.

In our proof we wish to clearly distinguish between the regularization and the particle discretization error. To this end, we introduce the auxiliary problem

$$\frac{\partial \omega^\varepsilon}{\partial t} + \text{div}(\mathbf{u}^\varepsilon \omega^\varepsilon) = 0, \qquad (2.6.1)$$

$$\mathbf{u}^\varepsilon = \mathbf{K}_\varepsilon \star_p \omega^\varepsilon, \qquad (2.6.2)$$

$$\omega^\varepsilon(\cdot, 0) = \omega_0^\varepsilon. \qquad (2.6.3)$$

We first prove Lemma 2.6.2.

Lemma 2.6.2. *Under the assumptions of Theorem 2.6.1 we have for* $t \in [0, T]$

$$\|\omega^\varepsilon\|_{r-1,\infty} \leq C(T), \qquad (2.6.4)$$

$$\|(\mathbf{u} - \mathbf{u}^\varepsilon)(\cdot, t)\|_{0,p} \leq C(T)\varepsilon^r. \qquad (2.6.5)$$

Proof. Concerning relation (2.6.4), note that, for $\varepsilon = 0$, this is the classical regularity result for the Euler equation that merely results from the combination of the advection of the vorticity equation and the regularizing property of the convolution by $\mathbf{K}$ (see Section A.3). For $\varepsilon \neq 0$, it thus suffices to observe that $\mathbf{K}_\varepsilon$ has the same smoothing property as $\mathbf{K}$, uniformly with respect to ε, and that $\|\omega_0^\varepsilon\|_{r-1,\infty} \leq \|\omega_0\|_{r-1,\infty}$.

To prove relation (2.6.5) let us first focus on the initialization error. We claim that for all finite p

$$\left\|\omega_0 - \omega_0^\varepsilon\right\|_{-1,p} \leq C\varepsilon^r. \qquad (2.6.6)$$

Using the definition of ω_0^ε and the fact that ζ has integral 1, we get

$$\omega_0(\mathbf{x}) - \omega_0^\varepsilon(\mathbf{x}) = \sum_{\mathbf{k}\in\mathbf{Z}^2} \int_\Omega [\omega_0(\mathbf{x}) - \omega_0(\mathbf{y})]\zeta_\varepsilon(\mathbf{y} - \mathbf{x} + \mathbf{k}L)\,d\mathbf{y}. \qquad (2.6.7)$$

We now develop $\omega_0(\mathbf{x}) - \omega_0(\mathbf{y})$ by the Taylor formula with the integral remainder to get, after cancellations due to moment properties (2.3.1) of ζ,

$$\omega_0(\mathbf{x}) - \omega_0^\varepsilon(\mathbf{x})$$
$$= \sum_{\mathbf{k}\in\mathbf{Z}^2} \int_0^1 \int_\Omega \{(\mathbf{y} - \mathbf{x})\nabla\omega_0[\mathbf{x} + t(\mathbf{y} - \mathbf{x})]\}^r \zeta_\varepsilon(\mathbf{y} - \mathbf{x} + \mathbf{k}L)\,d\mathbf{y}\,dt,$$

where we have used the notation $[\mathbf{x}\nabla f]^m = \sum_{\alpha+\beta=m} C_m^\alpha x_1^\alpha (\partial^\alpha f/\partial x_1^\alpha) x_2^\beta (\partial^\beta f/\partial x_2^\beta)$.

If we now insert this expansion into Eq. (2.6.7) and integrate it against a test function ϕ, we obtain

$$\int_\Omega \left[\omega_0(\mathbf{x}) - \omega_0^\varepsilon(\mathbf{x})\right]\phi(\mathbf{x})\,d\mathbf{x}$$
$$= \sum_k \int_\Omega \int_\Omega \int_0^1 [\mathbf{z}\nabla\omega_0(\mathbf{x} + t\mathbf{z})]^r\,\phi(\mathbf{x})\zeta_\varepsilon(\mathbf{z} + \mathbf{k}L)\,d\mathbf{x}\,d\mathbf{z}\,dt. \qquad (2.6.8)$$

We then integrate by parts once with respect to $\mathbf{x}$ all the derivatives appearing when developing $[\mathbf{z}\nabla\omega_0(\mathbf{x} + t\mathbf{z})]^r$ to get

$$\left|\int_\Omega \left[\omega_0(\mathbf{x}) - \omega_0^\varepsilon(\mathbf{x})\right]\phi(\mathbf{x})\,d\mathbf{x}\right| \le C\|\omega_0\|_{r,p}|\phi|_{1,p^*}\sum_k \int_\Omega |\mathbf{z}|^r|\zeta_\varepsilon(\mathbf{z} + \mathbf{k}L)|d\mathbf{z}.$$
$$(2.6.9)$$

where p^* is such that $1/p + 1/p^* = 1$.

It remains now to use the change of variable $\mathbf{z}' = \mathbf{z}/\varepsilon$ in the above integral to obtain

$$\left|\int_\Omega \left[\omega_0(\mathbf{x}) - \omega_0^\varepsilon(\mathbf{x})\right]\phi(\mathbf{x})\,d\mathbf{x}\right| \le C\varepsilon^r|\phi|_{1,p^*}\int_{\mathbf{R}^2} |\mathbf{z}|^r|\zeta(\mathbf{z})|\,d\mathbf{z}, \qquad (2.6.10)$$

which proves our claim.

Let us next set $e = \omega - \omega^\varepsilon$. Subtracting Eqs. (2.6.1)–(2.6.3) from the Euler equations satisfied by $(\mathbf{u}, \omega)$ we get

$$\frac{\partial e}{\partial t} + \text{div}(\mathbf{u}^\varepsilon e) = \text{div}[(\mathbf{u}^\varepsilon - \mathbf{u})\omega], \qquad (2.6.11)$$

$$e(\cdot, 0) = \omega_0 - \omega_0^\varepsilon. \qquad (2.6.12)$$

Since, by relation (2.6.4), the derivatives of $\mathbf{u}^\varepsilon$ are bounded uniformly with respect to ε, we can apply Theorem A.2.7 [estimate (A.2.12) with $m = 1$, $A = 1$] to obtain

$$\|e(\cdot, t)\|_{-1,p} \le C\left[\|e_0\|_{-1,p} + \int_0^t \|(\mathbf{u} - \mathbf{u}^\varepsilon)(\cdot, s)\|_{0,p}\,ds\right]. \qquad (2.6.13)$$

Now we can write

$$\mathbf{u} - \mathbf{u}^\varepsilon = (\mathbf{K} - \mathbf{K}_\varepsilon) \star_p \omega + \mathbf{K}_\varepsilon \star_p e = \mathbf{K} \star_p (\omega - \zeta_\varepsilon \star_p \omega) + \mathbf{K}_\varepsilon \star_p e.$$

From Calderon's theorem [estimate (A.3.6)], and the arguments already used for the proof of relation (2.6.6) above we get that:

$$\|\mathbf{K} \star_p (\omega - \zeta_\varepsilon \star_p \omega)\|_{0,p} \le C \|\omega - \zeta_\varepsilon \star_p \omega\|_{-1,p} \le C\varepsilon^r \|\omega\|_{r-1,\infty},$$

and thus

$$\|\mathbf{u} - \mathbf{u}^\varepsilon\|_{0,p} \le C(\varepsilon^r + \|e\|_{-1,p}). \tag{2.6.14}$$

Combining relations (2.6.6), (2.6.13), and (2.6.14) yields

$$\|e(\cdot, t)\|_{-1,p} \le C \left[\varepsilon^r + \int_0^t \|(e)(\cdot, s)\|_{-1,p} \, ds \right].$$

Our estimate (2.6.5) then follows by Gronwall's theorem. □

We can now give the proof of Theorem 2.6.1.

Proof of Theorem 2.6.1. Let $h > 0$ and fix m large enough so that $ms > r$. We are going to prove that

$$\|\mathbf{u}^\varepsilon - \mathbf{u}^h\|_{0,p} = O(h^m \varepsilon^{1-m}). \tag{2.6.15}$$

Together with relation (2.6.5) and the overlapping condition $h \le \varepsilon^{1+s}$, this will prove our result.

Since $r > 1$, in view of relation (2.6.4) we know that, for $t \le T$, $\|\mathbf{u}^\varepsilon\|_{1,\infty}$ is bounded uniformly with respect to ε; moreover, using the smoothing properties of the kernel $\mathbf{K}_\varepsilon$ [estimate (A.3.7)], we can write $|\mathbf{u}^\varepsilon(\cdot, t)|_{k,\infty} = O(\varepsilon^{1-k})$. In other words, there exists a constant C^* such that

$$\|\mathbf{u}^\varepsilon(\cdot, t)\|_{1,\infty} + \sum_{k=2}^m \varepsilon^{k-1} |\mathbf{u}^\varepsilon(\cdot, t)|_{k,\infty} \le C^*.$$

We then introduce the time T_h^* defined as

$$T_h^* = \max \left[t > 0; \|\mathbf{u}^h(\cdot, t)\|_{1,\infty} + \sum_{k=2}^m \varepsilon^{k-1} |\mathbf{u}^h(\cdot, t)|_{k,\infty} \le 2C^* \right]. \tag{2.6.16}$$

If we denote by e^h the vorticity error $\omega^\varepsilon - \omega^h$, we readily see that e^h satisfies

$$\frac{\partial e^h}{\partial t} + \text{div}(\mathbf{u}^h e^h) = \text{div}[(\mathbf{u}^h - \mathbf{u}^\varepsilon)\omega^\varepsilon], \tag{2.6.17}$$

$$e^h(\cdot, 0) = \omega_0^\varepsilon - \omega_0^h. \tag{2.6.18}$$

We can therefore write $e^h = \alpha^h + \beta^h$, where

$$\frac{\partial \alpha^h}{\partial t} + \mathrm{div}(\mathbf{u}^h \alpha^h) = 0, \quad \alpha_0^h = \alpha^h(\cdot, 0) = \omega_0^\varepsilon - \omega_0^h, \qquad (2.6.19)$$

$$\frac{\partial \beta^h}{\partial t} + \mathrm{div}(\mathbf{u}^h \beta^h) = \mathrm{div}[(\mathbf{u}^h - \mathbf{u}^\varepsilon)\omega^\varepsilon], \quad \beta^h(\cdot, 0) = 0. \qquad (2.6.20)$$

The term α^h reflects the influence of the particle discretization, whereas the control of β^h is related to the stability of the method.

Let us first analyze α^h.

If ϕ is a smooth test function, in view of Lemma A.1.2 we can write

$$|\langle \alpha_0^h, \phi \rangle| = \left| \int_\Omega \omega_0^\varepsilon \phi \, d\mathbf{x} - \sum_p h^2 \omega_0^\varepsilon(\mathbf{x}_p)\phi(\mathbf{x}_p) \right| \le C h^m |\omega_0^\varepsilon \phi|_{m,1}.$$

If we develop the derivatives of $\omega_0^\varepsilon \phi$, we deduce that

$$|\langle \alpha_0^h, \phi \rangle| \le C h^m \sum_{k=0}^m |\phi|_{k,p^*} |\omega_0^\varepsilon|_{m-k,p},$$

where, as usual, we have denoted by p^* the conjugate exponent to p. Next, recall that ω_0^ε has been obtained through a regularization of ω_0 by the cutoff function ζ_ε and that, by assumption, ω_0 is at least in $W^{1,\infty}(\Omega)$, whence $|\omega_0^\varepsilon|_{k,p^*} \le C\varepsilon^{-k}$, for $k \le m$, and $|\omega_0^\varepsilon|_{m,p^*} \le C\varepsilon^{1-m}$. Thus

$$|\langle \alpha_0^h, \phi \rangle| \le C h^m \left(\sum_{k=1}^m \varepsilon^{-m+k} |\phi|_{k,p^*} + \varepsilon^{1-m} \|\phi\|_{0,p^*} \right).$$

Thanks to Lemma A.2.8, this allows us to split α_0^h into m components as follows:

$$\alpha_0^h = \sum_{k=0}^m \bar{\alpha}_k^h, \quad \bar{\alpha}_k^h \in W^{-k,p}(\Omega);$$

$$\begin{cases} \|\bar{\alpha}_0^h\|_{0,p} \le C h^m \varepsilon^{-m+1} \\ \|\bar{\alpha}_k^h\|_{-k,p} \le C h^m \varepsilon^{-m+k} \text{ for } 1 \le k \le m \end{cases}.$$

Since problem (2.6.19) is linear, this decomposition in turn leads to a decomposition of α in components α_k^h solutions of

$$\frac{\partial \alpha_k^h}{\partial t} + \mathrm{div}\left(\mathbf{u}^h \alpha_k^h\right) = \mathrm{div}[(\mathbf{u}^h - \mathbf{u}^\varepsilon)\omega^\varepsilon], \quad \alpha_k^h(\cdot, 0) = \bar{\alpha}_k^h.$$

In view of the definition of T_h^* in Eq. (2.6.16), we can use Theorem A.2.7 [estimate (A.2.12)] with $A = \varepsilon^{-1}$, $\mathbf{a} = \mathbf{u}^h$, and $\mathbf{a}_0 = 0$ and write each of these components as a sum of smoother components:

$$\alpha_k^h = \sum_{0 \le l \le k} \alpha_{k,l}^h, \quad \alpha_{k,l}^h \in W^{-l,p}(\Omega),$$

with

$$\|\alpha_{0,0}^h\|_{0,p} \le Ch^m \varepsilon^{-m+1},$$

$$\varepsilon^{k-1} \|\alpha_{k,0}^h\|_{0,p} + \sum_{1 \le l \le k} \varepsilon^{k-l} \|\alpha_{k,l}^h\|_{-l,p} \le C \|\alpha_k^h\|_{-k,p} \le Ch^m \varepsilon^{-m+k},$$

$$\text{for } 1 \le k \le m. \quad (2.6.21)$$

The above decomposition gives us a precise account for the consistency of the particle discretization.

Let us now turn to the stability analysis. From Eq. (2.6.20) and Theorem A.2.7 we infer that, for $t \le T_h^*$,

$$\|\beta^h(\cdot, t)\|_{-1,p} \le C \int_0^t \|(\mathbf{u}^\varepsilon - \mathbf{u}^h)(\cdot, s)\|_{0,p} \|\omega^\varepsilon(\cdot, s)\|_{0,\infty} \, ds. \quad (2.6.22)$$

By relation (2.6.4), $\|\omega^\varepsilon\|_{0,\infty}$ is bounded independently of ε. Moreover,

$$\mathbf{u}^\varepsilon - \mathbf{u}^h = \mathbf{K}_{\varepsilon *_p} e^h = \mathbf{K}_{\varepsilon *_p} \left(\beta^h + \sum_{0 \le l \le k, 0 \le k \le m} \alpha_{k,l}^h \right),$$

and we can use the smoothing properties of the kernel $\mathbf{K}_\varepsilon$ [estimate (A.3.7)] to write, for $t \le T_h^*$,

$$\|\mathbf{u}^\varepsilon - \mathbf{u}^h\|_{0,p} \le C \left[\|\beta^h\|_{-1,p} + \sum_{k,l} \|\alpha_{k,l}\|_{-l,p} \max(1, \varepsilon^{1-l}) \right]. \quad (2.6.23)$$

Now straightforward calculations starting from relations (2.6.21) precisely lead to

$$\sum_{k,l} \|\alpha_{k,l}\|_{-l,p} \max(1, \varepsilon^{1-l}) \le Ch^m \varepsilon^{1-m}.$$

Therefore, combining relations (2.6.22) and (2.6.23) yields, still for $t \le T_h^*$,

$$\|(\mathbf{u}^\varepsilon - \mathbf{u}^h)(\cdot, t)\|_{0,p} \le C \left[h^m \varepsilon^{1-m} + \int_0^t \|(\mathbf{u}^\varepsilon - \mathbf{u}^h)(\cdot, s)\|_{0,p} \, ds \right]. \quad (2.6.24)$$

By Gronwall's theorem, this proves that Eq. (2.6.15) holds up to time T_h^*.

It remains now to prove that actually $T_h^* = T$. For that we differentiate our estimates at time T_h^*. From relation (2.6.22) we deduce that

$$\|\beta^h\|_{-1,p} + \sum_{k,l} \|\alpha_{k,l}\|_{-l,p} \min(1, \varepsilon^{1-l}) \leq Ch^m \varepsilon^{1-m}.$$

Our cutoff function is C^∞ so that we can differentiate the velocities as much as we wish. Moreover, recall that $h = \varepsilon^{1+s}$ and we have chosen m such that $ms > r$. Thus

$$\|\mathbf{u}^\varepsilon - \mathbf{u}^h\|_{k,p} \leq Ch^m \varepsilon^{1-m-k} \leq C\varepsilon^{r-k}.$$

We now use the fact that, for $p > 2$, $W^{k+1,p}(\Omega) \subset W^{k,\infty}(\Omega) \subset W^{k,p}(\Omega)$, and, by interpolation [estimate (A.3.5)], we deduce from the preceding estimate that, up to time T_h^*,

$$\|\mathbf{u}^\varepsilon - \mathbf{u}^h\|_{k,\infty} \leq C\varepsilon^{r-a-k}$$

for all positive a. This yields

$$\|\mathbf{u}^h(\cdot, t)\|_{1,\infty} + \sum_{k=2}^m \varepsilon^{k-1} |\mathbf{u}^h(\cdot, t)|_{k,\infty}$$

$$\leq \|\mathbf{u}^\varepsilon(\cdot, t)\|_{1,\infty} + \sum_{k=2}^m \varepsilon^{k-1} |\mathbf{u}^\varepsilon(\cdot, t)|_{k,\infty} + C\varepsilon^{r-1-a}.$$

Finally, from our choice of the constant C^*, we obtain, after choosing $a < 1/2$ and ε small enough,

$$\|\mathbf{u}^h(\cdot, T_h^*)\|_{1,\infty} + \sum_{k=2}^m \varepsilon^{k-1} |\mathbf{u}^h(\cdot, T_h^*)|_{k,\infty} \leq 3/2C^*.$$

If we had $T_h^* < T$, by time continuity, the above estimate could allow us to find a larger time t that satisfies the estimate required in the definition of T_h^*, in contradiction with the fact that T_h^* has been chosen to be as large as possible. Therefore $T_h^* = T$, and our proof is completed.   □

Let us now briefly indicate how the same error estimates can be obtained for the Euler equations in the whole space or in domains with solid boundaries.

In the first case, the only technical detail that has to be carefully studied originates in the fact that, because our domain is not bounded anymore, Sobolev estimates have to take into account the fact that a function can decay at infinity less rapidly than its derivatives [see estimate (A.3.4) and the comment following Theorem A.3.1]. One thus needs to handle separately components of the error that are derivatives and components that are not. If we look for error estimates for the velocity in L^p, these later components have to be controlled in L^q, instead of L^p, where q is the Sobolev conjugate exponent to p. Except for this slight modification, the proof goes along the same line as that of the one given above (see Ref. 60 for the details in the case of smooth data).

For the case of a bounded domain with solid walls, let us assume no-through-flow boundary conditions and that $\omega_0 \in W_0^{r-1,\infty}(\Omega)$ (this is the situation indicated at the beginning of Section 2.5). The vorticity remains in the same space for all time and the velocity is then given by the system of equations (2.5.1) and (2.5.2). It is easy to check that the vorticity remains in $W_0^{r-1,\infty}(\Omega)$ for all time. The mollification consists of the convolution with a cutoff function after the vorticity is extended by zero outside Ω. This procedure leads to the same estimates for the regularization error as in the periodic or unbounded cases. To obtain optimal (in terms of the regularity of the flow) error estimates we need to initialize the particles on a smooth vorticity field, as was done in the periodic case, with the additional constraint that it must vanish at the boundary. This can be done through a classical regularization–truncation technique. One then obtains ω_0^ε infinitely smooth, vanishing as well as all its derivatives on $\partial\Omega$ and satisfying $\|\omega_0^\varepsilon - \omega_0\|_{-1,p} \leq C\varepsilon^r$. Estimates (2.6.4) and (2.6.5) are still valid, and the convergence proof can proceed as in the periodic case.

So far, only semidiscrete vortex schemes have been analyzed. A convergence proof for fully discrete schemes, that is, those in which the particle paths are obtained through a time-advancing method, is given in Ref. 8. The remarkable feature of this result is that, whatever method is used, no stability constraint on the time-step conditions the convergence of vortex methods. Unlike grid-based methods, vortex methods do not suffer any Courant-Friedrichs-Levy condition relating the size of the time step to the minimal grid size. For high Reynolds number flows, if one wishes to resolve the viscous scales, this may lead to important computational savings. The convergence rate of the method is $O(\varepsilon^r + \Delta t^s)$, if an sth-order time-stepping method is used. The unconditional stability of vortex methods can easily been understood in light of the convergence proof just given. Let us for example consider the case of a forward Euler time discretization. It is not difficult to realize that the vortex method based on this time discretization of the particle trajectories is a continuous-in-time discretization of a modified Euler equation in which the velocity, instead of being computed at each time, would be frozen at the beginning of every time step. Clearly this modified Euler equation enjoys the same stability and regularity properties as the original system. If one denotes by $\mathbf{u}^{\Delta t}$ its solution and by $\mathbf{u}^{\Delta t,h}$ the velocity for the fully discrete vortex scheme, one thus has

$$\mathbf{u}^{\Delta t} - \mathbf{u}^{\Delta t,h} = O(\varepsilon^r).$$

On the other hand,

$$\mathbf{u}^{\Delta t} - \mathbf{u} = O(\Delta t),$$

which gives the expected convergence result.

Some comments on the result of Theorem 2.6.1 itself are now in order. First, when one looks at its proof, there is a clear distinction between the roles played by the first-order and the higher-order derivatives of the velocity. While there is a compensating effect that makes it unnecessary to control uniformly with respect to the smoothing parameter high-order derivatives, the lack of control of the first derivatives would introduce an $\exp(-C/\varepsilon)$ term in front of all other error terms, which would be catastrophic. Convergence for flows that do not have this minimal smoothness is still an essentially open question. We will briefly survey in Subsection 2.6.2 below some partial answers that have been given to this question. Note that, even for smooth flow, the exponential constant in the error estimates reflects a possible deterioration of the accuracy of the method as time goes on. This important fact, which we have already observed in Section 2.4, is acknowledged in all the numerical studies of vortex methods. We will discuss it in more detail and see how to improve this aspect of vortex methods in Chapter 7.

Let us finally comment on the regularization step introduced at the initialization of the method. Theorem 2.6.1 can also be understood as a convergence result for an infinitely smooth initial vorticity field but one in which the strength of the derivatives of the vorticity is explicitly taken into account. Assume for example that, through a vorticity generation at some boundary in a viscous flow, we have to deal with an initial vorticity field in which small scales related to the viscosity are present, resulting in high values for the derivatives. More precisely, let us assume that

$$|\mathbf{u}|_{1,\infty} \leq C, \quad |\mathbf{u}|_{k,2} \leq C\nu^{\frac{-k+1}{2}} \quad \text{for } k > 1$$

(this scale law is discussed in the context of two-dimensional turbulence in Ref. 100). In other words, we are in a situation, important in practice, in which diffusion plays the role of the regularization artificially introduced in our scheme. The above proof shows that if we choose $\varepsilon \leq C\nu^{1/2}$, the error estimate for the velocity in L^2 is

$$\|\mathbf{u}_\nu - \mathbf{u}^h\|_{0,2} \leq C\varepsilon^r \nu^{-r/2},$$

where we have denoted by $\mathbf{u}_\nu$ the exact solution for a given viscosity, r is the order of our cutoff, and C is a constant independent of ν (of course here we take into account the viscosity for the scales but forget it in the equations of the fluid, so we actually compare numerical and exact solutions of the Euler equations). This means that, if we consider ν as a small parameter, convergence in the treatment of the advection part of the flow (not speaking about the diffusion) will be achieved if $\varepsilon \ll C\nu^{1/2}$, as could have been expected.

Let us also mention some recent results [56] in the same spirit concerning the convergence of particle methods in the linear case, with optimal convergence rates. In this analysis, particles are initialized on a smoothed field but with a mollification scale equal to h instead of ε. In this case the cutoff functions have to satisfy discrete rather than continuous moment properties, and the method is very much reminiscent of regridding strategies, which will be discussed in Section 7.2.

2.6.2. Other Convergence Results

As we have seen, the key ingredient in the convergence proof is the uniform control of first-order derivatives of the velocity. We wish here to address briefly some limit cases when this control cannot be guaranteed.

The lack of control of the flow derivatives can happen for two reasons:

- the continuous flow field itself is not C^1,
- the overlapping condition is not satisfied, which does not allow us to recover uniform control of the computed velocities.

In the first situation, there is one case in which it is possible to bring the analysis back to the smooth case. This is when the lack of estimates for the derivatives of the velocity comes from oscillations in the initial vorticity field. More precisely, following Ref. 77, assume that

$$\omega_0(\mathbf{x}) = f(\mathbf{x}; \mathbf{x}/\delta),$$

where f is a smooth function defined in $[0, 1]^2$ and δ is a small oscillation parameter. Then it is possible to prove the following homogenization result:

$$\mathbf{u} \to \bar{\mathbf{u}} \text{ in } L^2,$$

as $\delta \to 0$, where $\bar{\mathbf{u}}$ denotes the velocity field corresponding to the mean initial vorticity field $\bar{\omega}_0(\mathbf{x}) = \int_{[0,1]^2} f(\mathbf{x}; \mathbf{y}) \, d\mathbf{y}$.

This means that oscillations cancel in the computation of the velocity. E and Hou [77] were able to prove through sharp estimates of the error that the same cancellation property is true for the vortex method. Although there is clearly no uniform bound with respect to δ of the derivatives of the velocity, the approximate velocity field converges strongly to the exact one independently of the oscillation parameter δ, except for some exceptional values of δ (to avoid these exceptional cases, the authors use the notion of convergence that is essentially independent of δ).

If one now tries to attack the problem of lack of control on the derivatives of the velocity from a more general point of view, a natural idea is to try to obtain *a priori* estimates on the smoothed approximate vorticity $\omega_\varepsilon^h = \omega^h * \zeta_\varepsilon$.

If the initial vorticity is in some L^p for $p > 1$ and thus the exact vorticity remains in the same L^p and if one is able to prove that ω_ε^h is bounded in L^p independently of ε and h, then standard compactness arguments will allow one to prove easily the strong convergence of the velocity field.

This is the point of view developed in Ref. 46, in which $p = \infty$. To prove uniform bounds for the vorticity it is crucial to control local accumulations of particles. Assume that ζ has support in the ball of radius 1 and denote by $J^h(t)$ the maximal number of particles in a ball of radius ε; then one clearly has

$$\left\| \omega_\varepsilon^h \right\|_{L^\infty} \leq Ch^2\varepsilon^{-2}J^h(t).$$

The coefficient J^h can be evaluated in terms of the minimal distance between particles, which itself, because of the incompressibility of the flow, depends on the maximal distance $l(t)$ between particles initially at distance h from each other:

$$J^h(t) \leq C\varepsilon^2 h^{-2}[1 + Cl(t)/\varepsilon]^2.$$

If the flow was Lipschitz-continuous, then $l(t)$ would remain of the order of h; however, in our situation the best one can expect is a quasi-Lipschitz approximate flow (that is, Lipschitz up to a logarithmic correction), and thus

$$l(t) \leq Ch^{-t\alpha_h},$$

where α_h denotes a tentative L^∞ bound for the vorticity.

If one assumes an overlapping condition of the type $h \leq C\varepsilon^{1+s}$, a boot-strap argument allows one then to conclude that an L^∞ bound actually holds for the vorticity, up to time T^* such that $(1 + s)\exp(-CT^*) = 1$, for some constant

C. That $T^* \to 0$ if $s \to 0$ translates the fact that the deterioration in the control of the particle dispersion requires more and more overlapping as time goes on.

The same basic strategy (namely trying to obtain enough *a priori* estimates to prove strong convergence on the velocity) has been more recently used in Ref. 37 under weaker regularity assumptions ($\omega_0 \in L^p$ for some $p > 4/3$). The proof makes more extensive use of the incompressibility of the flow. The basic remark is the following: Assume that $p = 2$ and $\mathbf{a}$ is an incompressible velocity field and denote by $\xi \in \Xi$ all possible random initializations (see Section A.1 for a more clear-cut definition), in the unit square. Then one can control the difference between the smoothed approximate and exact vorticities advected by $\mathbf{a}$ in the following sense:

$$\int_{\Xi} \int \left| \frac{1}{N} \sum_{i=1}^{N} \omega_0(\xi_i) \zeta_\varepsilon[\mathbf{x} - \mathbf{X}_a(t; \xi_i, 0)] \right.$$
$$\left. - \int \omega_0(\mathbf{y}) \zeta_\varepsilon[\mathbf{x} - \mathbf{X}_a(t; \mathbf{y}, 0)] \, d\mathbf{y} \right|^2 \, d\mathbf{x} \, d\xi \le \frac{C}{N\varepsilon^2}, \qquad (2.6.25)$$

where $\mathbf{X}_a$ denotes the characteristics associated with the flow field $\mathbf{a}$. Observe that this estimate, which easily follows from Lemma A.1.3, does not require any smoothness for the velocity field $\mathbf{a}$ (provided it is divergence free). It yields an L^2 estimate for the smoothed particle approximation for almost all initializations as soon as $N \gg \varepsilon^{-2}$, a condition that is similar to the usual overlapping condition.

To be brought back to the above linear analysis, one needs to consider a time discretization of the Euler equation, with advection of the particles given by the flow field at the beginning of each time step. Every new time step eventually requires a new initialization of particles, something that can be done by resampling the smoothed particle solution at fresh random locations. The proof of the convergence of the overall algorithm follows from a precise control of the exceptional set of initializations, for which estimate (2.6.25) fails, in order to prove that it is possible to find sequences of initializations that will provide L^2 estimates at every time step.

One can guess that the random resampling involved at the beginning of every time step induces a numerical diffusion. It is actually possible to prove that, for an appropriate choice of the discretization parameters, the algorithm provides a convergent approximation to the Navier–Stokes equations (see Chapter 5 for a discussion of resampling schemes in the context of vortex schemes for the Navier–Stokes equations).

Let us now turn to the second aspect of convergence mentioned above, namely the case in which the overlapping condition is not satisfied. The extreme case of this situation is the one in which $\varepsilon = 0$. In the resulting method, often called the point-vortex method, to compute the velocity of a given particle, one exactly adds the contribution of all other particles (the self-velocity created by the particle under consideration itself is, however, forgotten).

This method has been proved in Ref. 95 (see also [109]) to converge quadratically to the exact solution, provided it is smooth enough. The structure of the proof resorts to techniques used in the early analysis [25]. It consists of using discrete norms on the particle locations. More precisely let us denote by X_i and X_i^h exact and computed locations, respectively, of the particles (starting from a deterministic initialization on a mesh of size h) and set, for $p > 1$,

$$\mathbf{e}_i = \mathbf{X}_i - \mathbf{X}_i^h, \quad \|\mathbf{e}\|_{l^p} = \left(\sum_i h^2 |\mathbf{e}_i|^p \right)^{1/p}.$$

To prove the consistency of the method, one first finds out how much the exact particles fail to solve the differential system giving the locations of the numerical particles; one thus writes

$$\frac{d\mathbf{X}_i}{dt} = \sum_j \mathbf{K}(\mathbf{X}_i - \mathbf{X}_j)\alpha_j + \rho_i,$$

where ρ_i stands for the discretization error. But we have

$$\mathbf{u}(\mathbf{X}_i) = \mathbf{K} \star \omega(\mathbf{X}_i) = \int \mathbf{K}[\mathbf{x} - \mathbf{X}(t; \mathbf{y}, 0)]\omega_0(\mathbf{y})\, d\mathbf{y}, \qquad (2.6.26)$$

so that ρ_i amounts to a quadrature error of the kind analyzed in Lemma A.1.2. Because $\mathbf{K}$ is not smooth it seems that only first-order accuracy can be obtained; however, because $\mathbf{K}$ is odd (a property first exploited by Beale [19]) one can subtract the singularity of $\mathbf{K}$ and show second-order accuracy. Thus $\rho_i = O(h^2)$.

Next, to show the stability of the method, one subtracts the equations of motions of exact and numerical particles and gets

$$\frac{d\mathbf{e}_i}{dt} = \sum_{j \neq i} \left[\mathbf{K}(\mathbf{X}_i^h - \mathbf{X}_j^h) - \mathbf{K}(\mathbf{X}_i - \mathbf{X}_j) \right] \alpha_j h^2 + \rho_i. \qquad (2.6.27)$$

Let us now assume for a while that the numerical particles remain well separated in the sense that they satisfy, like the exact particles, the property

$$i \neq j \Rightarrow \left| \mathbf{X}_i^h(t) - \mathbf{X}_j^h(t) \right| \geq C^* h \qquad (2.6.28)$$

for all time $t \in [0, T]$ and some constant C^*. This allows us to replace within the sum on the right-hand side of Eq. (2.6.27) the kernel $\mathbf{K}$ by the convolution

$\mathbf{K}_h$ of $\mathbf{K}$ with a cutoff $\zeta_h(\mathbf{x}) = h^{-2}\zeta(\mathbf{x}/h)$, where ζ has support in the ball of radius C^*. Then we can use the stability estimates derived for the regular smoothed vortex methods to get

$$\left\| \sum_{j \neq i} [\mathbf{K}(\mathbf{X}_i^h - \mathbf{X}_j^h) - \mathbf{K}(\mathbf{X}_i - \mathbf{X}_j)] \right\|_{l_p} \leq C \|e\|_{l^p}$$

(the norms above are discrete L^p norms; we can obtain this estimate by writing a Taylor expansion of the left-hand side with an integral remainder and then using Calderon's theorem).

Together with Eq. (2.6.27) this yields $\|e\|_{l^p} = O(h^2)$. Finally we note that, since by the smoothness of the flow the exact particles are well separated, this in turn implies that Eq. (2.6.28) is actually satisfied. A boot-strap argument allows us to conclude.

2.6.3. Convergence of the Variable-Blob Method

As we have seen in Section 2.2 it maybe useful to compute particle velocities on the basis of blobs with nonuniform core size, with formulas of the type

$$\mathbf{u}^h(\mathbf{x}) = \sum_p \alpha_p \mathbf{K}_{\varepsilon(\mathbf{x}_p)}(\mathbf{x} - \mathbf{x}_p^h).$$

The convergence analysis corresponding to this situation has been given in Refs. 106 and 107. In 106, the core size is assumed to be given through an area-conserving mapping. This allows us to handle to some extent shear flows, in which blobs could be elongated along the strain direction, but not more general flows, in particular, wakes in which blob volumes should be allowed to increase as particles move downstream. Here we outline the proof given in Ref. 107, in which blobs can vary in a more general fashion. More precisely let us assume that there exists a smooth function f such that

$$\varepsilon(\mathbf{x}) = \varepsilon f(\mathbf{x}) \text{ with } 0 < C < f(\mathbf{x}) < C'. \tag{2.6.29}$$

In view of the preceding discussions, the particle discretization error is clearly controlled as in the case of uniform blobs. We thus need to focus on only the regularization error, which can be expressed as

$$E = \int \phi(\mathbf{y})\zeta_{\varepsilon(\mathbf{y})}(\mathbf{x} - \mathbf{y})\,d\mathbf{y} - \phi(\mathbf{x}),$$

where ϕ is a smooth test function. For simplicity, let us assume that the cutoff ζ has a compact support of size unity and is of order r in the sense of Eqs. (2.3.1).

We first set

$$\mathbf{z} = \frac{\mathbf{x} - \mathbf{y}}{\varepsilon(\mathbf{y})}.$$

In view of Eq. (2.6.29) we can write

$$\frac{\partial z_i}{y_j} = \frac{1}{\varepsilon(\mathbf{y})} \left[\delta_{ij} - \varepsilon z_i \frac{\partial f}{\partial y_j} \right],$$

and a Taylor expansion of f around $\mathbf{x}$ yields

$$\frac{\partial z_i}{y_j} = \frac{1}{\varepsilon(\mathbf{y})} \left(\delta_{ij} + \sum_{k=1}^{r} \varepsilon^k \mathbf{z}^k a_k \right) + O(\varepsilon^r).$$

This first shows that, for ε small enough, $\mathbf{y} \to \mathbf{z}$ defines a one-to-one smooth mapping and that the Jacobian of this mapping can be evaluated as $\varepsilon(\mathbf{y})^2$ $[1 + \sum_{k=1}^{r} \varepsilon^k \mathbf{z}^k b_k] + O(\varepsilon^r)$. As a result

$$\int \zeta_{\varepsilon(\mathbf{y})}(\mathbf{x} - \mathbf{y}) \, d\mathbf{y} = 1 + O(\varepsilon^r),$$

and we can rewrite E as

$$E = \int [\phi(\mathbf{y}) - \phi(\mathbf{x})] \zeta_{\varepsilon(\mathbf{y})}(\mathbf{x} - \mathbf{y}) \, d\mathbf{y} + O(\varepsilon^r).$$

If ϕ is of class C^r, its Taylor expansion gives

$$E = \sum_{|\alpha| \leq r} c_\alpha \|\phi\|_{r,\infty} \int \mathbf{z}^\alpha \zeta(\mathbf{z}) \varepsilon(\mathbf{y})^{|\alpha|} \, d\mathbf{z} + O(\varepsilon^r).$$

On expanding f again around $\mathbf{x}$, one finally obtains

$$E = \sum_{|\alpha| \leq r-1} d_\alpha \|\phi\|_{r,\infty} \varepsilon^{|\alpha|} \int \mathbf{z}^\alpha \zeta(\mathbf{z}) \, d\mathbf{z} + O(\varepsilon^r),$$

and, in view of Eqs. (2.3.1), $E = O(\varepsilon^r)$. The vortex method with smoothly varying blob size has therefore the same convergence properties as those of uniform blobs. We will give in Chapter 5 numerical illustrations of variable-blob techniques in the context of the Navier–Stokes equations.

2.6.4. Conservation Properties

Besides theoretical convergence, another important criterion for the efficiency of a numerical method is its behavior with respect to the natural invariants of the flow. For the two-dimensional Euler equations, such invariants are the total circulation,

$$I_0 = \int \omega(\mathbf{x}, t)\, d\mathbf{x},$$

the linear and angular impulses,

$$I_1 = \int \mathbf{x}\omega(\mathbf{x}, t)\, d\mathbf{x}, \quad I_2 = \int |\mathbf{x}|^2\omega(\mathbf{x}, t)\, d\mathbf{x},$$

and the energy

$$E = \int |\mathbf{u}(\mathbf{x}, t)|^2\, d\mathbf{x}.$$

We can obtain the numerical equivalent of the first three invariants by replacing ω with ω^h and taking the duality of this measure with the functions 1, $\mathbf{x}$, or $|\mathbf{x}|^2$. We obtain

$$I_0^h(t) = \sum_p \alpha_p,$$

$$I_1^h(t) = \sum_p \mathbf{x}_p^h(t)\alpha_p,$$

$$I_2^h(t) = \sum_p |\mathbf{x}_p^h(t)|^2\alpha_p.$$

Note that the definition of these quantities does not rely on vortex blobs, but only on particles of vorticity. The definition of energy requires more care. As a matter of fact it can be easily checked that, in general unbounded flows, the velocity decays at infinity like $|\mathbf{x}|^{-1}$ and therefore cannot have finite energy, unless the circulation is 0, in which case the velocity decays like $|\mathbf{x}|^{-2}$. However, it is possible to write an always well-defined energy as

$$E' = \int \psi(\mathbf{x}, t)\omega(\mathbf{x}, t)\, d\mathbf{x} = \int\int G(\mathbf{x} - \mathbf{x}')\omega(\mathbf{x}, t)\omega(\mathbf{x}', t)\, d\mathbf{x}\, d\mathbf{x}'.$$

In the above expression, ψ is the stream function associated with the incompressible velocity $\mathbf{u}$. It is easy to check that E' is always finite, provided the vorticity decays fast enough at infinity, is constant in time and, finally, as shown by an immediate integration by parts, is equal to the energy defined in the usual way when the velocity decays like $|\mathbf{x}|^{-2}$ at infinity.

We naturally obtain the equivalent of E' in a vortex method by replacing G with G_ε and ω with ω^h. This yields

$$E'^h(t) = \sum_{p,p'} \alpha_p\alpha_{p'} G_\varepsilon\left[\mathbf{x}_p^h(t) - \mathbf{x}_{p'}^h(t)\right].$$

Theorem 2.6.3. *Assume that ζ has radial symmetry. Then all four invariants I_i^h, $i \in [0, 2]$ and E^h are conserved by the vortex method.*

Proof. Let us first check the assertion for I_1^h (I_0^h is obviously conserved). We have

$$\frac{dI_1^h}{dt} = \sum_p \alpha_p \frac{dx_p^h}{dt} = \sum_p \alpha_p u^h\left(x_p^h, t\right),$$

that is,

$$\frac{dI_1^h}{dt} = \sum_p \alpha_p \alpha_{p'} K_\varepsilon\left(x_p^h - x_{p'}^h\right).$$

But, since ζ is even, K_ε is odd, and exchanging the indices p and p' in the above sum turns it into its opposite. So it must be zero, which shows that $I_1^{\prime h}$ is constant.

Let us turn to I_2^h. We have

$$\frac{dI_2^h}{dt} = \sum_p \alpha_p x_p^h \cdot \frac{dx_p^h}{dt} = \sum_{p,p'} \alpha_p \alpha_{p'} x_p^h \cdot K_\varepsilon\left(x_p^h - x_{p'}^h\right).$$

Writing $x_p^h = \frac{1}{2}(x_p^h + x_{p'}^h) + \frac{1}{2}(x_p^h - x_{p'}^h)$ allows us to split this sum into two pieces. The first one is readily seen to cancel out, by a symmetry argument similar to the one just used. The second one is given as

$$\sum_{p,p'} \alpha_p \alpha_{p'} \left(x_p^h - x_{p'}^h\right) \cdot K_\varepsilon\left(x_p^h - x_{p'}^h\right).$$

We claim that each term of this sum is zero. Indeed, since ζ has radial symmetry, we have $G_\varepsilon(x) = G_\varepsilon(|x|)$, and thus

$$K_\varepsilon(x) = \operatorname{curl} G_\varepsilon(x) = \left(\frac{x_2}{|x|}, -\frac{x_1}{|x|}\right) \frac{\partial G_\varepsilon(x)}{\partial r}(|x|)$$

is orthogonal to x.

The fact that $E^{\prime h}$ is also conserved can be checked with the same techniques. However, it is enough to recognize that the system of ODEs governing the motions of particles is Hamiltonian, with Hamiltonian $E^{\prime h}$ [considered as a function of $(x_p^h)_p$]. As a matter of fact, we can write for each particle $x_p^h = (x_p^{h,1}, x_p^{h,2})$

$$\frac{dx_p^{h,1}}{dt} = \frac{1}{\alpha_p} \frac{\partial E^{\prime h}}{\partial x_p^{h,2}}, \qquad \frac{dx_p^{h,2}}{dt} = -\frac{1}{\alpha_p} \frac{\partial E^{\prime h}}{\partial x_p^{h,1}}.$$

The energy E'^h is therefore automatically conserved along the trajectories of the system. □

As mentioned in the discussion of vortex sheet calculation at the beginning of this chapter, the conservation of these invariants is a clear indication that, unlike most grid-based methods, vortex methods are naturally free of numerical dissipation resulting from spatial discretization.

Time discretization of the equations governing the particle trajectories is of course a potential factor of discrepancy in the conservation properties of the vortex method. However, it is observed in practice that reasonably high-order schemes (second-order Adams–Bashforth or Runge–Kutta schemes) allow us to satisfy energy and moment conservations rather well, even for large time steps. As an illustration, let us consider again the case of a circular patch subject to high distortion. Figure 2.7 shows the time evolution of energy and angular impulse for two values of the time step ($\Delta t = 2$, $\Delta t = 1$) with a fourth-order Runge–Kutta

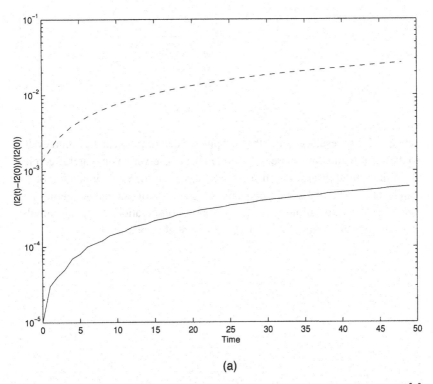

(a)

Figure 2.7. (a) Angular impulse, (b) kinetic energy for a circular patch $\omega(\mathbf{x}) = (1-|\mathbf{x}|^2)^3$ and a time step $\Delta t = 2$ (dashed curve) and $\Delta t = 1$ (solid curve).

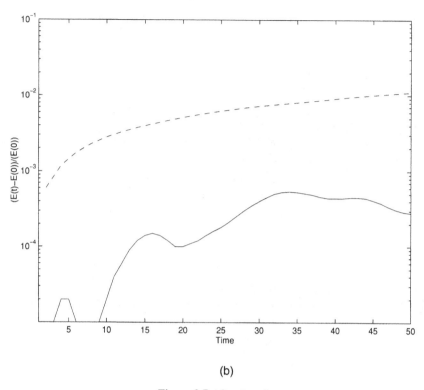

(b)

Figure 2.7. (*Continued*)

scheme. These results confirm that, despite the deterioration in the accuracy due
to the shear in the flow, there is no limit in the conservation of I_2 and E coming
from the spatial discretization (note that, based on the total circulation of the
patch, a time step of 1 corresponds in average to $1/8$ of a complete rotation). In
practical implementations of vortex methods, conservation of angular impulse
and energy can always be considered as essentially satisfied.

3

Three-Dimensional Vortex Methods
for Inviscid Flows

The need of a specific discussion of vortex schemes in the context of three-dimensional flows stems from the very nature of the vorticity equation that in three dimensions incorporate a stretching term. This term fundamentally affects the dynamics of the flow; it is in particular responsible for vorticity intensification mechanisms that make long-time inviscid calculations very difficult. Vorticity stretching is considered as the mechanism by which energy is being transferred between the large and the small scales in the flow. In order to resolve related phenomena, such as the energy cascade, an adequate treatment of diffusion is thus even more crucial than in two dimensions. However, the recipes for deriving diffusion algorithms are the same in two and three dimensions (they are discussed in Chapter 5), and we focus here on inviscid three-dimensional vortex schemes. Vorticity intensification in general is associated with a rapid stretching of Lagrangian elements, which makes it also crucial to maintain the regularity of the particle mesh; we refer to Chapter 7 for a general discussion of regridding techniques.

We will discuss here two classes of vortex methods that extend to three dimensions the two-dimensional schemes introduced in Chapter 2. In the first one, the vorticity is replaced by a set of points (particles), just as in two dimensions, but these particles carry vectors instead of scalars. The stretching term in the vorticity equation is accounted for by appropriate laws that modify the circulations of the particles. We call these methods vortex particle methods.

The second class of methods, the so-called vortex filament methods, is of a different nature. Here, the quantity carried by the Lagrangian computational elements is a scalar, as in two dimensions, but these elements are no longer discrete points but rather curves – filaments – that carry circulation. The vortex filament methods have in common with the two-dimensional particle schemes the fact that the weights of the computational elements remain constant in time.

55

A third class of methods, combining vortex particles and a grid-based Poisson solver for the calculation of particle velocities, will be described in Chapter 8.

Historically vortex filament methods have preceded vortex particle methods [71, 134, 136, 51]. The reason is probably that they have a more natural physical interpretation than vortex particle methods; in particular their definition is very reminiscent of Kelvin's circulation conservation theorem, which is one of the main properties of inviscid three-dimensional flows. As we will see, this is reflected by a very good behavior of the method with respect to the fundamental invariants of these flows.

Although there are in practice some strong links between the implementation of vortex filament and vortex particle methods, it seems that the latter have been only recently formalized [8]. From a numerical point of view, they can be considered as a more straightforward generalization of the two-dimensional schemes, with the appropriate modifications to handle vortex stretching.

For both classes of methods, we will follow in Sections 3.1 and 3.2 the same general outline: first we give the definitions, then we discuss their conservation properties, as this is where the differences between the physical nature of these methods are best understood, and we give some numerical illustrations. We also underline in Section 3.2 the links, from the computational point of view, between filament and particle methods. The convergence analysis for the methods of both types is done in Section 3.3. A specific discussion is devoted in Section 3.4 to the important issue of the divergence of the vorticity field when discretized by vortex particles.

3.1. Vortex Particle Methods

Let us first recall (see Chapter 1) the vorticity–velocity form of the Euler equations in three dimensions. The vorticity $\omega = \nabla \times \mathbf{u}$ is a vector and satisfies

$$\frac{\partial \omega}{\partial t} + \mathrm{div}(\mathbf{u} : \omega) - (\omega \cdot \nabla)\mathbf{u} = 0, \tag{3.1.1}$$

$$\omega(\cdot, 0) = \omega_0, \tag{3.1.2}$$

$$\mathbf{u} = \mathbf{K} \star \omega. \tag{3.1.3}$$

The notation $\mathbf{u} : \omega$ stands for the tensor of component $u_i \omega_j$ and div $(\mathbf{u} : \omega)$ is the vector of component $\partial u_i \omega_j / \partial x_i$. We assume here and throughout this chapter that the velocity vanishes at infinity or is periodic.

The kernel $\mathbf{K}$ is now a matrix-valued function such that

$$\mathbf{K} \star \omega(\mathbf{x}) = -\frac{1}{4\pi} \int \frac{\mathbf{x} - \mathbf{y}}{|\mathbf{x} - \mathbf{y}|^3} \times \omega(\mathbf{y}) \, d\mathbf{y} = \mathrm{curl}(G \star \omega).$$

In the above formulas $G(\mathbf{x}) = (4\pi|\mathbf{x}|)^{-1}$ is the elementary solution of the Poisson equation in three dimensions and $\times$ denotes the vector cross product. In Lagrangian form Eq. (3.1.1) can be rewritten as

$$\frac{D\omega}{dt} = (\omega \cdot \nabla)\mathbf{u}.$$

In this formulation, it is clear that one basic feature of this system is that the stretching term can produce local intensification and reorientation of the vorticity, resulting in particular in a loss of regularity for the solution in a finite time, depending on the strength of the initial vorticity.

3.1.1. Definitions

Equation (3.1.1) is a transport-deformation equation for the vorticity. Based on the discussion in Appendix A.2 (see in particular Corollary A.2.6 and the comment on systems that follows), a vortex particle method is obtained by setting

$$\omega^h(\mathbf{x}, t) = \sum_p \alpha_p^h(t)\delta\left[\mathbf{x} - \mathbf{x}_p^h(t)\right], \tag{3.1.4}$$

where the particles $\mathbf{x}_p^h$ follow the trajectories defined by the vector field $\mathbf{u}^h$ and α_p^h are vectors that solve the ODEs:

$$\frac{d\alpha_p^h}{dt} = \alpha_p^h \cdot \nabla\mathbf{u}^h\left(\mathbf{x}_p^h, t\right) = \left[\nabla\mathbf{u}^h\left(\mathbf{x}_p^h, t\right)\right]\left[\alpha_p^h\right]. \tag{3.1.5}$$

In the above system $[\nabla\mathbf{u}^h][\alpha_p^h]$ denotes the product of the matrix $(\partial u_i^h/\partial x_j)_{i,j}$ by the vector α_p^h.

As in two dimensions, to account for the nonlinear aspect of the flow equations, the vector field $\mathbf{u}^h$ is reconstructed from the vorticity field by application of the Biot–Savart law (3.1.3) to a mollified vorticity field. To this effect we choose a cutoff function ζ satisfying moment conditions up to the order r; then we set

$$\zeta_\varepsilon(\mathbf{x}) = \varepsilon^{-3}\zeta(\mathbf{x}/\varepsilon),$$

$$G_\varepsilon = G \star \zeta_\varepsilon, \quad \mathbf{K}_\varepsilon = \nabla G_\varepsilon.$$

Finally the velocity is computed as

$$\mathbf{u}^h = \mathbf{K}_\varepsilon \star \omega^h, \tag{3.1.6}$$

where the above convolution has to be understood in the sense of vector products, as in Eq. (3.1.3). The actual calculation of the regularized kernel follows the same lines as in two dimensions. If ζ is a spherically symmetric cutoff function, one obtains

$$\mathbf{K}_\varepsilon(\mathbf{x}) = -\frac{\mathbf{x}}{|\mathbf{x}|^3} f\left(\frac{\mathbf{x}}{\varepsilon}\right),$$

where

$$f(r) = \int_0^r \zeta(s)s^2\, ds.$$

The rules given in Section 2.3 to construct high-order cutoff functions apply in three dimensions as well. For example, combining two cubic Gaussians at scales 1 and $2^{-1/3}$ leads Beale and Majda [26] to the following fourth-order kernel:

$$\mathbf{K}_\varepsilon(x) = -\frac{\mathbf{x}}{4\pi|\mathbf{x}|^3}\left(1 + e^{-|\mathbf{x}|^3/\varepsilon^3} - 4e^{-2|\mathbf{x}|^3/\varepsilon^3}\right).$$

The method is finally defined by Eqs. (3.1.4)–(3.1.6). The initialization of particles is done along the same lines as in two dimensions. Typically particles are initially located at the nodes of a uniform grid, and they are assigned a circulation $\alpha_p^h(0)$ equal to the local vorticity multiplied by h^3, where h is the grid size. Other choices can obviously be made to take advantage of the particular geometry of the initial vorticity field or to allow local grid refinements.

We now come to two variants of the method just presented. They are both based on the following algebraic identity:

$$[\nabla\mathbf{u}][\omega] - [\nabla\mathbf{u}]^T[\omega] = \omega \times \operatorname{curl}\mathbf{u} = 0.$$

Hence the Euler equations can be expressed with an equivalent formulation if we replace the stretching term $(\omega \cdot \nabla)\mathbf{u} = [\nabla\mathbf{u}][\omega]$ with $[\nabla\mathbf{u}]^T[\omega]$ or any algebraic combination of both forms. This leads to two new schemes in which Eq. (3.1.5) is replaced with either

$$\frac{d\alpha_p^h}{dt} = [\nabla\mathbf{u}^h]^T[\alpha_p^h] \qquad (3.1.7)$$

or

$$\frac{d\alpha_p^h}{dt} = \frac{1}{2}([\nabla\mathbf{u}^h] + [\nabla\mathbf{u}^h]^T)[\alpha_p^h]. \qquad (3.1.8)$$

It is important to observe that, although based on equivalent formulations of the equations, these schemes do not lead, after discretization, to the same numerical method, for with definitions (3.1.4)–(3.1.6), $\omega^h \neq \nabla \times \mathbf{u}^h$. However, one can already predict similar convergence results for the three schemes. Observe that the scheme of Eq. (3.1.8) allows some computational savings over the original scheme of Eq. (3.1.5) because of the symmetry of the kernel involved in the computations of the weights. We will show below that Eq. (3.1.7), often referred to as the transpose scheme, offers the advantage of conserving the total circulation [45, 197].

Another conservative vortex scheme can be derived based on the conservative form of the stretching term. If we rewrite the stretching term in Eq. (3.1.1) as div (ω : **u**), to derive a vortex scheme we need to expand this term as

$$\text{div}(\omega : \mathbf{u}) = (\omega \cdot \nabla)\mathbf{u} + \mathbf{u}(\text{div }\omega).$$

We thus have to evaluate div ω on the particles. A general recipe for differentiating a particle field is to regularize it, then to differentiate the resulting smooth field, and finally to evaluate it on the particles. However, proceeding this way and combining the result with a treatment of the term $(\omega \cdot \nabla)\mathbf{u}$ with derivatives of the velocity obtained by differentiating Eq. (3.1.6), as for the preceding schemes, would not lead to a conservative method. To overcome this difficulty one has to apply the same strategy in the computation of the velocity derivatives as for the derivatives of the vorticity in the term div ω and use the formula

$$\partial_i \mathbf{u}(\mathbf{x}) = \sum_p v_p \mathbf{u}_p \partial_i \zeta_\varepsilon (\mathbf{x} - \mathbf{x}_p),$$

where $\mathbf{x}_p$ and v_p are the locations and the volumes of particles, respectively, and $\mathbf{u}_p$ are their velocities obtained through the regularized Biot–Savart law. Assuming for simplicity that the cutoff ζ has spherical symmetry, we obtain the following vortex scheme (where, for clarity, we omit the superscript h):

$$\frac{d\omega_p}{dt} = \sum_{j,q} v_q \left(\mathbf{u}_q w_p^j + \mathbf{u}_p w_q^j \right) \frac{(x_p^j - x_q^j)}{|\mathbf{x}_p - \mathbf{x}_q|} \zeta_\varepsilon'(|\mathbf{x}_p - \mathbf{x}_q|). \tag{3.1.9}$$

In this formula ω_p denotes the local vorticity values (the circulations are given by $\alpha_p = v_p \omega_p$) and the superscript j refers to the components of the vorticity. A version of this scheme in the context of vortex-in-cell methods will be presented in Section 8.2.

3.1.2. Conservation of Circulation

The basic invariants of three-dimensional inviscid flows are the total circulation, the linear and the angular impulses, the energy, and the helicity. In this section we restrict ourselves to the circulation,

$$\Gamma = \int \omega(\mathbf{x}, t) \, d\mathbf{x}, \tag{3.1.10}$$

whose conservation turns out to be already a difficulty for vortex particle methods.

Let us first recall the reason why the circulation is conserved for the continuous equations: since $\operatorname{div} \omega \equiv 0$, the stretching term can be written in a conservative form, $\operatorname{div}(\omega : \mathbf{u})$, and thus does not contribute to the production of circulation. Unfortunately, in vortex particle methods the vorticity ω^h is no longer guaranteed to be divergence free: At time zero, the approximate vorticity field is consistent with only pointwise values of ω_0 on a discrete set of points. This is clearly not enough to enforce a condition on its derivatives. Henceforth there is no guarantee that the circulation will be conserved by the original scheme of Eq. (3.1.5).

The fact that the approximate vorticity field, either in the particle form ω^h or in its regularized form $\omega^h \star \zeta_\varepsilon$, is not divergence free is by itself a difficulty that has long been a major obstacle to vortex particle methods. In principle such a violation of the fundamental laws of fluid mechanics may result in a nonphysical topology of vorticity lines (in particular the possibility of having nonclosed lines). We will discuss this issue in more detail in Section 3.4.

Let us now turn to the so-called transpose scheme of Eq. (3.1.7). As we already mentioned, its main feature is that it does conserve circulation.

Proposition 3.1.1. *Assume that the cutoff function ζ is even. Then the vortex particle method defined by Eq. (3.1.7) is conservative in the sense that*

$$\frac{d}{dt} \sum_p \alpha_p^h(t) \equiv 0.$$

Proof. Let us check the above identity for each component k. For simplicity we will drop the superscript h. We can write

$$\left(\sum_p \frac{d\alpha_p}{dt} \right)_k = \sum_{p,i} \frac{\partial u_i}{\partial x_k} (\alpha_p)_i = \sum_{p,p',i} \left[\frac{\partial \mathbf{K}_\varepsilon}{\partial x_k} (\mathbf{x}_p - \mathbf{x}_{p'}) \times \alpha_{p'} \right]_i (\alpha_p)_i$$

$$= \sum_{p,p'} \left[\frac{\partial \mathbf{K}_\varepsilon}{\partial x_k} (\mathbf{x}_p - \mathbf{x}_{p'}) \times \alpha_{p'} \right] \cdot (\alpha_p)$$

$$= \sum_{p,p'} \left[\frac{\partial \mathbf{K}_\varepsilon}{\partial x_k} (\mathbf{x}_p - \mathbf{x}_{p'}), \alpha_{p'}, \alpha_p \right],$$

where the term inside the last sum represents a determinant. Since the derivatives of $\mathbf{K}_\varepsilon$ are even, if we exchange the role of p and p', the sum is transformed into its opposite. It must therefore be zero. □

Unfortunately it does not seem possible to check that other natural invariants are algebraically conserved by this method. We will also see that it has problems with accuracy and the solenoidal condition. As for the scheme of Eq. (3.1.9), it is clearly conservative by construction.

3.1.3. Numerical Results

Let us briefly summarize the numerical results reported in Ref. 23 by comparing the efficiency of the schemes of Eqs. (3.1.5) and (3.1.7). It is hard to find test problems in three dimensions, unlike in two dimensions, in which the exact vorticity is available and that does not reduce by symmetry to a two-dimensional flow. The flow suggested by Beale et al. is a vortex ring with swirl. Its precise form – in particular the specific values of the vorticity inside its core – is obtained through a variational approach (see Refs. 23 and 137 for details). This is a steady solution, up to a uniform translation along the rotation axis, but particle paths are helices so its vortex approximation is truly three dimensional. It appears that the most accurate results are obtained with the original scheme. However these tests were performed without any regridding of the particles and it is plausible that, when the techniques of Section 7.2 are used, all schemes yield about the same accuracy.

Vortex particle methods are particularly suitable for viscous flows, and numerical illustrations of these methods will be provided in the following chapters.

3.2. Vortex Filament Methods

A vortex filament is a vorticity field that is concentrated on a curve that is either closed or extending to infinity; its direction is the local tangent at each point of the curve, and its strength is the vorticity flux across a cross section of the filament. For a collection of vortex filaments, it is worthwhile to think in terms of centerlines of vortex tubes, with longitudinal walls parallel to the vorticity field and centered around each filament. The circulation of each filament is evaluated by the vorticity flux on cross sections of each vortex tube and, as we have seen in Chapter 1, this flux is actually independent of the chosen cross section.

3.2.1. Definitions

This section is devoted to a rephrasing in mathematical terms the definition of filaments and of Kelvin's theorem. This will be useful for a rigorous study of the conservation properties of filament methods, as well as for fully exploiting the notion of weak solutions to advection equations that is, as in two dimensions, the main tool in the convergence analysis.

Let us start with the definition of filaments.

Definition 3.2.1. *Given a smooth closed oriented curve $\mathcal{L}$ in $\mathbf{R}^3$, we will call a vortex filament with unit circulation supported by $\mathcal{L}$ the vector-valued measure $\mu_{\mathcal{L}}$ defined by*

$$\langle \mu_{\mathcal{L}}, \phi \rangle = \int_{\mathcal{L}} \frac{\partial \gamma}{\partial \xi} \cdot \phi[\gamma(\xi)] \, d\xi,$$

where $\gamma(\xi)$ defines an orientation-preserving parameterization of $\mathcal{L}$, ϕ is a continuous test function, and the quantity inside the integral is a scalar product in $\mathbf{R}^3$.

It is easy to check that this definition does not depend on the chosen parameterization, that vortex filaments are automatically divergence free (take ϕ to be a gradient in the definition and then observe that the quantity in the integral is a total derivative along the curve), and that the flux of the vorticity along any cross section is one.

The first distinctive feature of vortex filaments is that they are transported by the flow with unchanged circulation. Let us give the mathematical formulation of this well-known fact in the framework of weak solutions to advection equations.

Theorem 3.2.2 (Kelvin's circulation theorem). *Assume that $\mathbf{u}$ is a given smooth function and that ω_0 is a vortex filament along the curve $\mathcal{L}_0$ with circulation α (in other words $\omega = \alpha \mu_{\mathcal{L}_0}$). Then the unique measure solution to Eq. (3.1.1) is the vortex filament,*

$$\omega(\mathbf{x}, t) = \alpha \mu_{\mathcal{L}(t)},$$

where $\mathcal{L}(t)$ satisfies

$$\frac{d\mathcal{L}}{dt} = \mathbf{u}(\mathcal{L}, t), \quad \mathcal{L}(0) = \mathcal{L}_0.$$

Proof. Since we are dealing with a linear problem, we can of course assume that $\alpha = 1$. Following the definition of measure solutions (see Appendix A, Definition A.2.5, and the remark following Corollary A.2.6), we have to show that

$$\int_0^T \left\langle \mu_{\mathcal{L}(t)}, \left[-\frac{\partial \phi}{\partial t} - (\mathbf{u} \cdot \nabla)\phi - [\nabla \mathbf{u}]^T \phi \right] (\cdot, t) \right\rangle dt = \langle \mu_{\mathcal{L}_0}, \phi(\cdot, 0) \rangle.$$

If we denote by I the integral on the left-hand side above, from Definition 3.2.1 we get

$$I = \int_0^T \int_{\mathcal{L}(t)} \left[-\frac{\partial \phi}{\partial t} - (\mathbf{u} \cdot \nabla)\phi - [\nabla \mathbf{u}]^T \phi \right] [\gamma(\xi, t), t] \cdot \frac{\partial \gamma}{\partial \xi}(\xi, t) \, d\xi \, dt.$$

By differentiating with respect to ξ the system $\dot{\gamma} = \mathbf{u}(\gamma, t)$, we obtain

$$\frac{d}{dt}\left(\frac{\partial \gamma}{\partial \xi} \right) = [\nabla \mathbf{u}(\gamma, t)] \left[\frac{\partial \gamma}{\partial \xi} \right].$$

Thus

$$([\nabla \mathbf{u}]^T \phi)[\gamma(\xi, t), t] \cdot \frac{\partial \gamma}{\partial \xi}(\xi, t) = \phi[\gamma(\xi, t), t] \cdot \frac{d}{dt}\left(\frac{\partial \gamma}{\partial \xi} \right)(\xi, t).$$

Furthermore,

$$\left[-\frac{\partial \phi}{\partial t} - (\mathbf{u} \cdot \nabla)\phi \right] [\gamma(\xi, t), t] \cdot \frac{\partial \gamma}{\partial \xi}(\xi, t) = -\frac{d}{dt}\phi[\gamma(\xi, t), t] \cdot \frac{\partial \gamma}{\partial \xi}(\xi, t),$$

and therefore

$$I = -\int_{\mathcal{L}} d\xi \int_0^T \frac{d}{dt}\phi[\gamma(\xi, t)] \cdot \frac{\partial \gamma}{\partial \xi}(\xi, t) \, dt = \int_{\mathcal{L}} \phi[\gamma_0(\xi, 0] \cdot \frac{\partial \gamma_0}{\partial \xi}(\xi) \, d\xi,$$

where $\gamma_0 = \gamma(\cdot, 0)$ is a parameterization of $\mathcal{L}_0$ and we have used the fact that an admissible test function must satisfy $\phi(\cdot, T) = 0$. In view of the definition of a filament we have finally verified that

$$I = \langle \mu_{\mathcal{L}_0}, \phi(\cdot, 0) \rangle,$$

which ends the proof. □

We can then define vortex filament methods in a straightforward way by coupling the definition

$$\omega^h(\mathbf{x}, t) = \sum_p \alpha_p \mu_{\mathcal{L}_p^h(t)} \tag{3.2.1}$$

with the filament dynamics

$$\frac{d\mathcal{L}_p^h}{dt} = \mathbf{u}^h\left(\mathcal{L}_p^h, t\right) = \mathbf{K}_\varepsilon \star \omega^h. \tag{3.2.2}$$

The mollified kernel $\mathbf{K}_\varepsilon$ is computed as usual through the convolution with a cutoff function; if ζ has spherical symmetry, $\mathbf{K}_\varepsilon = \nabla G_\varepsilon$, and $G_\varepsilon = G \star \zeta_\varepsilon$, we obtain the following formula for advecting the filaments:

$$\frac{d\mathcal{L}_p^h}{dt} = -\sum_q \alpha_q \int_{\mathcal{L}_q} \frac{\rho\left[\frac{\gamma_p - \gamma_q(\xi)}{\varepsilon}\right]}{|(\gamma_p - \gamma_q)|^3}[\gamma_p - \gamma_q(\xi)] \times \frac{\partial \gamma_q}{\partial \xi}\, d\xi, \tag{3.2.3}$$

where

$$\rho(r) = \int_0^r \zeta(s)s^2\, ds.$$

In some sense, vortex filaments for the three-dimensional Euler equations are very similar to vortex particles in two dimensions: their weights are scalar and are passively transported by the flow; moreover, the number of degrees of freedom (one per filament) is comparable with that for a two-dimensional problem (one per particle). However, the way filaments are initialized requires more care than for two-dimensional particles, since the initial filaments must approximate not only the strength of the vorticity but also the direction of the initial vortex lines of the real flow. Jets or wakes in temporal evolution provide particular examples well adapted to this type of discretization (we will show below in Subsection 3.2.4 illustrations of such flows). In complex geometries this is not in general an easy thing to do, but in some important cases, physical insight can help (as, for instance, when vorticity is generated at the edge of an obstacle). Let us also mention that filament surgery is required when stretching starts producing small unphysical scales within the filaments. This issue will be discussed in more detail in Section 5.6, which is devoted to subgrid models.

Although vortex filaments have a clear physical meaning, one must be cautious not to push too far the physical interpretation. If we rewrite $\mathbf{u}^h$ as $\mathbf{K} \star (\omega^h \star \zeta_\varepsilon)$, we may view the vorticity field as a collection of vortex elements centered around each filament. However, these elements are merely numerical elements and not strictly speaking vortex tubes: In principle their core radius ε is a numerical parameter that is constant along the filament direction (it can, however, vary from one element to another, depending on the local refinement desired in the direction transverse to the filaments, and some authors have used core values that vary along filaments and in time) and vorticity is not parallel to the walls of these elements. As a matter of fact these elements have no individual meaning: They must be considered collectively. Furthermore, as we can expect from the numerical analysis in the two-dimensional case (which, as we will see below, carries on to three dimensions), their accuracy in representing the real vorticity field is subject to an overlapping condition.

3.2.2. Conservation Properties

The beauty of vortex filament methods is most apparent in the behavior of the method with respect to the natural invariants of three-dimensional inviscid flows.

First it is clear by its definition that the method conserves the circulation. This is linked to the fact that filaments, unlike vortex particles, are divergence-free particles. It is actually possible to go further and check that linear impulse, angular impulse, and energy are conserved as well, provided the cutoff is an even function.

Let us consider the linear impulse $\int \mathbf{x} \times \omega^h \, d\mathbf{x}$. For simplicity in the notations we drop again the superscript h. Replacing as usual integrals using the duality between measures and continuous functions and using the mathematical definition of a filament, we can write the linear impulse as

$$I(t) = \sum_p \alpha_p \int_{\mathcal{L}_p} \gamma_p(\xi, t) \times \frac{\partial \gamma_p}{\partial \xi}(\xi, t) \, d\xi.$$

We thus have

$$\frac{dI}{dt} = \sum_p \alpha_p \left[\int_{\mathcal{L}_p} \frac{d\gamma_p}{dt}(\xi, t) \times \frac{\partial \gamma_p}{\partial \xi}(\xi, t) \, d\xi \right.$$
$$\left. + \int_{\mathcal{L}_p} \gamma_p(\xi, t) \times \frac{d}{dt} \frac{\partial \gamma_p}{\partial \xi}(\xi, t) \, d\xi \right].$$

Observe now that

$$\frac{d}{dt} \frac{\partial \gamma_p}{\partial \xi} = \frac{\partial}{\partial \xi} \mathbf{u}[\gamma_p(\xi, t), t],$$

so that integrating by parts over each filament yields

$$\int_{\mathcal{L}_p} \gamma_p(\xi, t) \times \frac{d}{dt} \frac{\partial \gamma_p}{\partial \xi}(\xi, t) \, d\xi = \int_{\mathcal{L}_p} \gamma_p(\xi, t) \times \frac{\partial}{\partial \xi} \mathbf{u}[\gamma_p(\xi, t), t] \, d\xi$$
$$= -\int_{\mathcal{L}_p} \frac{\partial}{\partial \xi} \gamma_p(\xi, t) \times \mathbf{u}[\gamma_p(\xi, t), t] \, d\xi.$$

Henceforth we have to check that

$$A = \sum_p \int_{\mathcal{L}_p} \mathbf{u}[\gamma_p(\xi, t), t] \times \frac{\partial \gamma_p}{\partial \xi}(\xi, t)\, d\xi = 0.$$

Using the explicit form of $\mathbf{u} = \mathbf{K}_\varepsilon \star \omega$, we get

$$A = \sum_{p,q} \int_{\mathcal{L}_p}\int_{\mathcal{L}_q} \left[\mathbf{K}_\varepsilon(\gamma_p - \gamma_q) \times \frac{\partial \gamma_p}{\partial \xi_p}\right] \times \frac{\partial \gamma_q}{\partial \xi_q}\, d\xi_p\, d\xi_q,$$

where, to avoid confusion, we have distinguished between the integration variables along $\mathcal{L}_p$ and $\mathcal{L}_q$. Routine algebraic calculations then give

$$A = \sum_{p,q} \int_{\mathcal{L}_p}\int_{\mathcal{L}_q} \left[\mathbf{K}_\varepsilon(\gamma_p - \gamma_q) \cdot \frac{\partial \gamma_p}{\partial \xi_p}\right] \cdot \frac{\partial \gamma_q}{\partial \xi_q}$$
$$- \left(\frac{\partial \gamma_p}{\partial \xi_p} \cdot \frac{\partial \gamma_q}{\partial \xi_q}\right) \mathbf{K}_\varepsilon(\gamma_p - \gamma_q)\, d\xi_p\, d\xi_q.$$

The second integral in the above right-hand side is zero because $\mathbf{K}_\varepsilon$ is odd. As for the first one, we note that

$$\mathbf{K}_\varepsilon(\gamma_p - \gamma_q) \cdot \frac{\partial \gamma_p}{\partial \xi_p} = \frac{\partial}{\partial \xi_p} G_\varepsilon(\gamma_p - \gamma_q),$$

and thus

$$\int_{\mathcal{L}_p} \mathbf{K}_\varepsilon(\gamma_p - \gamma_q) \cdot \frac{\partial \gamma_p}{\partial \xi_p}\, d\xi_p = 0$$

for all ξ_q, which finally proves that $A = 0$.

The conservation of the angular impulse $\int \mathbf{x} \times (\mathbf{x} \times \omega)\, d\mathbf{x}$ follows from similar arguments so let us now turn to the kinetic energy. For the continuous equations we have

$$E = \int |\mathbf{u}(\mathbf{x}, t)|^2\, d\mathbf{x} = \int \psi(\mathbf{x}, t) \cdot \omega(\mathbf{x}, t)\, d\mathbf{x}.$$

Note that, unlike in the two-dimensional case, the integration by parts leading to the above identity is always valid, provided that the vorticity decreases fast enough at infinity. The reason is that the velocity then decays as $|\mathbf{x}|^{-2}$ and the boundary terms on a sphere at infinity have no contribution in this integration by part.

For the numerical solution we argue as in two dimensions. The energy is computed with the formula below, based on the particle vorticity and the regularized stream function:

$$E = \int \psi^\varepsilon(\mathbf{x}, t) \cdot \omega^h(\mathbf{x}, t) \, d\mathbf{x},$$

where

$$\psi^\varepsilon = G_\varepsilon \star \omega^h.$$

By the definition of ω^h this gives

$$E = \sum_{p,q} \alpha_p \alpha_q \int_{\mathcal{L}_p} \int_{\mathcal{L}_q} G_\varepsilon(\gamma_p - \gamma_q) \frac{\partial \gamma_p}{\partial \xi_p} \cdot \frac{\partial \gamma_q}{\partial \xi_q} \, d\xi_p \, d\xi_q.$$

By differentiation we obtain

$$\frac{dE}{dt} = \sum_{p,q} \alpha_p \alpha_q \int_{\mathcal{L}_p} \int_{\mathcal{L}_q} [\mathbf{u}(\gamma_p) - \mathbf{u}(\gamma_q)] \cdot \mathbf{K}_\varepsilon(\gamma_p - \gamma_q) \frac{\partial \gamma_p}{\partial \xi_p} \cdot \frac{\partial \gamma_q}{\partial \xi_q}$$

$$+ G_\varepsilon(\gamma_p - \gamma_q) \left[\frac{\partial}{\partial \xi_p} \mathbf{u}(\gamma_p) \cdot \frac{\partial \gamma_q}{\partial \xi_q} + \frac{\partial}{\partial \xi_q} \mathbf{u}(\gamma_q) \cdot \frac{\partial \gamma_p}{\partial \xi_p} \right] d\xi_p \, d\xi_q. \quad (3.2.4)$$

An integration by parts over filaments then gives

$$\int_{\mathcal{L}_p} G_\varepsilon(\gamma_p - \gamma_q) \frac{\partial}{\partial \xi_p} \mathbf{u}(\gamma_p) \cdot \frac{\partial \gamma_q}{\partial \xi_q} \, d\xi_p$$

$$= -\int_{\mathcal{L}_p} \frac{\partial}{\partial \xi_p} G_\varepsilon(\gamma_p - \gamma_q) \mathbf{u}(\gamma_p) \cdot \frac{\partial \gamma_q}{\partial \xi_q} \, d\xi_p$$

$$= -\int_{\mathcal{L}_p} \left[\frac{\partial \gamma_p}{\partial \xi_p} \cdot \mathbf{K}_\varepsilon(\gamma_p - \gamma_q) \right] \left[\mathbf{u}(\gamma_p) \cdot \frac{\partial \gamma_q}{\partial \xi_q} \right] d\xi_p$$

and a similar formula when p and q are exchanged. So Eq. (3.2.4) can be rewritten:

$$\frac{dE}{dt} = \sum_{p,q} \alpha_p \alpha_q \int_{\mathcal{L}_p} \int_{\mathcal{L}_q} \mathbf{u}(\gamma_p) \cdot \left\{ \mathbf{K}_\varepsilon(\gamma_p - \gamma_q) \left(\frac{\partial \gamma_p}{\partial \xi_p} \cdot \frac{\partial \gamma_q}{\partial \xi_q} \right) \right.$$

$$\left. - \frac{\partial \gamma_q}{\partial \xi_q} \left[\frac{\partial \gamma_p}{\partial \xi_p} \cdot \mathbf{K}_\varepsilon(\gamma_p - \gamma_q) \right] \right\} + \mathbf{u}(\gamma_q) \cdot \left\{ \mathbf{K}_\varepsilon(\gamma_p - \gamma_q) \left(\frac{\partial \gamma_p}{\partial \xi_p} \cdot \frac{\partial \gamma_q}{\partial \xi_q} \right) \right.$$

$$\left. - \frac{\partial \gamma_p}{\partial \xi_p} \left[\frac{\partial \gamma_q}{\partial \xi_q} \cdot \mathbf{K}_\varepsilon(\gamma_p - \gamma_q) \right] \right\} d\xi_p \, d\xi_q.$$

We now recognize in the two terms between braces formulas for double-vector products:

$$\frac{dE}{dt} = \sum_{p,q} \alpha_p \alpha_q \int_{\mathcal{L}_p} \int_{\mathcal{L}_q} \mathbf{u}(\boldsymbol{\gamma}_p) \cdot \left\{ \frac{\partial \boldsymbol{\gamma}_p}{\partial \xi_p} \times \left[\mathbf{K}_\varepsilon (\boldsymbol{\gamma}_p - \boldsymbol{\gamma}_q) \times \frac{\partial \boldsymbol{\gamma}_q}{\partial \xi_q} \right] \right\}$$

$$- \mathbf{u}(\boldsymbol{\gamma}_q) \cdot \left\{ \frac{\partial \boldsymbol{\gamma}_q}{\partial \xi_q} \times \left[\mathbf{K}_\varepsilon (\boldsymbol{\gamma}_p - \boldsymbol{\gamma}_q) \times \frac{\partial \boldsymbol{\gamma}_p}{\partial \xi_p} \right] \right\} d\xi_p \, d\xi_q.$$

Using the definition of the velocity, we can finally rewrite this as

$$\frac{dE}{dt} = \sum_p \alpha_p \int_{\mathcal{L}_p} \mathbf{u}(\boldsymbol{\gamma}_p) \cdot \left\{ \frac{\partial \boldsymbol{\gamma}_p}{\partial \xi_p} \times \mathbf{u}(\boldsymbol{\gamma}_p) \right\} d\xi_p$$

$$+ \sum_q \alpha_q \int_{\mathcal{L}_q} \mathbf{u}(\boldsymbol{\gamma}_q) \cdot \left\{ \frac{\partial \boldsymbol{\gamma}_q}{\partial \xi_q} \times \mathbf{u}(\boldsymbol{\gamma}_q) \right\} d\xi_q = 0.$$

3.2.3. Vortex Particles versus Vortex Filaments

In this section we discuss some practical considerations concerning the links between vortex particles and vortex filaments. In an effective implementation of a vortex filament method, it is obviously not possible to follow the continuous curves. One needs instead to track a finite number of markers along these curves and reconstruct the vortex lines at each time step, in particular to compute the tangent vectors in the line integrals involved in the velocity computation of Eq. (3.2.3).

These markers are clearly good candidates to be considered as vortex particles, and one may wonder about the possible links that result from this remark between vortex filament methods and vortex particles methods as derived in Section 3.1. Let us first clarify precisely the way markers can actually be used. Each marker (indicated by a parameter ξ_m along a given filament $\mathcal{L}_p$) moves from one time step to the next with its local velocity. Then markers have to be linked to each other through some interpolation procedure to reconstruct a curve. This in particular means that, with each marker, one can associate an elementary vector line element dl_m parallel to the filament direction. These ingredients are then mixed together to give a velocity field through the following natural quadrature rule:

$$\int_{\mathcal{L}_p} \mathbf{K}_\varepsilon [\mathbf{x} - \boldsymbol{\gamma}_p(\xi)] \times \frac{\partial \boldsymbol{\gamma}_p}{\partial \xi_p} d\xi_p \simeq \sum_m \mathbf{K}_\varepsilon [\mathbf{x} - \boldsymbol{\gamma}_p(\xi_m)] \times dl_m.$$

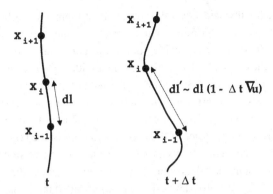

Figure 3.1. Stretching and markers along filaments.

This brings us back to a vortex particle method in which the circulations of the particles are given by $\alpha_p dl_m$. The most natural (and crude) choice is to take for dl_m the vector linking two successive markers along $\mathcal{L}_p$, which implicitly means that we make a piecewise linear curve reconstruction. This is actually the filament vortex method pioneered by Rehbach [171] and extensively used since then.

To analyze this analogy further let us observe that if $\mathbf{X}_m$ and $\mathbf{X}_{m+1}$ are two successive markers on a filament, with $dl_m = (\mathbf{X}_{m+1} - \mathbf{X}_m)/\Delta\xi$, one has, to the leading order (see Figure 3.1 for an illustration),

$$\frac{d}{dt}(\mathbf{X}_{m+1} - \mathbf{X}_m) = [\mathbf{u}(\mathbf{X}_{m+1}) - \mathbf{u}(\mathbf{X}_m)] \simeq [(\mathbf{X}_{m+1} - \mathbf{X}_m) \cdot \nabla]\mathbf{u}(\mathbf{X}_m).$$

As a result the particle strengths are, to the leading order, subject to the same evolution rule as in Section 3.1. Vortex filament methods can thus be viewed as vortex particles methods in which, the stretching has never to be explicitly taken into account to update the circulations, but is instead implicitly recovered through the modification by advection of the elementary line elements.

Of course the above discussion is based on first-order approximations of the stretching. A more accurate procedure would require more careful reconstruction of filaments at each time step (some of the numerical results given below are actually based on cubic interpolation). In this case, it is readily seen that the vortex filament method becomes similar to a vortex particle method, in which the rules to update the weights of the particles would use velocity derivatives computed through finite-difference formulas along the filaments, with a finite-difference stencil and order related to the interpolation method used to track the filaments.

In summary, for inviscid flows, vortex filament methods can be viewed as "clever" vortex particle methods, which take advantage of the connectivity of the particle distribution to enable substantial computational savings. The connectivity of the elements also facilitates the regridding/refinement strategies that may become necessary. Since the locations of markers are monitored at all times, it is easy to insert a new particle between two successive markers whenever their distance exceeds a certain value. This critical distance is in general taken to be of the order of the core radius used in the regularization of the Biot–Savart kernel. This allows us to compensate for the vorticity stretching parallel to the filaments and thus control the resolution in these directions. Knio and Ghoniem [118] have also devised in a method called the transport element method a way to monitor distances between neighboring filaments and insert new elements in the direction transverse to the filaments.

As we have already mentioned, the drawback of the filament method is that it relies on the knowledge of the topology of the vorticity at the initial stage and whenever vorticity is created. Moreover, the Lagrangian character of filaments is a purely inviscid property of three-dimensional flows: In the presence of diffusion, points in neighboring filaments will exchange vorticity through diffusion at a rate that depends on their distance, and the concept of a unique circulation along a given material filament does not hold anymore. As a result, there is no clear-cut way to simulate diffusion with a filament method, although filament surgery or core-spreading techniques are sometimes used to provide some dissipation mechanisms. It follows from these remarks that filament methods are not readily suitable for wall-bounded viscous flows.

In closing, let us remark that similar comments hold for the two-dimensional contour dynamics method [75, 167]. These methods consist of tracking level curves of the scalar vorticity. These curves can be seen as filaments for the gradient of the vorticity, which itself turns out to be the solution of a transport-deformation equation similar to the three-dimensional equation for the vorticity. The only difference between the two methods, besides the dimension of the discrete system, lies in the singularity of the kernels (logarithmic for contour dynamics, r^{-2} for filament methods) used to compute the velocities.

3.2.4. Numerical Illustrations

We first present some numerical results due to Knio and Ghoniem [118, 119] that can be seen as the three-dimensional counterpart of Krasny's calculations in Section 2.1. The flow under consideration is a vortex shear layer, periodic in the streamwise and spanwise directions, undergoing perturbations in both directions. Vortices are initially distributed uniformly along filaments on a few

material surfaces parallel to the shear layer plane. A typical grid resolution to
start these calculation involves $20 \times 14 \times 5$ points in the streamwise, spanwise,
and transverse directions, respectively. The initial perturbation is introduced by
moving the filaments through sine waves with small amplitudes in two directions.

Figure 3.2(a) shows vortices' locations in spanwise and streamwise cross
sections. One can observe the formation of roll-up in the streamwise direction,
typical of Kelvin–Helmholtz instability, while the three-dimensional character
of the flow is manifested in the mushroomlike structures in the spanwise direction, as a result of the braid distortion. Figure 3.2(b) shows the intensification of
the streamwise vorticity in the braids of the shear layer. A regridding strategy as
outlined above is crucial in order to obtain this kind of result. At the time shown
in these figures, the initial number of vortices has been multiplied by a factor

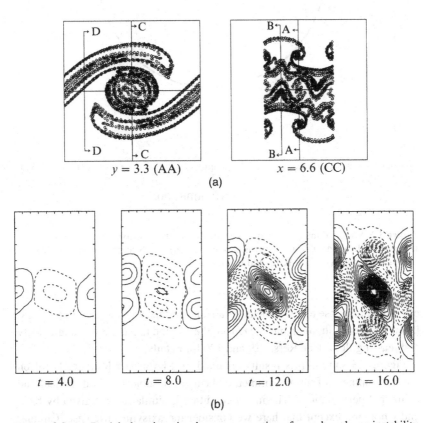

Figure 3.2. (a) Particle locations in plane cross sections for a shear layer instability
(x and y are the streamwise and the spanwise coordinates, respectively), (b) time evolution of the streamwise vorticity in the cross section D (courtesy of A. Ghoniem).

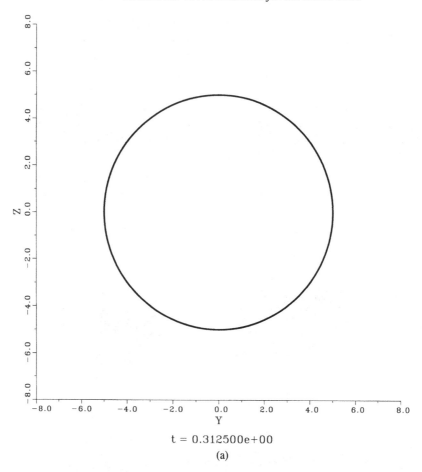

t = 0.312500e+00

(a)

Figure 3.3. Vortex filament simulations of the instability of an axisymmetric shear layer subject to helical and azimuthal perturbations. (a), (b), (c), Streamwise and (d), (e), (f), side views (courtesy of J. Martin and E. Meiburg).

of $\sim$10, in response to a significant strain and surface area increase in the shear layer. Knio and Ghoniem have extended the method to handle variable-density and reacting flows (see Refs. 119 and 120 for details).

Figure 3.3 summarizes results obtained by Martin and Meiburg [146] on the development of an axisymmetric shear layer subject to both helical and azimuthal perturbations. The initial condition is similar to that given by Knio and Ghoniem, except that here we consider the axisymmetric case. Circular filaments are uniformly initialized on a ring, with uniform circulation, and then slightly displaced to generate the perturbation. Periodic boundary conditions

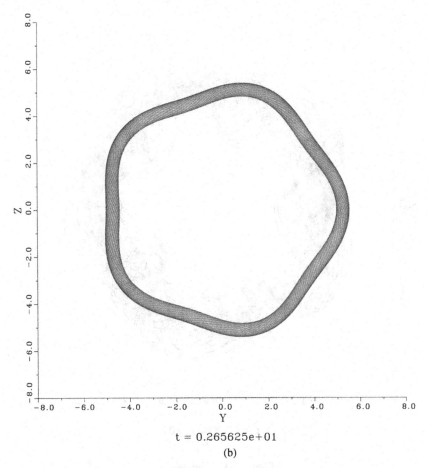

t = 0.265625e+01

(b)

Figure 3.3. (*Continued*)

are enforced by use of image filaments in the computation of the velocity. At the end of the calculation one can observe strong streamwise vorticity concentrated in the braids connecting the main rings. In this computation a cubic spline interpolation procedure is used to compute the local tangents to the filaments and to insert new elements when needed in order to maintain at all times a good resolution along the filaments.

Our last example is borrowed from Ref. 179. It concerns the phenomenon of vortex breakdown in a swirling axial jet. The vortex breakdown comes from the formation of a bulge that contains a recirculation zone. As the flow develops, this bubble becomes unstable and its asymmetry results in a wakelike behavior in the downstream zone. The vortex filament simulation shown in Figure 3.4

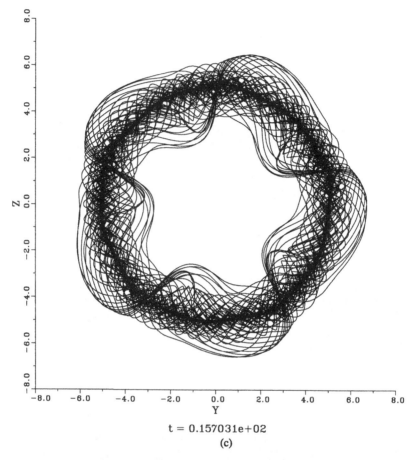

t = 0.157031e+02

(c)

Figure 3.3. (*Continued*)

enables a very sharp description of the tilting of the filaments, which is at the origin of the breakdown. In this simulation, filaments are laid on three coaxial sleeves. New elements are created upstream on every filament as soon as the preceding element has moved downstream more than a prescribed resolution, which is taken as a fraction of the core radius. The position of the new elements in the upstream plane is determined so as to ensure the connectivity of the filaments. A dissipation model is used, which consists of a continuous increase of the core radius of the elements, in the spirit of the core-spreading method (see Subsection 5.6.2) and a clipping strategy allows to remove all elements whose core radius exceed 10 times their initial value.

In conclusion, these calculations show that vortex filament methods are a useful tool for direct numerical simulations of inviscid flows. By concentrating

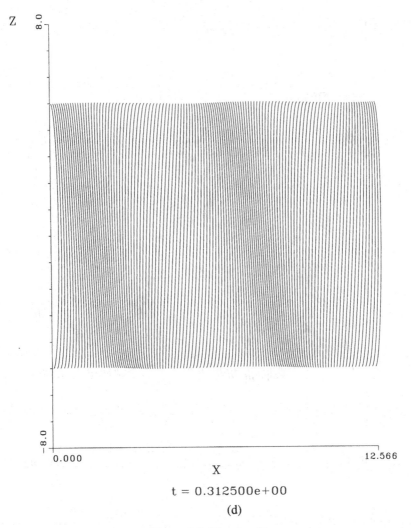

Z

8.0

-8.0

0.000   12.566

X

t = 0.312500e+00

(d)

Figure 3.3. (*Continued*)

on the vortical zone of the flow and allowing refinement at a minimal dissipation cost, they allow to investigate fine structures and the changes in the flow topology that are characteristic of three-dimensional vortex dynamics.

3.3. Convergence Results

Historically, the first convergence result for vortex methods in three dimensions has been given by Beale and Majda in one of the very first numerical analysis work on vortex methods [25]. This analysis is based on a reformulation of

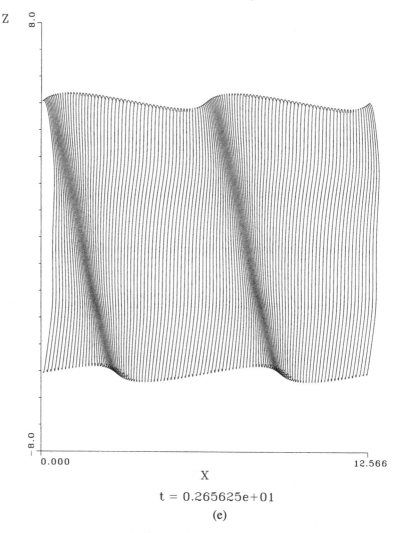

Z

8.0

−8.0

0.000   12.566

X

t = 0.265625e+01

(e)

Figure 3.3. (*Continued*)

the method of Eq. (3.1.5) in terms of the derivatives of the flow map. As a matter of fact, a comparison of Eq. (3.1.5) with Eq. (1.1.10) shows that vorticity and material lines follow the same dynamics. Therefore Eq. (3.1.5) can be equivalently rewritten as

$$\alpha_p = v_p \omega_p, \quad \omega_p^h(t) = [D_x \mathbf{X}(\mathbf{x}; t, 0)] [\omega_p^h(0)], \qquad (3.3.1)$$

where $\mathbf{X}$ denotes the characteristic of the flow and, as usual, v_p denotes the

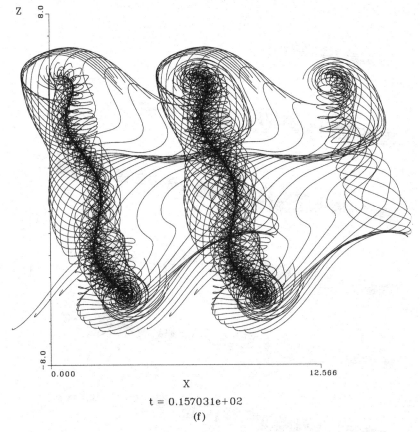

Z

8.0

−8.0

0.000   12.566

X

t = 0.157031e+02

(f)

Figure 3.3. (*Continued*)

volume of the particles. The numerical method suggested in Ref. 25 is to approximate the values of $[D_x\mathbf{X}(\mathbf{x}_p; t, 0)]$ by means of finite differences based on the pointwise values of the flow map (that is, actual particle locations) corresponding to grid points close to the original particle location $\mathbf{x}_p$.

The proof given in Ref. 25 relies heavily on the use of discrete norms associated with the grid made by the particles at their initial locations. The overall rate of convergence is conditioned both by the order and by the smoothness of the cutoff, as in two dimensions, and by the order of accuracy of the finite-difference scheme used in the approximation of the derivatives of the flow map. Greengard [89] gave a convergence proof of a method in the same spirit, but using derivatives of the flow map only along the direction parallel to the vorticity. This method is closely related to vortex filament methods, and we postpone its discussion to the end of this section.

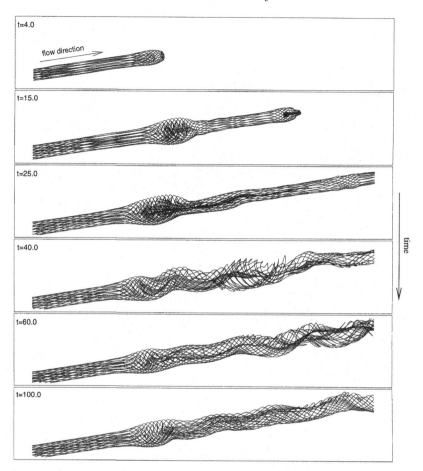

Figure 3.4. Vortex filament simulation of the vortex breakdown in a swirling jet (courtesy of J.-C. Saghbini and A. Ghoniem).

3.3.1. Vortex Particle Methods

The first convergence proof concerning the three-dimensional vortex particle method as defined by Eqs. (3.1.4) and (3.1.5) is due to Beale [19]. This proof assumes that the vorticity is smooth and with compact support and that the cutoff is at least fourth order. A later proof [60] works in more general situations. We give below the essential steps of a simplified version of this proof [44, 45], which is merely an extension of the argument developed in Chapter 2.

As for the two-dimensional case, we state the convergence result in the case of periodic boundary conditions, so that the proof can focus on the key features of the method. The initialization step is done on the basis of a regularized initial vorticity field:

$$\alpha_p(0) = \omega_0^\varepsilon(\mathbf{x}_p)h^3; \quad \omega_0^h = \sum \alpha_p \delta(\mathbf{x} - \mathbf{x}_p),$$

where the initial particles are on a uniform mesh with mesh size h and $\omega_0^\varepsilon = \zeta_\varepsilon \star_p \omega_0$, where ζ is a cutoff function and $\star_p$ denotes the periodic convolution (we refer to Appendix A for the notation and the needed mathematical background).

Theorem 3.3.1. *Let $r \geq 2$ and $T > 0$. Assume that $\omega_0 \in W^{r-1,\infty}(\Omega) \cap W^{2,\infty}(\Omega)$, where Ω is a square periodic box, and that ζ is a cutoff function of order r. Assume further that there exists $s > 0$ such that $h \leq \varepsilon^{1+s}$; then there exists a time $T(\omega_0)$ such that, if h is small enough, the following estimate holds for $t \in [0, T(\omega_0)]$:*

$$\|(\mathbf{u} - \mathbf{u}^h)(\cdot, t)\|_{L^p(\Omega)} \leq C(T)\varepsilon^r.$$

Observe that the restriction on the time for which the error estimate holds only reflects the fact that the three-dimensional Euler equations cease to be well posed past a certain critical time, which depends on the strength of the initial vorticity [142].

Proceeding as in two dimensions, in order to establish this error estimate we first focus on the regularization error. We denote by $(\mathbf{u}_\varepsilon, \omega_\varepsilon)$ the solution to the problem,

$$\frac{\partial \omega^\varepsilon}{\partial t} + \operatorname{div}(\mathbf{u}^\varepsilon \omega^\varepsilon) - (\omega_\varepsilon \cdot \nabla)\mathbf{u}_\varepsilon = 0, \tag{3.3.2}$$

$$\omega^\varepsilon(\cdot, 0) = \omega_0^\varepsilon, \tag{3.3.3}$$

$$\mathbf{u}^\varepsilon = \mathbf{K}_\varepsilon \star_p \omega^\varepsilon, \tag{3.3.4}$$

and we prove estimates similar to relations (2.6.4) and (2.6.5). For a time $T(\omega_0)$ and $t \leq T(\omega_0)$ we first have

$$\|\omega^\varepsilon(\cdot, t)\|_{2,\infty} \leq C. \tag{3.3.5}$$

If we next set $\mathbf{e} = \omega - \omega_\varepsilon$, we obtain, by the same argument as in two dimensions,

$$\|\mathbf{e}(\cdot, 0)\|_{-1,p} \leq C\varepsilon^r. \tag{3.3.6}$$

By subtracting Eq. (3.3.2) from Eq. (3.1.1) we also have

$$\frac{\partial \mathbf{e}}{\partial t} + \operatorname{div}(\mathbf{u}^\varepsilon \mathbf{e}) - (\mathbf{e} \cdot \nabla)\mathbf{u}^\varepsilon = \operatorname{div}\left[(\mathbf{u}^\varepsilon - \mathbf{u})\boldsymbol{\omega}\right] - (\boldsymbol{\omega} \cdot \nabla)(\mathbf{u} - \mathbf{u}^\varepsilon). \quad (3.3.7)$$

We thus may apply Theorem A.2.7 to relations (3.3.6) and (3.3.7) with

$$T = T(\boldsymbol{\omega}_0), m = 1, A = 1, \mathbf{a} = \mathbf{u}^\varepsilon, \mathbf{a}_0 = [\nabla \mathbf{u}^\varepsilon]$$

to get

$$\|\mathbf{e}(\cdot, t)\|_{-1,p} \leq C \left[\|\mathbf{e}_0\|_{-1,p} + \int_0^t \|(\mathbf{u} - \mathbf{u}^\varepsilon)(\cdot, s)\|_{0,p} \, ds \right.$$
$$\left. + \int_0^t \|(\boldsymbol{\omega} \cdot \nabla)(\mathbf{u} - \mathbf{u}^\varepsilon)(\cdot, s)\|_{-1,p} \, ds \right]. \quad (3.3.8)$$

But we have

$$\|(\boldsymbol{\omega} \cdot \nabla)(\mathbf{u} - \mathbf{u}^\varepsilon)\|_{-1,p} \leq \|\boldsymbol{\omega}\|_{1,\infty} \|\mathbf{u} - \mathbf{u}^\varepsilon\|_{0,p},$$

so that we end up with

$$\|\mathbf{e}(\cdot, t)\|_{-1,p} \leq C \left[\varepsilon^r + \int_0^t \|(\mathbf{u} - \mathbf{u}^\varepsilon)(\cdot, s)\|_{0,p} \, ds \right].$$

The proof then proceeds as in two dimensions: We write $\mathbf{u} - \mathbf{u}^\varepsilon = (\mathbf{K} - \mathbf{K}_\varepsilon) \star_p \boldsymbol{\omega} + \mathbf{K}_\varepsilon \star_p \mathbf{e}$, and then we apply Calderon's theorem [estimate (A.3.6)] to obtain

$$\|(\mathbf{u} - \mathbf{u}^\varepsilon)(\cdot, s)\|_{0,p} \leq C[\varepsilon^r + \|\mathbf{e}(\cdot, t)\|_{-1,p}].$$

We thus have

$$\|\mathbf{e}(\cdot, t)\|_{-1,p} \leq C \left[\varepsilon^r + \int_0^t \|(\mathbf{e})(\cdot, s)\|_{-1,p} \, ds \right]$$

and, by Gronwall's theorem,

$$\|\mathbf{e}(\cdot, t)\|_{-1,p} \leq C\varepsilon^r.$$

We now turn to the evaluation of $\mathbf{u}^\varepsilon - \mathbf{u}^h$. We first choose m such that $ms > r+1$. Let us show that

$$\|\mathbf{u}^\varepsilon - \mathbf{u}^h\|_{0,p} = O(h^m \varepsilon^{-m}). \quad (3.3.9)$$

It results from relations (3.3.4), (3.3.5), and (A.3.7) that we can find a constant C^* depending only on ω_0 such that, for $t \le T(\omega_0)$,

$$\|\mathbf{u}^\varepsilon(\cdot, t)\|_{2,\infty} + \sum_{k=3}^{m} \varepsilon^{k-2} |\mathbf{u}^\varepsilon(\cdot, t)|_{k,\infty} \le C^*.$$

We then introduce the time $T_h^* \le T(\omega_0)$, defined as

$$T_h^* = \max \left[0 < t \le T(\omega_0); \|\mathbf{u}^h(\cdot, t)\|_{2,\infty} + \sum_{k=3}^{m} \varepsilon^{k-2} |\mathbf{u}^h(\cdot, t)|_{k,\infty} \le 2C^* \right].$$

As in two dimensions, we split the error into two terms: $\mathbf{u}^\varepsilon - \mathbf{u}^h = \alpha^h + \beta^h$, where

$$\frac{\partial \alpha^h}{\partial t} + \text{div}(\mathbf{u}^h \alpha^h) - (\alpha^h \cdot \nabla)\mathbf{u}^h = 0, \quad \alpha^h(\cdot, 0) = \omega_0^\varepsilon - \omega_0^h, \quad (3.3.10)$$

and

$$\frac{\partial \beta^h}{\partial t} + \text{div}(\mathbf{u}^h \beta^h) - (\beta^h \cdot \nabla)\mathbf{u}^h$$
$$= \text{div}[(\mathbf{u}^h - \mathbf{u}^\varepsilon)\omega^\varepsilon] - (\omega^\varepsilon \cdot \nabla)(\mathbf{u}^h - \mathbf{u}^\varepsilon), \quad \beta^h(\cdot, 0) = 0. \quad (3.3.11)$$

The only difference with the two-dimensional case, besides the zero-order term in the advection equations on the left-hand sides above, is the additional term, which is due to the stretching, on the right-hand side of Eq. (3.3.11). This term, however, can easily be handled in our functional framework. By definition of the space $W^{-1,p}(\Omega)$,

$$\|(\omega^\varepsilon \cdot \nabla)(\mathbf{u}^h - \mathbf{u}^\varepsilon)\|_{-1,p} \le |\omega^\varepsilon|_{1,\infty} \|(\mathbf{u}^h - \mathbf{u}^\varepsilon)\|_{0,p}.$$

We then apply Theorem A.2.7 with $m = 1$ to bound α^h and with arbitrary m to bound β^h. This yields exactly the same estimates as for the corresponding error terms in two dimensions. We finally recover that $T^* = T(\omega_0)$ by differentiating $\mathbf{u} - \mathbf{u}_\varepsilon^h$ (note that one can obtain a slightly better control of the derivatives of $\mathbf{u}$, in particular a uniform bound on the second derivatives, because, on the one hand we required the initial vorticity field to be more smooth than in two dimensions, and, on the other hand, we chose a bigger value of m).

In closing let us mention that partial convergence results exist concerning the three-dimensional point-vortex method, that is, when no mollification is used for the velocity calculations. In Ref. 108 the method analyzed uses a finite-difference-based stretching evaluation, as in Ref. 25. The totally grid-free method is studied in Ref. 65 (note, however, that this reference contains a flaw in the consistency proof, and the convergence of the grid-free point-vortex method remains an open problem).

3.3.2. Vortex Filament Methods

Vortex filament methods are, like vortex particle methods, based on explicit weak solutions to the vorticity equation. The structure of the convergence proof will thus remain the same as that in Subsection 3.3.1 and one needs to focus on only the initialization error. To analyze this error in weak norms, we consider a test function ϕ; in view of Eq. (3.2.1) and Definition 3.2.1, we have

$$\left\langle \omega_0^h - \omega_0, \phi \right\rangle = \sum_p \alpha_p \int_{L_p} \phi[\gamma_p(\xi)] \cdot \frac{\partial \gamma_p}{\partial \xi} d\xi - \int \omega_0(\mathbf{x})\phi(\mathbf{x}) d\mathbf{x}.$$

(3.3.12)

In order to be brought back to a situation in which we can use optimal quadrature estimates, following Ref. 89, let us assume that the initial vortex lines can be smoothly mapped to parallel circles, in the following sense: there exists a smooth, one-to-one mapping Ξ from $[0, 1]^2 \times [0, 1]_p$ onto the support of ω_0 (where the index p above means that the opposite sides $[0, 1] \times \{0\}$ and $[0, 1] \times \{1\}$ are identified by periodicity) and a smooth scalar function λ defined on $[0, 1]^2$ such that

$$\omega_0[\Xi(\mathbf{y}, s)] = \lambda(\mathbf{y})\frac{\partial \Xi}{\partial s}(\mathbf{y}, s).$$

(3.3.13)

With these definitions it is clear that the vortex lines associated with ω_0 are the curves

$$s \longrightarrow \Xi(\mathbf{y}, s).$$

Let us now observe that, for ω_0 to be divergence free, the Jacobian determinant of the transform Ξ has to be independent of s. Indeed we must have, for all scalar function ψ,

$$\int \omega_0(\mathbf{x}) \cdot \nabla\psi(\mathbf{x}) d\mathbf{x} = 0.$$

Using the change of variables Ξ together with Eq. (3.3.13) yields

$$\int \lambda(\mathbf{y})\frac{\partial \Xi}{\partial s}(\mathbf{y}, s) \cdot \nabla\psi[\Xi(\mathbf{y}, s)]|\det[D_{\mathbf{y},s}\Xi(\mathbf{y}, s)]| \, d\mathbf{y} \, ds = 0,$$

or, equivalently,

$$\int \lambda(\mathbf{y}) \frac{\partial}{\partial s} \psi[\Xi(\mathbf{y}, s)] |\det[D_{\mathbf{y},s} \Xi(\mathbf{y}, s)]| \, d\mathbf{y} \, ds = 0.$$

After integration by parts with respect to the s variable, we obtain

$$\int \lambda(\mathbf{y}) \psi[\Xi(\mathbf{y}, s)] \frac{\partial}{\partial s} |\det[D_{\mathbf{y},s} \Xi(\mathbf{y}, s)]| \, d\mathbf{y} \, ds = 0.$$

Since this must be true for all function ψ, the s derivative of the Jacobian determinant must vanish and our claim is proved.

We can now set

$$\mu(\mathbf{y}) = \lambda(\mathbf{y}) |\det[D_{\mathbf{y},s} \Xi(\mathbf{y}, s)]| \qquad (3.3.14)$$

and we are ready to define our initialization. We select a regular array of vortex lines,

$$\mathcal{L}_p: \quad s \longrightarrow \Xi(\mathbf{p}h, s) ; \quad \mathbf{p} = (p_1, p_2) \in \mathbf{Z}^2,$$

and we assign them a circulation equal to μh^2:

$$\alpha_p = \mu(\mathbf{p}h) h^2.$$

Let us now come back to Eq. (3.3.12). We compute the integral on the right-hand side with the change of variables Ξ and use Eqs. (3.3.13) and (3.3.14) to obtain

$$\langle \omega_0 - \omega_0^h, \phi \rangle = \int \mu(\mathbf{y}) \frac{\partial \Xi}{\partial s} (\mathbf{y}, s) \phi[\Xi(\mathbf{y}, s)] \, d\mathbf{y} \, ds$$
$$- \int \sum_{\mathbf{p}} \mu(\mathbf{p}h) \frac{\partial \Xi}{\partial s} (\mathbf{p}h, s) \phi[\Xi(\mathbf{p}h, s)] \, ds.$$

This clearly shows that the initialization error amounts to a quadrature error with respect to the variable $\mathbf{y}$ (that is, in the local coordinates given by Ξ, the direction transverse to filaments), whereas integration with respect to the filament direction variable s is performed exactly.

If ω_0 is smooth enough we thus have (Lemma A.1.2)

$$|\langle \omega_0 - \omega_0^h, \phi \rangle| \leq C h^m \int \|\mu \frac{\partial \Xi}{\partial s}(\cdot, s) \phi[\Xi(\cdot, s)]\|_{m,1} \, ds$$
$$\leq C h^m \|\phi\|_{m,1}.$$

The convergence analysis can now proceed along the same lines as in Subsection 3.3.1 and yields, for a finite time only, depending on the initial condition,

$$\|\mathbf{u} - \mathbf{u}^h\|_{0,p} \leq C\varepsilon^r,$$

provided the cutoff is of order r and the overlapping condition $h \ll \varepsilon$ is satisfied. If the initial vorticity field is not smooth, the construction of a filament method based on a regularized initial vorticity field is more questionable than that for vortex particles, as it would require use of filament locations adapted to these vorticity fields. Note finally that, in view of the remark made in Section 3.2 concerning the analogy of filament method with the two-dimensional contour dynamics method, the above argument also shows the convergence of these methods.

3.4. The Problem of the Vorticity Divergence for Vortex Particle Methods

As we already said, the difficulty of vortex particle methods to reproduce a divergence-free vorticity field has its origin in the initialization stage of the method itself. However, one may advocate that, since the approximation of functions by particles is a consistent operation, this divergence can be made small enough by using a sufficient number of points. The issue is then to analyze to what extent a divergence in the initial field, however small it is, is amplified by the convection and stretching in the flow.

Let us first focus on the original particle scheme of Eq. (3.1.5) and show that this method actually performs rather well, at least in the early stages of the evolution of the particles. We first recall that the particle circulations resulting from this scheme satisfy

$$\omega_p^h(t) = [D_x \mathbf{X}^h(\mathbf{x}; t, 0)] [\omega_p^h(0)]. \tag{3.4.1}$$

As a result, if the vorticity at a given particle initially points toward a neighbor, then, to the leading order, this property will be preserved for all time. In other words, the connectivity of vortex lines is maintained by the method, at least as long as the stretching does not force too many particles apart from each other. Another way to understand this feature is to observe that, since the method is based on an explicit (weak) solution to the vorticity equation,

$$\frac{\partial \omega^h}{\partial t} + \text{div}(\mathbf{u}^h : \omega^h) - (\omega^h \cdot \nabla)\mathbf{u}^h = 0. \tag{3.4.2}$$

Straightforward differentiations yield

$$\frac{\partial(\text{div } \omega^h)}{\partial t} + \text{div}[\mathbf{u}^h(\text{div } \omega^h)] = 0. \tag{3.4.3}$$

This means that the scheme does not further amplify the violation of the divergence-free constraint on the vorticity. It will not seriously disrupt vortex lines if they are connected at time zero.

The transpose scheme of Eq. (3.1.7), although conservative, unfortunately does not share this property, which may explain why, at least in some cases, it does not perform as well as the original scheme. As for the other conservative scheme, Eq. (3.1.9), that we presented in Section 3.1, it is based on the conservative form of the Euler equation. By differentiation this gives

$$\frac{\partial}{\partial t}(\text{div } \omega^h) = 0,$$

which is even better than Eq. (3.4.3).

Let us now review a few tools that can be used to improve vortex particle schemes further from the point of view of the vorticity divergence.

3.4.1. Helmholtz Decomposition

The first class of methods consists of projecting the vorticity field onto a divergence-free field, through a Helmholtz decomposition. As any vector field, the vorticity field can be split into a gradient and a rotational part:

$$\omega = \text{grad } \pi + \text{curl } \phi. \tag{3.4.4}$$

A divergence-free vorticity field is recovered by subtracting the gradient part of ω. This is actually equivalent to using a Lagrange multiplier in the vorticity equation to enforce the divergence-free constraint on the vorticity (similar to the pressure in the velocity–pressure formulation) and to use a transport-projection algorithm.

The "pressure"-like term π is obtained by taking the divergence of Eq. (3.4.4). Assume that we are dealing with a grid-free vortex particle method without solid boundaries and that ω vanishes at infinity. It is then natural to write π as

$$\pi = -G_\varepsilon \star \text{div } \omega,$$

where, as usually, G_ε is a regularized version of G. The corrected particle vorticity is therefore

$$\omega_1' = \omega + \nabla G_\varepsilon \star \text{div } \omega. \tag{3.4.5}$$

Another possibility is to compute ϕ by taking the curl of Eq. (3.4.4). This leads to a vorticity field

$$\omega_2' = \nabla \times [\nabla \times (G_\varepsilon \star \omega)] . \tag{3.4.6}$$

One may observe that the second formula also can be written as

$$\omega_2' = \nabla \times \mathbf{u}_\varepsilon, \tag{3.4.7}$$

where $\mathbf{u}_\varepsilon$ is the regularized velocity field, a formula due to Novikov.

It is important to note that Eqs. (3.4.5) and (3.4.6) do not give the same fields: if div $\omega = 0$, the fields ω_1' and ω coincide, whereas ω_2' and ω differ. Formula (3.4.6) actually induces a dissipation resulting from the smoothing involved in the reconstruction of the vorticity. On the other hand, it is readily seen that formula (3.4.5) does not guarantee that ω_2', unlike ω_1', is divergence free. This difficulty, related to the coexistence of particles and regularized fields in vortex methods, is very similar to the ones encountered in remeshing or assignment techniques (See Chapter 7 and Section 8.1). A possible way to overcome it is to process the weights of the particles before applying formula (3.4.5).

More precisely, if the vorticity ω is given by particles with circulations α, one has to find weights β such that, denoting by $\star_d$ the discrete convolution on the particle locations,

$$\beta \star_d \zeta_\varepsilon = \nabla G_\varepsilon \star_d \text{ div } \omega$$

on all particles, then add β to α, according to Eq. (3.4.5). Solution procedures to solve this linear system are analyzed in Refs. 200 and 201. A natural approach, similar to Beale's method described in Section 7.1 [formula (7.1.3)], leads to an iterative procedure. However, it was found that this procedure becomes ill conditioned as the overlapping of the particles increases, which may be seen as a consequence of the fact that the linear system gets closer to an ill-posed deconvolution problem. The convergence of the iterations is also affected by the lack of smoothness of the vorticity, which is another difficulty, as three-dimensional flows, unlike their two-dimensional counterparts, tend to develop singularities spontaneously.

If this linear system is solved just after the time particles are remeshed on regular locations, one may consider inverting once for all the matrix obtained for these locations. We refer to Ref. 201 for a discussion of these various strategies.

As a numerical illustration of this relaxation procedure, we use the following example from Ref. 199. The initial configuration for this simulation consists of two vortex rings crossing each other. Although not physical, this is an enlightening test problem as each filament of a given ring gets wrapped around the other ring. In Figure 3.5 we follow the particles lying on one filament in each ring,

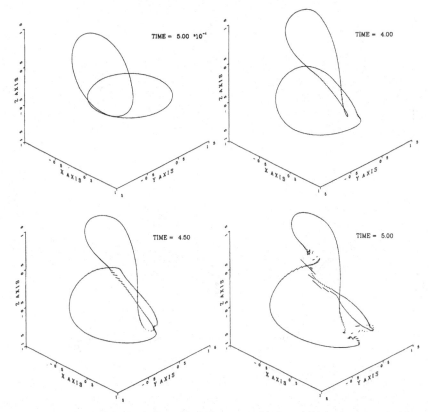

Figure 3.5. Evolution of two filaments with the original scheme.

with the unmodified scheme of Eq. (3.1.5). We observe that, for early times, the filament topology is well preserved. Figure 3.6 shows the same sequence, but with the following remeshing procedure: When particles go too far apart from each other, new particles are inserted, with circulations assigned according to Eq. (3.4.5), that is,

$$\sum_q \alpha_q^{\text{new}} \zeta_\varepsilon(\mathbf{x}_p - \mathbf{x}_q)$$

$$= \sum_q \alpha_q^{\text{old}} \zeta_\varepsilon(\mathbf{x}_p - \mathbf{x}_q) + \text{grad}\{\alpha_q^{\text{old}} \cdot \text{grad}[G_\varepsilon(\mathbf{x}_p - \mathbf{x}_q)]\}. \quad (3.4.8)$$

The improvement, at least until time 4.0, is clear. This is confirmed by the behavior of the diagnostics for linear and angular impulses, which are better conserved in the remeshed calculations (see Ref. 199 for details and further discussions).

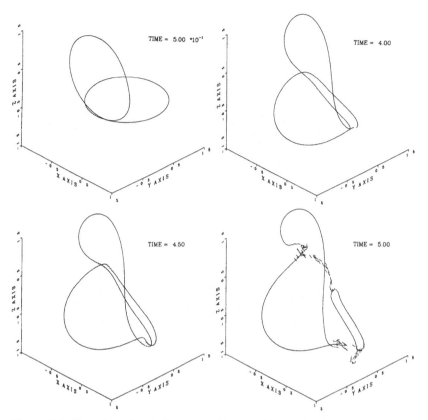

Figure 3.6. Evolution of two filaments with remeshing and relaxation scheme of Eq. (3.4.8).

3.4.2. Dissipation Techniques

The first technique we describe has been introduced in the field of Maxwell equations. In this context the preservation of a divergence-free magnetic field is crucial to ensure the conservation of charge. The relevant idea is then to dissipate the vorticity divergence by modifying the vorticity equation into

$$\frac{\partial \omega}{\partial t} + \text{div}(\mathbf{u} : \omega) - (\omega \cdot \nabla)\mathbf{u} = \sigma \, \text{grad} \, (\text{div} \, \omega). \qquad (3.4.9)$$

Taking the divergence of this equation indeed yields

$$\frac{\partial (\text{div} \, \omega)}{\partial t} + (\mathbf{u} \cdot \nabla)(\text{div} \, \omega) = \sigma \, \Delta(\text{div} \, \omega).$$

In particular,

$$\frac{d}{dt}\|\text{div } \omega\|_{0,2}^2 = -\sigma\|\nabla(\text{div } \omega)\|_{0,2}^2,$$

which confirms that this equation does dissipate vorticity divergence. One also deduces from Eq. (3.4.9) that

$$\frac{d}{dt}\|\omega\|_{0,2}^2 = -\sigma\|\text{div } \omega\|_{0,2}^2.$$

This shows that the algorithm is globally dissipative. In a Navier–Stokes scheme, with molecular viscosity ν, for this dissipation to be negligible compared with the molecular one, one has to make sure that

$$\|\text{div } \omega\|_{0,2}^2 \ll \|\nabla\omega\|_{0,2}^2.$$

The implementation of Eq. (3.4.9) in a vortex method can be done through a particle strength exchange scheme (see Subsection 5.4.4).

In closing, let us mention that another point of view on the divergence problem in vortex particle methods may be to consider that the divergence of the vorticity is only one particular manifestation of unphysical small scales produced in the course of the computations. If one adopts this point of view, it is natural to rely on turbulent viscosity models as natural tools to address this issue. We again refer to Chapter 5 on diffusion schemes, and more specifically to Section 5.6 for a discussion of turbulent viscosity models in vortex methods. Experience also shows that plain regridding strategies, when used with smooth, high-order interpolation functions (e.g., the M_4' scheme see Section 7.2), provide subgrid dissipation, of the hyperviscosity type, which essentially remedies the problem with the divergence.

4

Inviscid Boundary Conditions

Vortex methods were initially conceived as a tool to simulate the inviscid dynamics of vortical flows. The vorticity carried by the fluid elements is conserved in inviscid flows and simulating the flow amounts to the computation of the velocity field. In bounded domains the velocity field is constrained by the conditions imposed by the type and the motions of the boundaries. For an inviscid flow, it is not possible to enforce boundary conditions for all three velocity components as we have lost the highest-order viscous term from the set of governing Navier–Stokes equations. Usually for inviscid flows past solid bodies we impose conditions on the velocity component locally normal to the boundary.

The description of an inviscid flow can be facilitated when the velocity field is decomposed into two components that have a kinematic siginificance. In this decomposition, a rotational component accounts for the velocity field due to the vorticity in the flow whereas a potential component is used in order to enforce the boundary conditions and to ensure the compatibility of the velocity and the vorticity field in the presence of boundaries. This is the well-known Helmholtz decomposition.

Alternatively the evolution of the inviscid flow can be described in terms of an extended vorticity field. The enforcement of a boundary condition for the velocity components normal to the boundary does not constrain the wall-parallel velocity components. This allows for velocity discontinuities across the interface that may be viewed as velocity gradients over an infinitesimal region across the boundary. In turn, this tangential velocity discontinuity can be equivalently described by a vortex sheet distribution. The strength of this vortex sheet is adjusted so as to enforce the boundary conditions at the interface. In the same context angular motions of solid bodies in the flow may be equivalently described in terms of volume distribution of vorticity at the interior of the body. This allows for a complete description of the flow field in terms of an extended vorticity field. A compatibility condition between the components of the extended vorticity field is necessary in order to ensure the conservation of flow invariants while enforcing the kinematic boundary conditions.

From the engineering perspective the model of inviscid flow has been implemented for several decades, as an approximation of high Reynolds number nonseparated flows past surfaces such as ship hulls and airfoils. Hess in 1975 pioneered the computation of potential flows past airplane hulls by using boundary element techniques to discretize the governing integral equations, leading to the well-known panel methods. In order to calculate more realistic airloads, the next level of engineering approximation considers the inviscid evolution of the vorticity field generated by the aerodynamic configurations. This approximation has to account for the proper generation and shedding of the vorticity from the surface of the body into the wake. In steady lifting flows this mechanism is expressed in terms of the Kutta condition. In unsteady incompressible flows various engineering approximations have been used in order to *a priori* define the location and the injection rate of vorticity in the flow from the surface of the body. These shedding models provide a mechanism of relating the kinematic and the dynamic constituents of the flow and are well suited to the description of the flow in terms of an extended vorticity field. The problem of vorticity generation in a viscous flow is further addressed in Chapter 6.

In Section 4.1 we describe the Helmholtz decomposition of the velocity field and we consider the solution of Laplace's equation along with the kinematic boundary conditions. Our emphasis is on the integral solutions of the equations and their discretization by use of boundary element methods. In Section 4.2 we describe the alternative technique of enforcing the kinematic boundary conditions by means of an extended vorticity field. We introduce the concept of the surface vortex sheet and we present Poincaré's identity relating the values of the velocity field with its values at the boundary. We analyze the resulting integral equations and discuss their numerical approximation by using panel methods. Furthermore we emphasize the properties of these integral equations that allow us eventually to link the inviscid and the viscous descriptions of the flow field. In Section 4.3 we discuss the Kutta condition and vortex shedding mechanisms used in engineering approximations. In Section 4.4 we discuss the difficulties of smooth vortex methods associated with the modification of the vorticity integrals in the presence of boundaries in inviscid flows.

4.1. Kinematic Boundary Conditions

In the following sections we focus on the enforcement of the kinematic boundary conditions for flows around an impermeable solid body. Note that the results presented herein can be readily extended to flows with other boundary conditions, such as free-surface flows [15, 141].

We consider an impermeable solid body immersed in an incompressible fluid extending to infinity and we denote by $\mathbf{u}_b(t)$ the unsteady velocity of translation and by $\Omega_b(t)$ the unsteady rotational velocity of the body. The body surface may be viewed as an interface separating two volumes. We denote by V_i the volume interior to the fluid flow domain (exterior to the body) and by V_e the volume exterior to the fluid domain (and interior to the body). In the following sections the subscripts i and e refer to quantities in these domains. We distinguish both sides of the surface S as S_i (related to the volume V_i) and S_e (related to the volume V_e) and a unit vector $\mathbf{n}$ uniquely defined at every point of S and in the direction from S_e toward S_i.

4.2. Kinematics I: The Helmholtz Decomposition

In the Helmholtz decomposition the velocity field is decomposed into a rotational component and a solenoidal component as

$$\mathbf{u} = \mathbf{u}_\omega + \mathbf{u}_\phi, \tag{4.2.1}$$

where $\mathbf{u}_\omega$ accounts for the vortical part of the flow, and it is such that

$$\omega = \nabla \times \mathbf{u} = \nabla \times \mathbf{u}_\omega. \tag{4.2.2}$$

Moreover, we can impose the additional constraint of incompressibility on $\mathbf{u}_\omega$:

$$\nabla \cdot \mathbf{u}_\omega = 0. \tag{4.2.3}$$

Combining Eqs. (4.2.2) and (4.2.3), we obtain a vector Poisson equation for the velocity $\mathbf{u}_\omega$ in terms of the vorticity of the flow field:

$$\Delta \mathbf{u}_\omega = -\nabla \times \omega. \tag{4.2.4}$$

A solution to this equation for example is given by the Biot–Savart law,

$$\mathbf{u}_\omega = \int_{V_i} \mathbf{K}(\mathbf{x} - \mathbf{y})\, \omega(\mathbf{y})\, d\mathbf{y}, \tag{4.2.5}$$

where $\mathbf{K}(\mathbf{x})$ is the Biot–Savart kernel.

Having assigned the rotational part of the flow to the $\mathbf{u}_\omega$ velocity component we have that

$$\nabla \times \mathbf{u}_\phi = 0. \tag{4.2.6}$$

The second velocity component may now be equivalently described in terms of a potential as

$$\mathbf{u}_\phi = \nabla \Phi. \tag{4.2.7}$$

The condition of incompressibility ($\nabla \cdot \mathbf{u}_\phi = 0$) results in Laplace's equation for the potential:

$$\Delta \Phi = 0. \tag{4.2.8}$$

Clearly the decomposition Eq. (4.2.1) is not unique as any potential flow field may be added to it. We obtain a unique solution by enforcing the kinematic boundary conditions of the flow. For the potential flow past an impermeable solid body translating with a velocity $\mathbf{u}_b$, Eq. (4.2.8) is supplemented with the boundary conditions

$$\mathbf{u} \cdot \mathbf{n} = (\mathbf{u}_\omega + \nabla \Phi) \cdot \mathbf{n} = \mathbf{u}_b \cdot \mathbf{n}, \tag{4.2.9}$$

or, equivalently,

$$\nabla \Phi \cdot \mathbf{n} = (\mathbf{u}_b - \mathbf{u}_\omega) \cdot \mathbf{n}. \tag{4.2.10}$$

Hence the velocity field of the flow can be expressed as

$$\mathbf{u} = \int_{V_i} \mathbf{K}(\mathbf{x} - \mathbf{y})\, \omega(\mathbf{y})\, d\mathbf{y} + \nabla \Phi, \tag{4.2.11}$$

where Φ satisfies Laplace's equation and the boundary condition (4.2.10).

4.2.1. Laplace's Equation for Hydrodynamics

Laplace's equation is the fundamental equation in potential theory with applications ranging from electrodynamics to hydrodynamics.

Numerical solutions of Laplace's equation can be classified into two broad categories. Grid-based methods (e.g., finite differences, finite elements) rely on a discretization of the Laplacian operator over the whole domain, which results in a linear system of equations for the unknown values of the potential. For simple geometries the matrices involved are banded and the equation can be solved easily with a number of matrix inversion techniques. On the other hand, this requires the discretization of the entire domain. Moreover, for complex geometries the resulting linear system of equations does not always have a simple structure, and special inversion techniques are necessary in order to get efficient computations. This is a subject of ongoing research efforts, and efficient numerical schemes, such as multigrid, can be devised that reduce the computational cost.

Alternatively, the solution of Laplace's equation can be obtained by means of the equivalent Green's function formulation. Enforcing the boundary conditions results in a set of integral equations for the unknown distribution of discontinuities (singularities) of the potential and/or its derivatives at the boundary of the domain. These integral equations can then be discretized by use of various quadrature techniques, resulting in a set of algebraic equations for the unknown singularity strengths at the boundary. Complex geometries are relatively easy to handle as only the boundary of the domain needs to be discretized.

Integral equation techniques reduce by an order of magnitude the dimensionality of the problem by requiring the solution for the unknown values of potential discontinuities at only the boundary of the domain. Unlike the case of grid-based methods, the matrices here are dense, even for simple geometries. However, it is possible to construct fast inversion techniques that render the computational cost proportional to the number of boundary elements, leading to an optimal technique for the numerical solution of Laplace's and Poisson's equations.

A number of integral solutions to Laplace's equation for various geometries can be obtained by taking advantage of the linearity of the equation. In general, these solutions can be itemized as follows:

- *Linear superposition of elementary solutions.* This technique is based on the observation that the velocity vectors do not cross the flow streamlines. Hence a linear superposition of elementary solutions of Laplace's equation (such as sources/sinks) that produce streamlines that coincide with a body surface provides the solution for the potential flow problem around this body. This technique has been used extensively in the past for the potential flow solution about axisymetric bodies (e.g., Rankine's ovoid).

- *The method of images.* In this method, in order to enforce boundary conditions, we consider an odd or even extension (image) of the infinite-domain Green's function across the boundary. A simple example is the flow induced by a vortex over a flat surface, which can be equivalently described by the system of a dipole, whose axis coincides with the boundary of the domain. For two-dimensional geometries this technique can be extended to more complicated geometries by use of the conformal mapping technique.

- *The method of singularity distributions.* This is the most general technique for the integral solution of Laplace's equation. In fact the first two techniques may be viewed as its particular cases. The solution of Laplace's equation is expressed in terms of a continuous superposition of its Green's functions with weighting factors (strength), the discontinuities of the potential, and its derivatives at the boundary of the domain. The weighting factors are determined from the solution of a linear system of equations so as to enforce the no-through-flow boundary conditions.

The method of singularity distributions is based on the Poincaré identity that relates the value of a scalar field Φ to its discontinuities at the boundary of the domain. For the potential of an incompressible flow ($\Delta\Phi = 0$) the Poincaré identity gives

$$\Phi(\mathbf{x}) = \int_S [\mathbf{n}(\mathbf{x}') \cdot \nabla(\Phi_i - \Phi_e)(\mathbf{x}') G(\mathbf{x} - \mathbf{x}')$$
$$-(\Phi_i - \Phi_e)(\mathbf{x}') \mathbf{n}(\mathbf{x}') \cdot \nabla G(\mathbf{x} - \mathbf{x}')] dS(\mathbf{x}'), \qquad (4.2.12)$$

where the integrals associated with the value of the potential at infinity are zero and

$$G(\mathbf{x} - \mathbf{x}') = \begin{cases} \frac{1}{4\pi} \frac{1}{|\mathbf{x} - \mathbf{x}'|} & \text{in three dimensions} \\ -\frac{1}{2\pi} \log(|\mathbf{x} - \mathbf{x}'|) & \text{in two dimensions} \end{cases}. \qquad (4.2.13)$$

The discontinuities involved in the above boundary integral equations are called single- and double-layer potentials. We define the double-layer potential μ (or doublet) as

$$\mu = \Phi_e - \Phi_i \qquad (4.2.14)$$

and the single-layer potential σ (or source) as

$$\sigma = \frac{\partial \Phi_e}{\partial \mathbf{n}} - \frac{\partial \Phi_i}{\partial \mathbf{n}}. \qquad (4.2.15)$$

In order to enforce the boundary condition of no-through flow we have to evaluate the wall normal derivative of the potential ($\partial\Phi/\partial\mathbf{n}$) at the boundary of the domain. This can be calculated by means of a limiting process with Eq. (4.2.12), resulting in

$$\frac{\partial \Phi}{\partial \mathbf{n}}(\mathbf{x}_b) = \pm\frac{1}{2}\sigma(\mathbf{x}_b) + \int_S [\sigma(\mathbf{x}'_b) G(\mathbf{x}_b - \mathbf{x}'_b)$$
$$- \mu(\mathbf{x}'_b) \mathbf{n} \cdot \nabla G(\mathbf{x}_b - \mathbf{x}'_b)] dS(\mathbf{x}'_b) \qquad (4.2.16)$$

The + sign is used when we approach the interface from the side S_e, and the − sign is used when we approach the interface from the side S_i.

Enforcing the no-through-flow boundary conditions results in a system of integral equations for the unknown singularity strengths at the boundary of the domain. For example, requiring that the potential be continuous across the interface implies that $\mu = 0$, resulting in an integral equation for the source distribution on the surface of the body:

$$(\mathbf{u}_b - \mathbf{u}_\omega) \cdot \mathbf{n}(\mathbf{x}_b) = -\frac{1}{2}\sigma(\mathbf{x}_b) + \int_S \sigma\left(\mathbf{x}_b'\right) G\left(\mathbf{x}_b - \mathbf{x}_b'\right) dS\left(\mathbf{x}_b'\right). \quad (4.2.17)$$

The integrals in the above equations can be calculated numerically with a quadrature, resulting in a set of algebraic equations for the unknown strength of the singularity distribution. Usually this is achieved in terms of the well-known panel methods that are further discussed in Section 4.3.

4.3. Kinematics II: The Ψ–ω and the u–ω Formulations

The kinematic boundary conditions can be enforced without considering a decomposition into a potential and a vortical component for the velocity field of the domain. Instead, we consider an extended vorticity field that consists of the vorticity field in the fluid, a vortex sheet distribution on the boundary, and a uniform vorticity distribution inside the body corresponding to the solid-body rotation (see Fig. 4.1).

The velocity and the extended vorticity of the flow field can be described with the aid of a single vector streamfunction Ψ such that $\mathbf{u} = \nabla \times \Psi$. In two dimensions the no-through-flow boundary conditions can be expressed in terms of the streamfunction, and the whole problem reduces to the solution of a Poisson equation. This method is applicable in three dimensions as well, but the cross coupling of the streamfunction components resulting from the kinematic boundary conditions makes the method less easy to implement for complex geometries.

The use of the streamfunction can finally be bypassed by direct consideration of the relationship between the velocity and the extended vorticity field of the flow by means of the Poincaré identity. In this formulation the kinematic boundary conditions can be enforced by an adjustment of the strength of the vortex sheet distribution on the surface of the body.

4.3.1. The Extended Vorticity Field

Applying the condition of no-through flow results in a flow velocity whose tangential component at the body surface is in general different from the tangential velocity component in the interior of the body. This velocity difference across the interface may be viewed as an infinitesimally thin vorticity distribution (vortex sheet) along the surface of the body, defined as

$$\boldsymbol{\gamma}(\mathbf{x}) = \mathbf{n}(\mathbf{x}) \times [\mathbf{u}(\mathbf{x}_i) - \mathbf{u}(\mathbf{x}_e)]. \tag{4.3.1}$$

The physical character of the surface vortex sheet has been discussed already in Chapter 2, but it can be further elucidated through the following limiting process. We may consider the surfaces S_e and S_i as separated by a small distance ε. We suppose now that a vortex ω_γ is continuously distributed between S_e and S_i and is continuously normal to the local vector $\mathbf{n}$. This vortex induces velocities $\mathbf{u}_e$ and $\mathbf{u}_i$ on the surfaces S_e and S_i, respectively. In the limit of $\varepsilon \to 0$ the quantity $\varepsilon\omega_\gamma$ has a finite limit that we denote by $\boldsymbol{\gamma}$, defined as a vortex sheet. Using Kelvin's theorem on a loop around the surfaces S_i and S_e, we can easily show that

$$\boldsymbol{\gamma}(\mathbf{x}) \times \mathbf{n}(\mathbf{x}) = \mathbf{u}(\mathbf{x}_i) - \mathbf{u}(\mathbf{x}_e). \tag{4.3.2}$$

This vortex sheet is used to enforce the no-through-flow boundary conditions and at the same time corresponds to the fluid velocity on the boundary.

In a similar spirit, a solid-body rotation can be equivalently described in terms of a uniform vorticity field distributed in the interior of the body. The vorticity exterior to the body, the surface vortex sheet, and the solid-body rotation

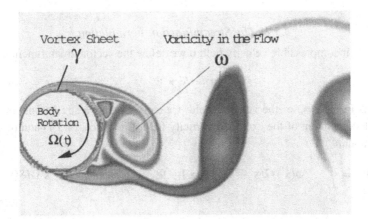

Figure 4.1. Definition of an extended vorticity field for flow past a translating and rotating cylinder.

constitute an extended vorticity field (Figure 4.1). We denote by $\mathbf{u}_\omega$ the velocity field induced by the vorticity of the flow and the solid-body rotation and by $\mathbf{u}_\gamma$ the velocity field induced by the surface vortex sheet.

Hence the velocity field throughout the domain can be represented as

$$\mathbf{u} = \mathbf{u}_\omega + \mathbf{u}_\gamma + \mathbf{u}_\infty, \tag{4.3.3}$$

where $\mathbf{u}_\infty$ denotes the velocity at infinity. For simplicity in this section we define

$$\mathbf{u}_{\text{ext}} = \mathbf{u}_\omega + \mathbf{u}_\infty. \tag{4.3.4}$$

The velocity and the vorticity field of each constituent are related by a Poisson equation with the form

$$\Delta \mathbf{u}_\alpha = \nabla \times \omega_\alpha, \tag{4.3.5}$$

where the index α can take the values ω and γ. Solving Poisson's equation for each constituent of the extended vorticity field, we obtain that

$$
\begin{aligned}
\mathbf{u}_\omega &= \nabla \times \int_{V_e \cup V_i} \omega(\mathbf{x}') G(\mathbf{x} - \mathbf{x}') \, dV(\mathbf{x}'), \\
\mathbf{u}_\gamma &= \nabla \times \int_S \gamma(\mathbf{x}') G(\mathbf{x} - \mathbf{x}') \, dS(\mathbf{x}'),
\end{aligned}
\tag{4.3.6}
$$

where $G(\mathbf{x} - \mathbf{x}')$ is given in Eq. (4.2.13). The strength of the vortex sheet must be adjusted so that the total velocity field of the flow satisfies the no-through-flow boundary condition.

4.3.2. The Streamfunction Formulation

For an incompressible velocity field $\mathbf{u}$ we define the vector streamfunction $\boldsymbol{\Psi}$:

$$\mathbf{u} = \nabla \times \boldsymbol{\Psi}. \tag{4.3.7}$$

As in the case of the velocity field, we can associate a streamfunction for each constituent of the extended vorticity field. From Eqs. (4.3.7) and (4.3.6) we obtain

$$\boldsymbol{\Psi}_\omega = \int_{V_e \cup V_i} \omega(\mathbf{x}') G(\mathbf{x} - \mathbf{x}') \, dV(\mathbf{x}'), \quad \boldsymbol{\Psi}_\gamma = \int_S \gamma(\mathbf{x}') G(\mathbf{x} - \mathbf{x}') \, dS(\mathbf{x}'). \tag{4.3.8}$$

We define also as $\boldsymbol{\Psi}_\infty$ the streamfunction corresponding to the velocity field at infinity, and for simplicity we define

$$\Psi_{\text{ext}} = \Psi_\omega + \Psi_\infty. \tag{4.3.9}$$

The vorticity field and the streamfunction are related as

$$\omega = \nabla \times \mathbf{u} = \nabla \times \nabla \times \Psi \tag{4.3.10}$$

$$= \nabla(\nabla \cdot \Psi) - \nabla^2 \Psi. \tag{4.3.11}$$

By requesting that the vector streamfunction be solenoidal ($\nabla \cdot \Psi = 0$), we obtain a Poisson equation for the streamfunction and the vorticity field:

$$\nabla^2 \Psi = -\omega. \tag{4.3.12}$$

Applying the Poincaré identity to each component of the streamfunction yields

$$\Psi(\mathbf{x}) = \Psi_{\text{ext}} + \int_S \frac{\partial(\Psi_i - \Psi_e)}{\partial \mathbf{n}}(\mathbf{x}')\,G(\mathbf{x} - \mathbf{x}')$$

$$- (\Psi_i - \Psi_e)(\mathbf{x}')\,\frac{\partial G}{\partial \mathbf{n}}(\mathbf{x} - \mathbf{x}')\,dS(\mathbf{x}'). \tag{4.3.13}$$

The above equation can be expressed in terms of singularity distributions on the boundary of the domain. We define the source distribution σ as

$$\sigma = \Psi_i - \Psi_e \tag{4.3.14}$$

and the vortex sheet γ as

$$\gamma = \frac{\partial \Psi_i}{\partial \mathbf{n}} - \frac{\partial \Psi_e}{\partial \mathbf{n}}, \tag{4.3.15}$$

so that Eq. (4.3.13) can be expressed as

$$\Psi(\mathbf{x}) = \Psi_{\text{ext}} + \int_S \left[\gamma(\mathbf{x}')\,G(\mathbf{x} - \mathbf{x}') - \sigma(\mathbf{x}')\,\frac{\partial G}{\partial \mathbf{n}}(\mathbf{x} - \mathbf{x}') \right] dS(\mathbf{x}'). \tag{4.3.16}$$

Note that the definition of the source distribution is consistent with the physical concept of defining the numerical difference in Ψ between two streamlines as the volume flow rate between the streamlines. Similarly the vortex sheet can be viewed as a velocity gradient across the interface.

For two-dimensional and axisymmetric flows, there is only one nonzero vorticity and streamfunction component, and the kinematic boundary conditions can be easily expressed in terms of the streamfunction. For example, requesting that the wall normal velocity at the surface of a stationary body be zero is equivalent to requesting that

$$\frac{\partial \Psi}{\partial \mathbf{s}} = 0 \quad \text{or} \quad \Psi_i = \Psi_e = 0 \tag{4.3.17}$$

on the surface of the body.

In this case, by considering the limiting form of the expressions in Eq. (4.3.16) and imposing the constraint that the streamfunction be continuous across the interface, we obtain an integral equation for the unknown strength of the vortex sheet that in two dimensions is given by

$$\frac{1}{2\pi} \int_S \gamma(\mathbf{x}') \log |\mathbf{x} - \mathbf{x}'| \, d\mathbf{x}' = \Psi_{\text{ext}}, \tag{4.3.18}$$

where Ψ_{ext} is the streamfunction that accounts for the vorticity field and the free stream on the exterior of the body.

In three dimensions this component-by-component analysis is not always possible as the kinematic boundary conditions involve a cross coupling of the various components of the streamfunction at the boundary $[\mathbf{u} \cdot \mathbf{n} = (\nabla \times \mathbf{\Psi}) \cdot \mathbf{n}]$. However, for simple geometries such as flat walls, suitable boundary conditions can be derived.

4.3.3. The Vorticity–Velocity Formulation

The kinematic boundary conditions can be enforced by direct exploitation of the relationship between the velocity and the extended vorticity field of the flow. This method is directly applicable to two- and three-dimensional flows and allows for a more physical description of the kinematic and viscous boundary conditions as they are related to the extended vorticity field.

The basis of the method is the following theorem, which describes the vector Poincaré identity for the velocity field $\mathbf{u}$ (with $\nabla \cdot \mathbf{u} = 0$) of the flow:

$$\mathbf{u}(\mathbf{x}) = \mathbf{u}_{\text{ext}} + \nabla \times \int_S [\mathbf{n} \times (\mathbf{u}_e - \mathbf{u}_i)](\mathbf{x}') G(\mathbf{x} - \mathbf{x}') \, dS(\mathbf{x}') \tag{4.3.19}$$

$$+ \nabla \left\{ \int_S [\mathbf{n} \cdot (\mathbf{u}_e - \mathbf{u}_i)](\mathbf{x}') G(\mathbf{x} - \mathbf{x}') \, dS(\mathbf{x}') \right\}, \tag{4.3.20}$$

where $\mathbf{u}_{\text{ext}} = \mathbf{u}_\omega + \mathbf{u}_\infty$.

If we require that the normal velocity component be continuous across the interface ($\mathbf{n} \cdot \mathbf{u}_e = \mathbf{n} \cdot \mathbf{u}_i$) we obtain that

$$\mathbf{u}_{\text{ext}} + \nabla \int_S \gamma(\mathbf{x}') G(\mathbf{x} - \mathbf{x}') dS(\mathbf{x}') = \mathbf{u}(\mathbf{x}). \qquad (4.3.21)$$

In order to enforce the boundary conditions we obtain first the limiting expressions of Eq. (4.3.21) for points on the surface of the body.

$$\mathbf{u}(\mathbf{x}_0) = \pm \frac{\gamma \times \mathbf{n}}{2} + \int_S \gamma(\mathbf{x}') \times \nabla G(\mathbf{x} - \mathbf{x}') dS(\mathbf{x}') \qquad (4.3.22)$$

where the $+$ or the $-$ sign appears when we approach the surface from the inside or the outside, respectively, of the flow domain.

The strength of the surface vortex sheet is of course unknown, and it would need to be determined by the kinematic boundary condition ($\mathbf{n} \cdot \mathbf{u}_e = \mathbf{n} \cdot \mathbf{u}_i$).

4.3.4. Boundary Integral Equations

We determine the strength of the vortex sheet by enforcing the kinematic boundary conditions. When we require that the wall normal velocity component be continuous across the interface the boundary condition of no-through flow can be imposed in terms of a Fredholm integral equation of the first or the second kind. Fredholm integral equations of the first kind, when discretized, usually lead to ill-conditioned systems of equations. We can obtain a well-conditioned system of equations by formulating the problem as a Fredholm integral equation of the second kind.

- We can obtain a Fredholm integral equation of the first kind for the unknown strength of the vortex sheet by enforcing that $\mathbf{u} \cdot \mathbf{n} = \mathbf{u}_b \cdot \mathbf{n}$ on the exterior of the body:

$$(\mathbf{u}_{\text{ext}} - \mathbf{u}_b) \cdot \mathbf{n} = \int_S \gamma(\mathbf{x}') \times \nabla G(\mathbf{x} - \mathbf{x}') \cdot \mathbf{n}(\mathbf{x}) dS(\mathbf{x}'). \qquad (4.3.23)$$

- The boundary condition of no-through flow can be expressed alternatively in terms of the tangential velocity component in the interior of the body, leading to a Fredholm integral equation of the second kind [121]

$$(\mathbf{u}_{\text{ext}} - \mathbf{u}_b) \cdot \mathbf{t} = + \frac{\gamma(\mathbf{x}) \times \mathbf{n}}{2} \cdot \mathbf{t}(\mathbf{x}) + \int_S \gamma(\mathbf{x}')$$
$$\times \nabla G(\mathbf{x} - \mathbf{x}') \cdot \mathbf{t}(\mathbf{x}) dS(\mathbf{x}'). \qquad (4.3.24)$$

However, Eq. (4.3.24) does not admit a unique solution in the case of multiply connected domains (e.g., any finite two-dimensional body, a torus in three dimensions, etc.). In order to obtain a unique solution and a well-conditioned system of equations, $m - 1$ constraints need be imposed (where m is the multiplicity of the domain). We may obtain such conditions by enforcing Kelvin's theorem along any curve that renders the domain simply connected. In engineering applications such an extra constraint is provided from the Kutta condition (see Subsection 4.4.6).

Both constraints may be viewed as compatibility conditions between the kinematic and the dynamic descriptions of the flow field. As Eq. (4.3.24) is often used in order to enforce the kinematic boundary conditions, we further discuss its uniqueness properties in the following sections.

4.3.5. Solution Uniqueness for Two-Dimensional Flows

In this subsection we are concerned with solid, nondeformable, impermeable boundaries of a body translating with velocity $\mathbf{u}_b(t)$ and rotating with angular velocity $\Omega_b(t)$ around its center of mass $\mathbf{x}_b$ in an incompressible flow. The surface of the body is denoted by S and its area by A_b, and we use the convention that $\mathbf{n}$ points outward from the body $\mathbf{s}$ is the direction of integration along the surface.

As discussed above, by enforcing the boundary condition of no-through flow, we can now write a boundary integral equation for the unknown strength of the scalar vortex sheet as

$$-\frac{\gamma(\mathbf{x})}{2} + \frac{1}{2\pi} \int_S \gamma(\mathbf{x}') \, \mathbf{n} \cdot \nabla \log|\mathbf{x} - \mathbf{x}'| \, d\mathbf{x}' = [(\mathbf{u}_{\text{ext}} - \mathbf{u}_b) \cdot \mathbf{s}]_i. \quad (4.3.25)$$

Equation (4.3.25) is a Fredholm integral equation of the second kind, the solution of which determines the strength (γ) of the free-surface vortex sheet when the right-hand side is given.

By setting

$$Q(\mathbf{x}, \mathbf{x}') = \frac{1}{\pi} \frac{\partial}{\partial n} [\log |\mathbf{x}(s) - \mathbf{x}(\mathbf{x}')|], \quad h = -2[(\mathbf{u}_{\text{ext}} - \mathbf{u}_b) \cdot \mathbf{s}]_i, \quad (4.3.26)$$

we may express Eq. (4.3.25) equivalently as

$$\gamma(\mathbf{x}) - \int Q(\mathbf{x}, \mathbf{x}') \, \gamma(\mathbf{x}') \, d\mathbf{x}' = h(\mathbf{x}). \quad (4.3.27)$$

Equation (4.3.27) is an integral equation of the second kind with a nontrivial homogeneous solution. According to the Fredholm alternative, a solution to

Eq. (4.3.27) exists if

$$\int_S h(\mathbf{x}') \, \xi(\mathbf{x}') \, d\mathbf{x}' = 0, \tag{4.3.28}$$

where ξ is the solution of the eigenvalue problem for the adjoint equation with $\lambda = 1$, i.e.,

$$\int_S Q(\mathbf{x}', \mathbf{x}) \, \xi(\mathbf{x}') \, d\mathbf{x}' = \lambda \xi(\mathbf{x}). \tag{4.3.29}$$

For the kernel $Q(\mathbf{x}, \mathbf{x}')$ considered, $\rho(\mathbf{x}) = \text{const.}$ is the solution for $\lambda = 1$ so the necessary condition is

$$\int_S h(\mathbf{x}') \, d\mathbf{x}' = 0. \tag{4.3.30}$$

This is indeed the case as the above condition is equivalent to

$$\int_S (\mathbf{u}_{\text{ext}} \cdot \mathbf{s})_i \, d\mathbf{x} = 0. \tag{4.3.31}$$

The solution of Eq. (4.3.27) is not unique because, if $f_0(\mathbf{x})$ is a solution, then an arbitrary number of solutions $\gamma(\mathbf{x})$ of the form

$$\gamma(\mathbf{x}) = \gamma_0(\mathbf{x}) + \alpha \rho(\mathbf{x}) \tag{4.3.32}$$

may be obtained, where α is an arbitrary constant and $\rho(\mathbf{x})$ is the solution of the following eigenvalue problem for $\lambda = 1$:

$$\int_S Q(\mathbf{x}, \mathbf{x}') \, \rho(\mathbf{x}') \, d\mathbf{x}' = \lambda \rho(\mathbf{x}). \tag{4.3.33}$$

We can obtain a unique solution by imposing one more additional constraint on the strength of the surface vortex sheet. This can be achieved by use of the momentum equation for the velocity field on both sides of the interface to obtain an evolution equation for the strength of the vortex sheet. It can be shown then that

$$\frac{d}{dt} \int_S \gamma(\mathbf{x}, t) \, d\mathbf{x} = -\frac{d\Gamma_\Omega}{dt}, \tag{4.3.34}$$

where

$$\Gamma_\Omega = \int_{V_e} \Omega(\mathbf{x}_b) \, d\mathbf{x}_b. \tag{4.3.35}$$

It is in the sense of Eq. (4.3.34) that the total vorticity is conserved in an inviscid flow.

Note that if there are multiple bodies present in the flow, the above condition needs to be satisfied for each individual body. For such geometries the kernel $Q(\mathbf{x}, \mathbf{x}')$ admits multiple eigenfunctions ρ. In particular, for a domain with multiplicity m one needs to supply $m - 1$ constraints for the strength of the vortex sheet. In calculations in which the vorticity is allowed to enter from the body into the fluid (simulating or modeling viscous effects) care must be exercised so that the vorticity field generated by each individual body is properly tracked.

Hence uniqueness for the strength of the vortex sheet can be enforced, or, equivalently, α is uniquely determined by

$$\alpha = -\Gamma_\Omega + \frac{\int_S \gamma_0(\mathbf{x})\, d\mathbf{x}}{\int_S \rho(\mathbf{x})\, d\mathbf{x}}. \tag{4.3.36}$$

Summarizing, we find that the governing set of equations is given by

$$\gamma(\mathbf{x}) - \frac{1}{\pi} \int_S \frac{\partial}{\partial \mathbf{n}} (\log |\mathbf{x} - \mathbf{x}'|)\, \gamma(\mathbf{x}')\, d\mathbf{x}' = -2\mathbf{u}_{\text{ext}} \cdot \mathbf{s}, \tag{4.3.37}$$

$$\frac{d}{dt} \int_S \gamma(\mathbf{x}, t)\, d\mathbf{x} = -\frac{d\Gamma_\Omega}{dt}. \tag{4.3.38}$$

These sets of equations may be solved using a panel method. Discretizing the body with M vortex panels results in a system of equations for the M unknown strengths, which may be expressed in matrix form:

$$Q f = h. \tag{4.3.39}$$

This system has $M + 1$ equations but with only M unknowns, the strengths of the panels. In order to solve this latter system of equations, several approaches are plausible: least-squares solution, introduction of a new unknown, elimination of one equation, etc. However, these approaches rely mainly on empirical criteria. An alternative way for the application of the inviscid boundary condition [121] is based on the spectral decomposition of the kernel $Q(\mathbf{x}, \mathbf{x}')$. This kernel may be decomposed as

$$Q(\mathbf{x}, \mathbf{x}') = \sum_{i=1}^{\infty} \lambda_i \rho_i(\mathbf{x}) \xi_i(\mathbf{x}'), \tag{4.3.40}$$

where λ_i and $\rho_i(\mathbf{x})$ are eigenvalues and eigenfunctions, respectively, given by the solution of Eq. (4.3.33) and $\xi_i(\mathbf{x})$ are eigenfunctions of the transposed kernel given by the solution of Eq. (4.3.29). These eigenfunctions are normalized so that Eqs. (4.3.29) and (4.3.33) are valid for the decomposed kernel defined in Eq. (4.3.40). Hence

$$\int_S \rho_i(\mathbf{x})\, \xi_j(\mathbf{x})\, ds = \begin{cases} 1, & \text{if } i = j \\ 0, & \text{otherwise} \end{cases}, \tag{4.3.41}$$

$$\xi_1(\mathbf{x}) = \text{const.} = 1/L, \tag{4.3.42}$$

where L is the perimeter of the shape considered. Note that the eigenfunction $\rho_1(\mathbf{x})$ corresponding to the eigenvalue $\lambda_1 = 1$ is a solution of Laplace's equation satisfying the inviscid boundary condition in the absence of any external flow field.

Now the condition imposed by Kelvin's theorem may be substituted by an equivalent one by multiplying both sides of Eq. (4.3.27) by $\lambda_1 \rho_1(\mathbf{x})\, \xi_1(\mathbf{x}')$, so that the new set of equations is

$$f(\mathbf{x}) - \int Q(\mathbf{x}, \mathbf{x}')\, f(\mathbf{x}')\, d\mathbf{x}' = h(\mathbf{x}), \tag{4.3.43}$$

$$\frac{\rho_1(\mathbf{x})}{L} \int f(\mathbf{x}')\, d\mathbf{x}' = -\frac{\rho_1(\mathbf{x})}{L} \Gamma_\Omega. \tag{4.3.44}$$

Adding Eq. (4.3.44) to Eq. (4.3.43) results in the final form of the integral equation that needs to be solved for the unknown strength of the vortex sheet:

$$\gamma(\mathbf{x}) - \int \left[Q(\mathbf{x}, \mathbf{x}') - \frac{\rho_1(\mathbf{x})}{L} \right] \gamma(\mathbf{x}')\, d\mathbf{x}' = h(\mathbf{x}) - \frac{\rho_1(\mathbf{x})}{L} \Gamma_\Omega. \tag{4.3.45}$$

In this form the new kernel is just the kernel given by Eqs. (4.3.40) but with the first term in the spectral decomposition eliminated. Hence the singularity associated with the kernel $Q(\mathbf{x}, \mathbf{x}')$ is annihilated and Eq. (4.3.45) is well posed. When Eq. (4.3.45) is solved by panel methods, the resulting system of equations is a well-conditioned one. Note that in the above formulation of the problem it is necessary to know *a priori* the form of the eigenfunction $\rho_1(\mathbf{x})$, which is not available in general for arbitrary shapes. For elliptic bodies (including the cylinder and the flat plate) this eigenfunction can be determined analytically and is given as a function of the eccentricity ε of the ellipse and the polar angle θ:

$$\rho_1(\theta) = [1 - \varepsilon \cos^2(\theta)]^{-1/2}. \tag{4.3.46}$$

For arbitrary configurations, we may obtain $\rho_1(\mathbf{x})$ by solving the eigenproblem

(4.3.33) by using panel methods. We then obtain the eigenfunction by using as collocation points the eigenvector of the resulting matrix. This adds to the computational cost of the method, but this eigenfunction needs to be computed only once for the considered shape. For the case of a cylinder, $\varepsilon = 0$, so the unknown function $\gamma(\mathbf{x})$ may be obtained directly as Eq. (4.3.45) reduces to

$$\gamma(\mathbf{x}) = h(\mathbf{x}) - \frac{1}{L}\Gamma_\Omega. \qquad (4.3.47)$$

4.3.6. The Kutta–Joukowski Condition

An extended potential flow model has been used extensively in the past decades as an engineering approximation of steady, lifting flows past streamlined bodies, such as airfoils at small angles of attack. In these flows the viscous effects are confined to a thin layer around the surface of the airfoil, and vorticity is shed smoothly from its trailing edge. Potential flow can serve as an approximation of this flow when it is complemented by an additional constraint that accounts for the generation of the lift force and produces a pattern in which the flow leaves smoothly at the trailing edge. As further discussed by Bachelor [18], when separation is avoided and vorticity is shed smoothly from the trailing edge of a lifting airfoil, a non-zero circulation is established for a contour surrounding the airfoil. This circulation is prescribed by the Kutta–Joukowski condition, which simply states that the proper circulation constant is precisely the value that causes the flow to leave smoothly at the trailing edge of the airfoil.

We can further elucidate the need for the Kutta–Joukowski condition in a potential flow model by considering the flow past an airfoil with a sharp trailing edge that satisfies Laplace's equation with the no-through-flow boundary condition. This can result in a flow pattern in which the streamlines go around the trailing edge (see Figure 4.2) and the velocity (or equivalently the suction) at the trailing edge of the airfoil becomes infinite. If we consider this inviscid flow model as an approximation to a viscous flow, it would correspond to a sharply reduced pressure and an inverse pressure gradient near the airfoil's trailing edge. In a steady flow this would lead to flow separation with streamlines that do not follow the shape of the airfoil, thus contradicting our initial assumptions. We can approximate a realistic flow pattern with the addition of a circulation constant, which brings this separation point to the trailing edge of the airfoil.

The use of this approximation has been strengthened by experimental results that have demonstrated that the lift experienced by an airfoil is practically

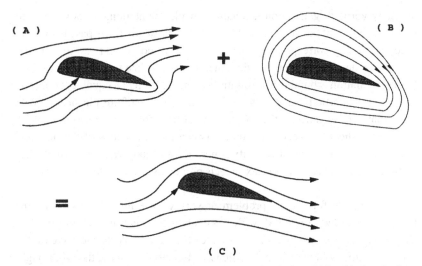

Figure 4.2. Enforcement of the Kutta condition: A circulation (B) is added to the flow pattern (A) to produce the flow pattern (C).

independent of the Reynolds number of the flow, once a critical value has been exceeded. Hence the governing processes can be represented by inertial considerations. However, in practice the trailing edge of the airfoil is not sharp, as it's hidden beneath viscous boundary layers. Moreover the pressure difference on the top and the bottom of the airfoil results in uneven boundary layer thicknesses. In reality the inviscid flow is not around the airfoil profile but around a rather different configuration.

In spite of all these difficulties, the Kutta condition is one of the major working assumptions in airfoil theory. Potential flow models that use the Kutta condition have been extensively used in the design of airfoils and have been able to provide a remarkably good estimate of the lift forces and pressure distributions for a large number of different airfoil configurations.

4.3.7. The Vortex Shedding Model for Inviscid Flow Calculations

In engineering calculations, in order to approximate the dynamics of high Reynolds number flows past lifting bodies, the inviscid approximation of potential flow is usually complemented by an unsteady vorticity generation mechanism to account for the injection of vorticity in the wake. A thorough review of such models can be found in Ref. 183. Unlike the case of a steady flow, here the goal is not to approximate a certain flow pattern, but to model the

unsteady vorticity generation mechanism while maintaining an otherwise inviscid evolution of the vorticity field. In general unsteady flows past airfoils are
accompanied by corresponding changes in the lift coefficient and the circulation about the airfoil. From Kelvin's theorem we realize that any change $\Gamma(t)$
in the circulation around the airfoil must be balanced by an equal and opposite
change in the vorticity shed in the wake. A description of this physical process
is facilitated if we consider the flow field in terms of the extended vorticity field.
The vortex sheet can serve as a link between the kinematics of the flow and
the dynamic phenomena of vorticity generation at a body surface. Vorticity creation can be achieved if the vortex sheet is properly shedded into the wake of the
airfoil.

We formulate further the problem of vortex shedding by considering the
flow in a bounded domain V_i with a boundary S, which is smooth except at
the trailing edge E, where it has a cusp (see Figure 4.3). When we solve for the
Euler equation with no external vorticity in V_i while enforcing the no-through-
flow boundary condition, a singularity exists at the sharp corner of the body.
This singularity can be easily derived through conformal mapping techniques
for profiles like Joukowski airfoils. In order to obtain a finite velocity at all body
locations, we can eliminate this singularity by creating a suitable vorticity field
in the domain. This vorticity must result from a shedding mechanism, which
we now describe.

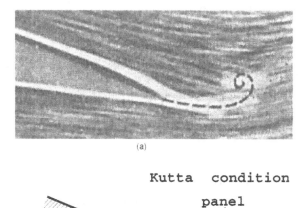

Figure 4.3. Shedding model for the trailing edge of an airfoil: (a) Flow at the trailing
edge of an impulsively started airfoil [164], (b) a point-vortex shedding model.

Let us start with the Euler equations written in terms of velocity, vorticity, and pressure as

$$\frac{\partial \mathbf{u}}{\partial t} - \mathbf{u} \times \omega + 1/2\nabla(|\mathbf{u}|^2) = -\nabla P. \qquad (4.3.48)$$

Taking the tangential component of the Euler equation and observing that $(\mathbf{u} \times \omega) \cdot \mathbf{s} = \omega \mathbf{u} \cdot \mathbf{n}$ and that $\mathbf{u} \cdot \mathbf{n} = 0$ on S, we can write, after integrating over S,

$$\frac{d}{dt} \int_S \mathbf{u} \cdot \mathbf{s}\, ds + \frac{1}{2}[|\mathbf{u}(E^+)|^2 - |\mathbf{u}(E^-)|^2] = 0. \qquad (4.3.49)$$

In the above equation we denote by $\mathbf{u}(E^+)$ and $\mathbf{u}(E^-)$ the values of the velocity on the upper and the lower part of the edge, respectively. Moreover, we have used the concept of the Kutta condition for steady lifting flows, setting the pressure to be equal at the two sides of the cusp.

Now if we recall that the circulation on S is equal to the integral of vorticity in V_i, we obtain

$$\frac{d}{dt} \int_{V_i} \omega\, dx = \frac{1}{2}[\mathbf{u}(E^+) - \mathbf{u}(E^-)] \cdot [\mathbf{u}(E^+) + \mathbf{u}(E^-)] \qquad (4.3.50)$$

This implies that vorticity must be created on S to ensure conservation of the circulation. Now the only point where this vorticity creation can take place is precisely the point E, since elsewhere the normal velocity is zero. Moreover, it is reasonable to set the velocity at this point to be the average velocity $\bar{\mathbf{u}} = 1/2[\mathbf{u}(E^+) + \mathbf{u}(E^-)]$. Hence Eq. (4.3.50) can be satisfied by introduction in the flow at each time step of an elementary vortex sheet with length $\delta t\, \bar{\mathbf{u}}$ and strength $[\mathbf{u}(E^+) - \mathbf{u}(E^-)]$. The effect of this vortex sheet is to extend the discontinuity from the surface of the airfoil to the region behind its trailing edge. This mechanism must be understood as a limiting procedure of the underlying viscous flow, in which vorticity would be created all over the body. The assumption behind the validity of this limiting behavior is that the vorticity does not enter the flow through diffusion, but only through convection at the shedding point. It is then convected downstream as a shear layer across which the velocity difference is that of the velocity field on the two sides of the trailing edge.

A natural procedure to simulate this shedding mechanism in the flow past a lifting airfoil consists of the following steps:

1. Discretize the boundary of the airfoil with M panels (see also Section 4.4).
2. Compute the surface vortex sheet in order to satisfy the no-through-flow boundary condition over M elements of the boundary, as indicated in Subsection 4.3.5.

3. Compute the tangential velocities at the end points E^+ and E^-; create a new vortex at the point $E + \Delta t/2[\mathbf{u}(E^+) + \mathbf{u}(E^-)]$ carrying a vorticity of magnitude $\Delta t[\mathbf{u}^2(E^+) - \mathbf{u}^2(E^-)]/2$.
4. Move the vorticities in the wake with the velocity induced by the extended vorticity field.

This shedding model allows us to handle vorticity creation by means of a quasi-inviscid method that fully takes advantage of the quasi-one-dimensional structure of the wake.

Nitsche and Krasny [158] have made three-dimensional calculations of this kind in order to simulate the vorticity generated at the orifice of a nozzle. In Figure 4.4 we present the results of these simulations which provide an excellent qualitative agreement with the results from related experiments.

4.4. Discretization of the Integral Equations

The integral equations that result from the enforcement of the kinematic boundary conditions can be discretized by use of various quadratures. A particular method that has been used extensively in aerodynamics is the panel method, pioneered by Hess in 1975.

In panel methods the boundary is discretized by use of a set of quadrilateral- or (for more flexibility) triangular-shaped elements and a distribution is assumed for the singularity strengths over each element. The integrals over each element (panel) can then be computed either by an exact integration or a local quadrature formula.

The integral equations are thus replaced by a set of algebraic equations, and the resulting matrix equations can be solved by means of direct and/or iterative matrix inversion techniques. These matrices are dense but fast summation methods (see Appendix B) can be implemented, resulting in a computational cost linearly proportional with the number of computational elements at the boundary.

4.4.1. Panel Methods

In order to outline the panel method, we consider the discretization of integral Eq. (4.3.25) that determines the strength of a vortex sheet distribution around a two-dimensional surface:

$$\gamma(\mathbf{x}) - \frac{1}{\pi} \int_S \frac{\partial}{\partial \mathbf{n}} (\log |\mathbf{x} - \mathbf{x}'|) \, \gamma(\mathbf{x}') \, d\mathbf{x}' = -2\mathbf{u} \cdot \mathbf{s}. \qquad (4.4.1)$$

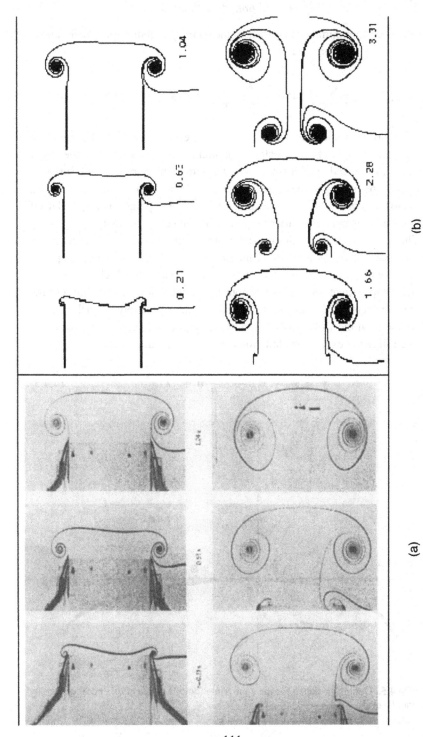

(a)

(b)

Figure 4.4. (a) Experimental visualizations and (b) simulations of the vortex generated by a jet emerging from a nozzle. (Courtesy of M. Nitsche)

111

When we approximate the surface of the body by a set of M panels, the above equation may be expressed as

$$\gamma(\mathbf{x}) - \frac{1}{\pi} \sum_{i=1}^{M} \int_{J_i} \gamma(\mathbf{x}') \times \frac{\mathbf{x} - \mathbf{x}'}{|\mathbf{x} - \mathbf{x}'|^2} |\mathbf{n} \, dl(\mathbf{x}')| = -2\mathbf{u} \cdot \mathbf{s}. \qquad (4.4.2)$$

The body points appearing in the integrals over each panel can be distinguished by two categories. The first category contains the points on the panel itself whereas the second contains points on the remaining portion of the surfaces. For points belonging to the surface outside each panel, the position vector can be set to the actual value and no singularity occurs. For points on the panel itself care must be exercised in order to evaluate the integrals correctly.

The key elements in the discretization are the approximation of the geometry of the body and the assumption of the singularity distribution over each panel. We denote by $\mathbf{x}_0$ a point on the surface of the body around which we construct a panel approximation, and we denote by $\mathbf{n}$ and $\mathbf{s}$ the wall normal and the tangential vector on the body surface at this point, respectively. Then the surface of the body in the vicinity of this point (see Figure 4.5) can be approximated by the following expansion along the local tangential direction:

$$\mathbf{y} = \mathbf{x}_0 + x_1 \mathbf{s} - \frac{1}{2} \frac{x_1^2}{R_0} \mathbf{n} + O\left(x_1^3\right), \qquad (4.4.3)$$

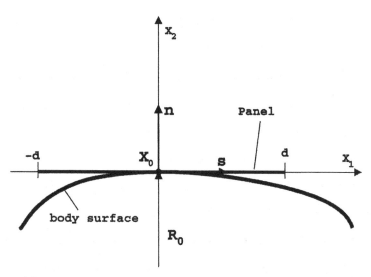

Figure 4.5. Definition sketch for the approximation of a segment of a body surface by a panel $(-d, d)$.

where R_0 denotes the local radius of curvature of the body. By using this expansion we can calculate the vector area element of each panel divided by the distance from the center of the panel as

$$\frac{dl_y}{|\mathbf{x} - \mathbf{x}'|^2} = \frac{dx_1}{r_p^2} \left[1 + \left(\frac{x_1^2}{2 R_0^2} \right) \right]. \qquad (4.4.4)$$

Similarly, an expansion of the vortex sheet distribution over each panel can be constructed as

$$\gamma(\mathbf{x}) = \gamma(\mathbf{x}_0) + x_1 \left. \frac{\partial \gamma}{\partial x_1} \right|_{\mathbf{x}_0} + \frac{1}{2} x_1^2 \frac{\partial^2 \gamma}{\partial x_1^2} + O\left(x_1^3\right). \qquad (4.4.5)$$

The accuracy of the method increases by an increase in the smoothness of the local approximation for the geometry and the singularity strengths. A Taylor series expansion in terms of the size of the panels has shown [102] that the order of the geometry approximation of the surface should be an order of magnitude larger than the order of the singularity strength approximation. Table 4.1 provides a summary of consistent approximations in terms of the complexity of the panel geometry and the respective location of the enforcement of the boundary condition and the local distribution of the singularity strength.

The articles by Hess [101, 102] and the book of Katz and Plotkin [113] are good resources for a variety of singularity/panel distributions. In general the set of integral equations is replaced by an equivalent system of linear equations of the form

$$A_{mn} s_n = r_m, \qquad (4.4.6)$$

where A_{mn} denotes the influence of a unit singularity strength from panel n at panel m, s_n is the unknown strength of the singularity distribution at panel n,

Table 4.1. *Consistent approximations for singularity strengths and panel geometry*

Order/Parameter	*0th Order*	*1st Order*	*2nd Order*
Singularity strength	Concentrated at a point – δ function	Constant across panel	Linear variation across panel
Panel geometry	Constant height	Internal or external facets	Parabolic or facet + curvature
Boundary condition	Single point on internal facet	Single point on actual surface	Average across panel

and r_m denotes the boundary condition as it is enforced on panel m. For constant geometries the above set of equations is inverted only once and is subsequently stored so that the enforcement of the boundary conditions is proportional to a matrix–vector multiply, thus scaling with the square of the boundary elements. However, this quadratic computational cost makes the method prohibitive for geometries consisting of more than a few thousand elements (as it is usually the case for three-dimensional configurations). This is another type of the classical N-body problem that we have encountered in the case of the computation of the Biot–Savart velocity field for vortex particles. This quadratic cost can be reduced if multipole expansions are used to approximate the potential or velocity field induced by the panels at distances that are larger than the characteristic length of the panel. In such cases an iterative method is implemented for the solution of the matrix equation. Moreover the matrix elements are never formed but the matrix–vector multiply is performed with the aid of the multipole expansions.

4.5. Accuracy Issues Related to the Regularization near the Boundary

The integral equations discussed in the previous sections contain terms involving the vorticity in the flow field. We discuss here issues relating to the regularization of the kernels of these integral equations in the presence of boundaries.

4.5.1. General Setting

The vorticity field is discretized by particles, and we wish to implement finite-size blobs in order to avoid singularities in the calculation of the velocity field and to allow for high-order quadrature approximations. The difficulty in implementing these smooth vortices can be illustrated by considering, for example, Eq. (4.2.5). In this equation the integral corresponding to the interior vorticity is restricted to the domain V_i. This implies that if a particle is close to the boundary and if we consider a spherical blob of the kind constructed in Chapter 2 around it, we have to ignore the part of the blob outside V_i. At the same time the basic feature of the blobs, namely that they contain the appropriate mass, implies that such a cutoff in a spherical blob will introduce immediately an $O(1)$ error in the reconstruction of a continuous vorticity field except, as we have seen in Section 2.6, if the vorticity vanishes sufficiently fast at the boundary.

One may argue that this error affects only the vorticity and is localized around the boundary. Alternatively, since the velocity is proportional to an integral of the vorticity, and, as the effect of regularization on the calculation of the velocity decays exponentially away from the blobs, this error will be of the order of $O(\varepsilon)$ around the boundary and will reduce to the regularization error away from the boundary [typically $O(\varepsilon^2)$]. However, it is important to realize that the computation of the velocity at the boundary directly interferes with the correct calculation of the boundary potentials that are required in formula (4.4.1) for example; therefore an $O(\varepsilon)$ error in the calculation of the velocity at the boundary may ultimately affect the calculation of the velocity well inside the domain. Moreover, if the calculation of the velocity at the boundary is done only within $O(\varepsilon)$ and if it is involved in a vorticity generation algorithm of a kind that we will discuss later on for the Navier–Stokes equations, then the $O(\varepsilon)$ error will accumulate at each time step and the whole algorithm will not be consistent unless $\varepsilon \ll \Delta t$. It is thus crucial to control the regularization difficulties at the boundary.

We discuss here two ways to remedy this situation. The first one is to compensate for the loss of vorticity resulting from chopping blobs at the boundary by locally increasing the strength of the cutoff with the constraint that the mass of the blobs remains equal to one.

Let ζ denote the smoothing function for a spherical blob. We denote by $\mathbf{x}$ a point in V_i and by r a positive number. One possibility is to consider the following blob shape at the point $\mathbf{x}$:

$$\zeta_\varepsilon(\mathbf{x}, \mathbf{y}) = \frac{1}{M(\mathbf{x})}\zeta[(\mathbf{x} - \mathbf{y})/\varepsilon],$$

where the normalization $M(\mathbf{x})$ is given by

$$M_\varepsilon(\mathbf{x}) = \int_\Omega \zeta\left(\frac{\mathbf{x} - \mathbf{y}}{\varepsilon}\right) d\mathbf{y}.$$

Unfortunately, in general these blobs do not lead to clear explicit formulas for their induced velocity field. Moreover it should be emphasized that they still do not provide a consistent approximation of the vorticity field as ε tends to 0. To illustrate this issue we reproduce here a simple example from Ref. 151 (more details on the construction of adapted blobs can be found in this reference). This is a one-dimensional example with $\Omega = [0, +\infty]$ and ζ is the top-hat function in the interval $[-1/2, 1/2]$. Straightforward calculations yield

$$M_\varepsilon(x) = \begin{cases} x + \varepsilon/2 & \text{if } 0 \leq x \leq \varepsilon/2 \\ \varepsilon & \text{if not} \end{cases}.$$

Now if we consider a uniform initial vorticity field $\omega \equiv 1$ and compare ω with its regularization we obtain

$$1 - \int_\Omega \zeta(x, y)\, dy = \begin{cases} 0 & \text{if } \varepsilon < x \\ 1 - x/\varepsilon + \log(x/\varepsilon) & \text{if } \varepsilon/2 \leq x \leq \varepsilon \ . \\ 1 - x/\varepsilon + \log(1/2) & \text{if } x \leq \varepsilon/2 \end{cases}$$

At $x = 0$ for example the error is $\log(1/2) = O(1)$. It thus seems that, compared with the usual spherical blobs, these blobs offer the advantage only of conserving the total vorticity, but they do not improve the pointwise accuracy of the computed velocity.

The second approach [194, 205] consists of using blobs with adapted shapes near the boundary. Flat vortex sheets and elliptical blobs with a minor axis smaller than the distance of the blob to the boundary are good examples of such shapes. However, the formulas giving the parameters of these shapes are not straightforward and it is not clear whether they can lead to efficient algorithms in general geometries.

In closing, we wish to state that these issues are to some extent much easier to deal with in the context of triangulated vortex methods (see Subsection 7.1.2) or vortex-in-cell methods (see Section 8.2). The advantages of these methods is that they do not require explicit regularization (hence avoiding the related difficulties) for the computation of the velocity as this regularization is implicitly contained in the reconstruction of a piecewise polynomial vorticity field on the underlying mesh.

4.5.2. Particular Cases

We describe here several particular cases in which the simplicity of the geometry allows simple and accurate procedures to overcome the difficulties at the boundary. To simplify the notation we will assume throughout this section that $\mathbf{u}_b \cdot \mathbf{n} = 0$.

In the general framework we assume that some smooth extension $(\tilde{\mathbf{u}}, \tilde{\omega})$ of the solution $(\mathbf{u}, \omega)$ of the original problem is available in some region (not necessarily large) outside V_i. The numerical algorithm can then be defined in the following way:

1. First initialize particles $\tilde{\mathbf{x}}_p$ with values of $\tilde{\omega}_0$ (this includes particles outside V_i).
2. At each time step define a collection of blobs,

$$\tilde{\omega}_\varepsilon = \sum_p \tilde{\alpha}_p \zeta_\varepsilon[\mathbf{x} - \tilde{\mathbf{x}}_p(t)]$$

where ζ_ε is a smoothing function with spherical symmetry.

3. Using, for example, integral formula (4.4.1), compute the velocity fields $\mathbf{u}_\varepsilon$ in V_i that satisfy

$$\nabla \times \mathbf{u}_\varepsilon = \boldsymbol{\omega}_\varepsilon \text{ in } V_i; \quad \mathbf{u}_\varepsilon \cdot \mathbf{n} = 0 \text{ on } S. \tag{4.5.1}$$

The only thing that is not defined so far in this algorithm is the velocity field of the particles lying outside V_i. To be more specific, let us focus on the two-dimensional case. The solution proposed in Ref. 203 is to extrapolate the velocity from values known inside the domain. If $\tilde{\mathbf{x}}_p \in \mathbf{R}^2 - V_i$, define $\tilde{\mathbf{x}}_p^0$, its nearest point on S, and the two points in V_i as

$$\tilde{\mathbf{x}}_p^1 = 2\tilde{\mathbf{x}}_p^0 - \tilde{\mathbf{x}}_p, \quad \tilde{\mathbf{x}}_p^2 = 3\tilde{\mathbf{x}}_p^0 - 2\tilde{\mathbf{x}}_p. \tag{4.5.2}$$

Finally move the points outside V_i with the velocity computed according to the rule

$$\tilde{\mathbf{u}}_\varepsilon(\tilde{\mathbf{x}}_p) = 3\mathbf{u}_\varepsilon\left(\tilde{\mathbf{x}}_p^1\right) - 2\mathbf{u}_\varepsilon\left(\tilde{\mathbf{x}}_p^2\right). \tag{4.5.3}$$

Clearly the accuracy of the whole scheme is conditioned only by criteria already defined for the blob shapes in the whole space problem and by the accuracy of the extrapolation steps of Eqs. (4.5.2) and (4.5.3). Owing to the fact that we assumed that $\tilde{\mathbf{u}}$ is smooth, this extrapolation can be proved to be of second order. It is possible to write improved extrapolation formulas.

It is important to note that, because of the no-through-flow boundary condition, particles outside the domain remain outside; therefore it is not neccesary to feed the neighborhood of V_i with fresh particles. Figure 4.6 shows how the contribution of particles outside V_i affects the calculation of the velocity in the domain in a way that clearly differs significantly from the previously seen schemes.

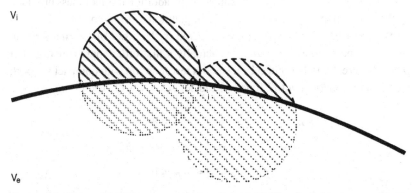

Figure 4.6. Vortex blobs for an extended vorticity field. The shaded area represents the parts of the blobs that contribute to the velocity evaluation in V_i.

The major practical difficulty here obviously lies in the initialization of particles outside the domain. This makes it necessary to have an explicit description of the extension of the initial vorticity. Moreover, the way to explicitly construct the particles at $\tilde{\mathbf{x}}_p^1$ and $\tilde{\mathbf{x}}_p$ is not obvious if the geometry is not simple (for example, there might be several neighbors of a given particle outside the domain if the boundary has corners).

Finally we describe a classical particular case in which all formulas are explicit and easy to implement. This is the case of the half plane or of geometries that can reduce to this case through conformal mapping. We assume here that

$$V_i = \{\mathbf{x} = (x_1, x_2), x_2 > 0\}.$$

We still wish to solve Eq. (4.5.1), with an additional prescribed behavior at infinity $\mathbf{u}(x_1, x_2) \to 0$ for $x_2 \to +\infty$, but now we observe that if $\mathbf{u}$ is a solution and if we make an odd extension ω across the boundary $x_2 = 0$, by defining

$$\bar{\omega}(x_1, x_2) = \begin{cases} \omega(x_1, x_2) & \text{if } x_2 > 0 \\ -\omega(x_1, x_2) & \text{if not} \end{cases},$$

then $\bar{\mathbf{u}} = \mathbf{K} \star \bar{\omega}$ coincides with $\mathbf{u}$ in V_i. Therefore we deal with a problem in the whole space and we can use smoothing kernels with spherical symmetry. In practice this consists of considering image particles across the boundary, carrying circulation with opposite sign. The accuracy in the regularization is now limited only by the smoothness of $\bar{\omega}$. Actually this function is not always differentiable at the boundary, and the regularization error on the vorticity might very well be large. However, we again observe that we compute velocities and not vorticities, and we can take advantage of the additional smoothness provided by the integration step involved in the computation of the velocity.

We repeat here the simple argument given in the proof of Lemma 2.6.2 to show that the regularized velocity is actually a first order approximation. Let $p > 1$ be a real number and ϕ be a smooth test function with compact support; let ζ be any spherical cutoff function and set $\bar{\omega}_\varepsilon = \bar{\omega} \star \zeta_\varepsilon$. Then

$$|\langle \bar{\omega}_\varepsilon - \bar{\omega}, \phi \rangle| = |\langle \bar{\omega}, \phi \star \zeta_\varepsilon - \phi \rangle|$$
$$\leq \|\bar{\omega}\|_{0, p*} \|\phi \star \zeta_\varepsilon - \phi\|_{0, p}$$
$$\leq C\varepsilon \|\bar{\omega}\|_{0, p*} |\phi|_{1, p}.$$

This estimate directly shows that, with the notations introduced in Appendix A,

$\|\bar{\omega}_\varepsilon - \bar{\omega}\|_{-1,p} = O(\varepsilon)$, and, in view of Eq. (A.3.6),

$$\|\mathbf{K} \star (\bar{\omega}_\varepsilon - \bar{\omega})\|_{0,p} \leq C\varepsilon.$$

This error estimate improves to second order away from a layer of width ε around the boundary.

It turns out that this method of regularization is more efficient than could be inferred from this estimate.

An important point to note is that the normal velocity given by this construction vanishes at the boundary, without any need to incorporate further single- or double-layer potentials that would propagate the boundary errors to the whole flow. Furthermore one can prove that the tangential velocity is computed with second-order accuracy. This is very important in view of the fact that this component of the velocity is generally used to incorporate vorticity in the flow when we are solving the Navier–Stokes equations. We summarize these properties in the following.

Proposition 4.5.1. *Assume that ω is of class C^1 with compact support in $V_i = \{\mathbf{x} = (x_1, x_2), x_2 \geq 0\}$. Let $\mathbf{u}, \bar{\omega}, \bar{\mathbf{u}}$ be defined as given above and set $\bar{\mathbf{u}}_\varepsilon = \bar{\mathbf{u}} \star \zeta_\varepsilon$, where ζ is a spherical cutoff function. Then*

  (i) $\|\bar{\mathbf{u}} - \bar{\mathbf{u}}_\varepsilon\|_{0,\infty} = O(\varepsilon)$,
  (ii) $\bar{\mathbf{u}}_\varepsilon \cdot \mathbf{n} = 0$ on S,
  (iii) $\|\mathbf{u} \cdot \mathbf{s} - \bar{\mathbf{u}}_\varepsilon \cdot \mathbf{s}\|_{L^\infty(S)} = O(\varepsilon^2)$.

Proof. The first assertion has already been proved, and the second one results from elementary symmetry considerations; so we turn immediately to the third one.

To simplify the notation we drop the overbars everywhere so ω will denote a function defined in $\mathbf{R}^2$ with a discontinuity on S and smooth on each side of $\partial\Omega$. We can always write $\omega(y_1, y_2) = \lambda(y_1)H(y_2) +$ smooth function in $\mathbf{R}^2$, where λ has compact support and H is the Heaviside function. The smooth part of the vorticity can be handled in the usual way. It will produce a regularization error of the order of $O(\varepsilon^2)$ and we can focus on the discontinuous part.

We can split the regularization between the x_1 and the x_2 components. The x_1 component, in which direction the vorticity is smooth, will again produce an order $O(\varepsilon^2)$ error. This enables us to focus on the case when $\lambda(y_1) \equiv 1$ in a ball of radius M and vanishes outside. We further simplify the calculations by assuming that ζ is the characteristic function of the square $[-1/2, +1/2]^2$. We can now write

$$\mathbf{u} \cdot \mathbf{s}(x_1, 0) - \bar{\mathbf{u}}_\varepsilon \cdot \mathbf{s}(x_1, 0) = \int \frac{y_2}{|x - y|^2} [\omega(\mathbf{y}) - \omega_\varepsilon(\mathbf{y})] \, dy_2$$

$$= \int dy_1 \int_{-\varepsilon}^0 (-1 + y_2/\varepsilon) \frac{y_2}{x_1^2 + y_1^2 + y_2^2} \, dy_2$$

$$+ \int dy_1 \int_0^\varepsilon (1 - y_2/\varepsilon) \frac{y_2}{x_1^2 + y_1^2 + y_2^2} \, dy_2$$

$$= \int_{-M}^M \log \frac{x_1^2 + y_1^2 + \varepsilon^2}{x_1^2 + y_1^2} \, dy_1 = O(\varepsilon^2).$$

Note that the terms of order ε that arise from the integrals of the terms $y_2^2/x_1^2 + y_1^2 + y_2^2$ cancel because of symmetry.   □

Let us now discuss how this particular case in which the calculations are simple and accurate can extend to more general situations. First, as already mentioned, it is possible to recover all cases that can be deduced from the half space through conformal mappings; this is the case of the domain exterior to a disk, for which the conformal mapping can be easily written in terms of complex coordinates. All polygonal domains, including domains where one or more of the vertices is located at infinity, can also be mapped to the half space through a conformal mapping that can be explicitly written in complex coordinates. In all these cases, the construction of image particles and the computation of the velocity are easy to implement.

More generally one may wonder about the accuracy of a procedure that would consist of locally replacing the computational domain by the half space limited by the tangent plane to a given point of the boundary. Assume that the boundary is smooth and there exists some conformal mapping F from V_i to the half space, but for which explicit formulas are not known. Assume that for each point $\mathbf{x}$ in an ε neighborhood of S, there exists a unique normal $\mathbf{n}(\mathbf{x})$ passing through this point and define $\bar{\mathbf{x}} = \mathbf{x} - 2d(\mathbf{x}, S)\nu(\mathbf{x})$. The exact procedure for determining the velocity with no-through flow at the boundary would be to consider images in the complex plane and then to use $\mathbf{x}' = F^{-1}[\bar{F}(\mathbf{x})]$ instead of $\bar{\mathbf{x}}$. However a Taylor series expansion of F around the point of the boundary $\mathbf{x} - d(\mathbf{x}, S)$ shows that the distance between $\bar{\mathbf{x}}$ and the true image $\mathbf{x}'$ is of the order of $O(\varepsilon^2)$. Therefore the construction of ω_ε based on the images $\bar{\mathbf{x}}$ introduces an additional $O(\varepsilon^2)$ error in the regularized vorticity, and one can expect the computed velocity to be an $O(\varepsilon^2)$ approximation of the exact velocity. Note, however, that the normal velocity component will not be exactly zero, unlike the case in which true images are used, and that the calculation of $\bar{\mathbf{x}}$ itself is not straightforward.

5

Viscous Vortex Methods

Vortex methods were originally conceived as a tool to model the evolution of unsteady, incompressible, high Reynolds number flows of engineering interest. Examples include bluff-body flows and turbulent mixing layers. Vortex methods simulate flows of this type by discretizing only the vorticity-carrying regions and tracking the computational elements in a Lagrangian frame. They provide automatic grid adaptivity and devote little computational effort to regions devoid of vorticity. Moreover the particle treatment of the convective terms is free of numerical dissipation.

Thirty years ago simulations using inviscid vortex methods predicted the linear growth in the mixing layer and were able to predict the Strouhal frequency in a variety of bluff-body flow simulations. In three dimensions, we have seen that inviscid calculations using the method of vortex filaments have provided us with insight into the evolution of jet and wake flows. However, the inviscid approximation of high Reynolds number flows has its limitations. In bluff-body flows viscous effects are responsible for the generation of vorticity at the boundaries, and a consistent approximation of viscous effects, including diffusion, is necessary at least in the neighborhood of the body. In three-dimensional flows, vortex stretching and the resultant transfer of energy to small scales produce complex patterns of vortex lines. The complexity increases with time, and viscous effects provide the only limit in the increase of complexity and the appropriate mechanism for energy dissipation. In this chapter we discuss the simulation of diffusion effects in the context of vortex methods. The design of particle schemes to handle viscous effects is compelled by the desire to maintain the Lagrangian character of vortex methods. The mechanisms of vorticity generation and the related numerical results are discussed in Chapter 6.

In a time-stepping scheme, a natural approach is to consider successively in substeps the inviscid and the viscous parts of the equations. This algorithm, known as viscous splitting, relates to the classical concept introduced by Prandtl, as early as 1904 [164], of distinguishing the viscous and inviscid phenomena in the flow.

Chorin [49] formulated the method of viscous splitting for vortex methods. For an unbounded flow, in a time-stepping algorithm the scheme proceeds as follows. In the first substep vortices move with the local velocity to satisfy the inviscid part of the equations. In three dimensions vortex stretching is considered at this substep by a change in the strength of the vorticity-carrying elements. Diffusion effects are taken into account at the following substep. Chorin envisioned the resolution of diffusion effects in the mean, exploiting the stochastic character of the diffusion equation. He introduced the well-known method of random walk, in which particles undergo a Brownian-like motion to simulate the effects of diffusion. Viscous splitting and random-walk methods are described in Sections 5.1 and 5.2, respectively.

Although physically appealing, the random-walk method suffers some limitations, essentially because of its lack of pointwise accuracy. In the 1980s other methods were proposed to allow more flexibility in the choice of numerical parameters and increase the accuracy with which diffusion effects are described. The first such scheme was introduced in 1983 [169] with the idea of modifying the strengths of the particles instead of their locations. It again required a viscous splitting step algorithm and relied on the Green's function solution of the diffusion equation for small time steps. The integral was discretized by use of the locations of the Lagrangian computational elements as quadrature points. It amounts to neighboring particles with Gaussian shapes exchanging their vorticity so as to simulate diffusion effects.

It is interesting to mention here the close link of this formulation with the description of the diffusive process envisioned by Lighthill [138]. Lighthill states that "vorticity is proportional to the angular momentum of a spherical particle about its mass center, and when it diffuses *from one particle to another* the relevant angular momentum is about quite a different point. The correct view relates diffusion of vorticity to that of momentum, through that of momentum gradient." This is a consistent definition with the redistribution of particle strengths as each component of vorticity is proportional to a momentum gradient. This observation serves as one more manifestation of the close link of physics and mathematical concepts in the development of vortex methods. These so-called resampling methods are discussed in Section 5.3.

Subsequently it was realized that the idea of redistribution of the particle strengths could be generalized in the framework of integral approximation of the diffusion equation. Moreover these observations lead to the realization that viscous splitting is not a prerequisite for these methods. Physically speaking, diffusion is acting in a continuous manner on the vorticity field while it is transported along the trajectories of the fluid elements. The result was a class of methods admitting high accuracy in both time and space discretization and able to handle a variable viscosity coefficient, which we would call, in general, particle strength exchange (PSE) methods. The PSE method is discussed and analyzed in Section 5.4.

Section 5.5 is devoted to two other variants on the theme of viscous redistribution. The first one is in the spirit of the free Lagrange methods, originally used for compressible flow calculations [86]. These techniques are close relatives of finite-element methods and are based on Voronoi diagrams for the generation of a well-triangulated space from the irregular particle locations. These techniques have been implemented in two-dimensional flows, but they introduce a significant degree of complexity into the method and may require considerable effort to be extended to three-dimensional flows. The second one consists of adaptively determining the amount of vorticity to be transferred between neighboring particles through discrete versions of the momentum properties used for the derivation of PSE schemes.

An immediate observation of the results obtained by means of the available viscous schemes is that vortex methods face the same problems as all standard numerical methods regarding the scales that can be resolved. Returning to the realm of engineering applications, we may consider then the applicability of viscous vortex methods to such simulations. As will be further elucidated, viscous vortex methods retain their adaptive properties and they can still be used as computational tools for the simulation of complex unsteady flows. However, as one cannot expect to resolve all the scales of the flow, turbulence models or large-eddy-simulation-(LES-) type formulations are necessary to apply viscous vortex methods to most problems of engineering interest.

As a matter of fact, it appears that earlier successes of two-dimensional viscous vortex methods (including the core expansion scheme) in reproducing flow quantities measured experimentally were due to the turbulence modeling implied by these techniques, rather than an accurate treatment of the diffusion. More recently, formal turbulence modeling and LES methodologies have been introduced. In the former category we discuss the method of hairpin removal, introduced by Feynman in 1957 [82] and Chorin in 1990 [52]. In the latter we consider recent results that rely on the formulation of the PSE scheme for flows of variable, nonlinear, spatial diffusivity.

In closing this introduction, let us mention that, in order to focus on the treatment of diffusion effects by vortex methods, we will deal in this chapter, except for its last section, with two-dimensional flows for which the Navier–Stokes equations become in a Lagrangian frame a diffusion-type equation. However, it must be pointed out that all the described numerical schemes can easily incorporate the treatment of vorticity stretching as described in Chapter 3. Some of our numerical illustrations will indeed concern three-dimensional simulations. Most often the numerical analysis carries on to three dimensions as well.

Except in Section 5.1 below, in which the implications in numerical accuracy of the viscous splitting is discussed for the full Navier–Stokes equations, we would prove the convergence of the schemes in only the linear case. The underlying claim is that if, on the one hand, one is able to prove stability and consistency in the approximation of the viscous terms and if, on the other hand, convergence is well understood for the nonlinear Euler equations (see the convergence results in Chapters 2 and 3), convergence for the full Navier–Stokes equations will follow.

5.1. Viscous Splitting of the Navier–Stokes Equations

The evolution of the vorticity field in a two-dimensional viscous flow is described by the Navier–Stokes equations which may be expressed in a velocity–vorticity $(\mathbf{u}, \omega)$ formulation as

$$\frac{\partial \omega}{\partial t} = -\mathbf{u} \cdot \nabla \omega + \nu \Delta \omega. \tag{5.1.1}$$

The velocity field is obtained by solving the Poisson equation $(\Delta \mathbf{u} = -\nabla \times \omega)$.

The evolution of the flow is considered in discrete time steps. In each time step the vorticity field is convected (according to $\mathbf{u} \cdot \nabla \omega$) and diffused (according to $\nu \Delta \omega$). The algorithm of viscous splitting consists of substeps in which the convective and the diffusive effects are considered successively. The algorithm may involve several substeps in order to increase the accuracy of the viscous splitting.

The two-step viscous splitting algorithm may then be expressed as

- convection:

$$\frac{\partial \omega}{\partial t} + \mathbf{u} \cdot \nabla \omega = 0;$$

- diffusion:

$$\frac{\partial \omega}{\partial t} = \nu \Delta \omega.$$

Equivalently the viscous splitting algorithm may be expressed in a Lagrangian frame as

- convection:

$$\frac{d\mathbf{x}_p}{dt} = \mathbf{u}(\mathbf{x}_p),$$

$$\frac{d\omega_p}{dt} = 0;$$

- diffusion:

$$\frac{d\mathbf{x}_p}{dt} = 0,$$

$$\frac{d\omega_p}{dt} = \nu\Delta\omega(\mathbf{x}_p),$$

where $\mathbf{x}_p$ and ω_p denote the locations and the vorticity, respectively, carried by the fluid elements. In the viscous splitting algorithm, in the first substep the fluid elements are advanced with the local flow velocity. Diffusion acts at these new locations to modify the vorticity field of the flow.

5.1.1. The Model of the Linear Convection–Diffusion Equation

Before analyzing the viscous splitting algorithm for the Navier–Stokes equation, we discuss a viscous splitting algorithm for the simpler case of a linear convection–diffusion equation.

We consider the viscous evolution of a scalar field $W(\mathbf{x}, t)$ convected by the known velocity field $\mathbf{c}(\mathbf{x}, t)$ according to the following linear convection–diffusion equation:

$$\frac{\partial W}{\partial t} + \mathbf{c} \cdot \nabla W = \nu\Delta W,$$

with initial condition $W(x)$. By denoting by A the convection $(\mathbf{c} \cdot \nabla)$ operator and by B the diffusion operator $(\nu\Delta)$,

$$\mathbf{c} \cdot \nabla \to A, \quad \nu\Delta \to B,$$

we may express the above equation as

$$\frac{dW}{dt} = AW + BW$$

[since our interest is ultimately in fully discrete schemes, we may prefer to view A and B as spatial discretizations of $(\mathbf{c} \cdot \nabla)$ and $(\nu \Delta)$, respectively]. Integrating the above equation in operator form, we may express the solution as

$$W(t) = W_0 e^{(A+B)t}.$$

We can construct the solution of the above equation by advancing in discrete time steps (δt). The equation

$$\frac{dW}{dt} = AW + BW$$

is solved at each interval $[n\delta t, (n+1)\delta t]$ with $W(n\delta t) = W^n$ as the initial condition. The solution may be expressed in operator form as

$$W^{n+1} = e^{(A+B)\delta t} W^n.$$

In a two-step viscous splitting algorithm, the equation is solved in the following two substeps:

- Substep 1 – convection:

$$\frac{dW}{dt} = AW.$$

When W^n is used as the initial condition, the solution of this substep may be expressed as

$$W^{n+\frac{1}{2}} = e^{A\delta t} W^n.$$

With $W^{n+\frac{1}{2}}$ as the initial condition, the diffusion equation is solved at the following substep.

- Substep 2 – diffusion:

$$\frac{dW}{dt} = BW.$$

The result of this two-step procedure is

$$W^{n+1} = e^{B\delta t} W^{n+\frac{1}{2}} = e^{B\delta t} e^{A\delta t} W^n.$$

In general

$$e^{B\delta t} e^{A\delta t} \neq e^{(A+B)\delta t}.$$

This may be clarified by a Taylor series expansion of the exponential forms for a small time step:

$$e^{B\delta t} e^{A\delta t} = \left(I + \delta t\, B + \frac{\delta t^2}{2} B^2 + \cdots \right) \cdot \left(I + \delta t\, A + \frac{\delta t^2}{2} A^2 + \cdots \right),$$

$$e^{(A+B)\delta t} = \left[I + \delta t\,(A+B) + \frac{\delta t^2}{2}(A+B)(A+B) + \cdots \right],$$

where I is the identity operator. It can be easily verified that the two expansions are equivalent only if the operators A and B commute, i.e., if $AB = BA$. In the present case, this would be the case for a velocity field independent of x. Generally speaking, the two-step viscous splitting algorithm is second-order accurate at each time step and first-order accurate overall.

Note that higher-order algorithms may be devised by considering further substeps. Higher-order schemes can be devised by considering the following three-substep procedure, which may be expressed in operator form as

$$W^{n+1} = \left(e^{B\delta t/2} e^{A\delta t} e^{B\delta t/2} \right) W^n.$$

A Taylor series expansion reveals that this procedure introduces errors of $O(\delta t)^3$ at each time step and $O(\delta t)^2$ overall.

We are reminded here that the above results hold for the idealized problem in which the velocity field is known *a priori*. In the case of the Navier–Stokes equations the nonlinear relationship of velocity and vorticity do not allow for this simplification.

We present below a rigorous analysis for the error introduced by viscous splitting in the fully nonlinear case in two dimensions.

5.1.2. Error Estimates for the Viscous Splitting of the Navier–Stokes Equations

In this section we make constant use of the notations and results of Appendix A. Let us also introduce some additional notation. For $\delta t > 0$ and $\omega_0 \in L^\infty(\mathbf{R}^2) \cap L^1(\mathbf{R}^2)$, we denote by $E(t)\omega_0$ the solution at time t to the Euler equation:

$$\frac{\partial \omega}{\partial t} + (\mathbf{u} \cdot \nabla)\omega = 0$$

with initial conditions ω_0. The velocity field is obtained from the solution of the Poisson equation $\Delta \mathbf{u} = -\nabla \times \omega$ by means of the Biot–Savart law as

$$\mathbf{u}(\mathbf{x}) = \int \mathbf{K}(\mathbf{x} - \mathbf{y})\omega(\mathbf{y})\, d\mathbf{y},$$

where, for two-dimensional flows,

$$\mathbf{K}(\mathbf{x} - \mathbf{y}) = -\frac{1}{2\pi} \frac{(\mathbf{x} - \mathbf{y})^{\perp}}{|\mathbf{x} - \mathbf{y}|^2}.$$

We denote by $H^{\nu}(t)\omega_0$ the solution at time t of the diffusion equation,

$$\frac{\partial \omega}{\partial t} = \nu \Delta \omega,$$

with initial vorticity ω_0. It is well known that the solution of the diffusion equation may be expressed in integral form as

$$\omega(\mathbf{x}, t) = \int \mathcal{G}(\mathbf{x} - \mathbf{y}, \nu t)\omega_0(\mathbf{y}) \, d\mathbf{y} = H^{\nu}(\delta t)\omega_0(\mathbf{x}), \tag{5.1.2}$$

where $\mathcal{G}$ (the "heat kernel") is the Green's function solution to the diffusion equation. In two dimensions

$$\mathcal{G}(\mathbf{x} - \mathbf{y}, s) = \frac{1}{4\pi s} e^{-|\mathbf{x} - \mathbf{y}|^2/4s}. \tag{5.1.3}$$

The viscous splitting of Navier–Stokes equation (5.1.1) produces the following sequence ω^n:

$$\omega^{n+1} = H^{\nu}(\delta t) E(\delta t) \omega^n, \tag{5.1.4}$$

where ω^n is sought as an approximation of the exact solution $\omega(\cdot, t_n)$, at time $t_n = n\delta t$. The convergence of this method is given by the following theorem:

Theorem 5.1.1 [24]. *If $\omega_0 \in W^{3,\infty}(\mathbf{R}^2) \cap W^{3,1}(\mathbf{R}^2)$ then the following estimate holds for all $p \in [1, +\infty]$:*

$$\|\omega^n - \omega(\cdot, t_n)\|_{0,p} \le C(T)\nu\delta t. \tag{5.1.5}$$

We reproduce here the essential steps in the proof given in Ref. 67. We denote by S^{ν} the operator that gives the exact solution ω of the Navier–Stokes equations, and we start from the following induction expression:

$$\omega_{n+1} - \omega(\cdot, t_{n+1}) = H^{\nu}(\delta t) E(\delta t)[\omega^n - \omega(\cdot, t_n)]$$
$$- [S^{\nu}(\delta t) - H^{\nu}(\delta t) E(\delta t)]\omega(\cdot, t_n). \tag{5.1.6}$$

We know (see Appendix A) that neither the heat equation nor the Euler equations increase L^p norms; hence

$$\|H^{\nu}(\delta t) E(\delta t)[\omega^n - \omega(\cdot, t_n)]\|_{0,p} \le \|\omega^n - \omega(\cdot, t_n)\|_{0,p}. \tag{5.1.7}$$

It remains now to prove that, for λ smooth enough,

$$\|[S^\nu(\delta t) - H^\nu(\delta t) E(\delta t)]\lambda\|_{0,p} \leq C\nu\delta t^2. \qquad (5.1.8)$$

Let us set $\mu(\cdot, t) = [S^\nu(t) - H^\nu(t) E(t)]\lambda$. From the definitions of S^ν, H^ν, and E, we get

$$\frac{\partial\mu}{\partial t} = [-(\mathbf{a}^\nu \cdot \nabla)S^\nu(t) + \nu\Delta S^\nu(t) - \nu\Delta H^\nu(t)E(t)$$
$$+ H^\nu(t)(\mathbf{a} \cdot \nabla)E(t)]\lambda,$$

where $\mathbf{a}^\nu(\cdot, t)$ and $\mathbf{a}(\cdot, t)$ stand for $\mathbf{K} \star [S^\nu(t)\lambda]$ and $\mathbf{K} \star [E(t)\lambda]$, respectively. We can rewrite the above equation as

$$\frac{\partial\mu}{\partial t} = \mathbf{a}^\nu \cdot \nabla\mu - \nu\Delta\mu + R_1^\nu + R_2^\nu, \qquad (5.1.9)$$

with

$$R_1^\nu(\cdot, t) = [(\mathbf{a} - \mathbf{a}^\nu) \cdot \nabla]H^\nu(t)E(t)\lambda,$$

$$R_2^\nu(\cdot, t) = H^\nu(t)(\mathbf{a} \cdot \nabla)E(t)\lambda - (\mathbf{a} \cdot \nabla)H^\nu(t)E(t)\lambda.$$

With zero initial conditions for μ, the stability for advection–diffusion equations requires control of the source terms R_1^ν and R_2^ν. Using either the maximum principle if $p = +\infty$ or multiplying Eq. (5.1.9) by $|\mu|^{p-2}\mu$, if p is finite, one easily obtains

$$\frac{d}{dt}\|\mu(\cdot, t)\|_{0,p} \leq \left\|(R_1^\nu + R_2^\nu)(\cdot, t)\right\|_{0,p}. \qquad (5.1.10)$$

We now turn to estimating R_1^ν and R_2^ν. For R_1^ν, we first write

$$\left\|R_1^\nu\right\|_{0,p} \leq \|(\mathbf{a} - \mathbf{a}^\nu)(\cdot, t)\|_{0,\infty}\|H^\nu(t)E(t)\lambda\|_{1,p}.$$

Next we observe that

$$\|H^\nu(t)E(t)\lambda\|_{1,p} \leq \|E(t)\lambda\|_{1,p} \leq C(\|\lambda\|_{1,1} + \|\lambda\|_{1,\infty}).$$

On the other hand, elliptic regularity properties of the convolution by $\mathbf{K}$ yield

$$\|(\mathbf{a} - \mathbf{a}^\nu)(\cdot, t)\|_{0,\infty} \leq C\left\{\|[S^\nu(t) - E(t)]\lambda\|_{0,\infty} + \|[S^\nu(t) - E(t)]\lambda\|_{0,1}\right\}.$$

But it is readily seen that

$$\|[S^\nu(t) - E(t)]\lambda\|_{0,\infty} + \|[S^\nu(t) - E(t)]\lambda\|_{0,1} \leq C\nu t(\|\lambda\|_{2,1} + \|\lambda\|_{2,\infty});$$

hence we get

$$\left\| R_1^\nu(\cdot, t) \right\|_{0,p} \le C \nu t (\|\lambda\|_{2,1} + \|\lambda\|_{2,\infty})^2. \tag{5.1.11}$$

For R_2^ν, we write $H^\nu = Id + \bar{H}^\nu$, where

$$\| \bar{H}^\nu(t) f \|_{k,p} \le \nu \max_{t' \le t} \| \Delta H^\nu(t') f \|_{k,p} \le C \nu t \| f \|_{k+2,p}.$$

The contributions of the identity obviously cancel in R_2^ν, and we are left with

$$\left\| R_2^\nu(\cdot, t) \right\|_{0,p} \le C \nu t (\|\lambda\|_{3,1} + \|\lambda\|_{3,\infty}).$$

Since $\mu(\cdot, 0) = 0$, when combined with relations (5.1.10) and (5.1.11), this proves relation (5.1.8). In view of Eqs. (5.1.6) and relation (5.1.7) we thus have

$$\|\omega_{n+1} - \omega(\cdot, t_{n+1})\|_{0,p} \le \|\omega_n - \omega(\cdot, t_n)\|_{0,p} + C \nu \delta t^2$$

and by induction our desired estimate (5.1.5).

5.2. Random-Walk Methods

The random-walk method was introduced by Chorin [49] for the simulation of slightly viscous flows by vortex methods. The method relies on the probabilistic interpretation of the Green's function integral solution to the diffusion equation and on the relationship between diffusion and random walk (Brownian motion) of vorticity-carrying particles.

The implementation of the method is inherently linked to the viscous splitting algorithm. The vorticity field established after the convection step is used as an initial condition for the diffusion equation. In order to simulate diffusion effects, the particles then undergo a Brownian motion. The positions of the particles are updated over one time step δt according to the following rule:

$$\mathbf{x}_p^{n+1} = \mathbf{x}_p^n + \boldsymbol{\xi}_p^n, \tag{5.2.1}$$

where $\boldsymbol{\xi}_p^n$ are random numbers drawn with a Gaussian probability distribution:

$$\mathcal{G}(\mathbf{y}, \varepsilon/2) = \frac{1}{\left(\sqrt{2\pi\varepsilon}\right)^d} e^{\frac{-y^2}{2\varepsilon}} \tag{5.2.2}$$

with zero mean and variance $\varepsilon = 2\nu\delta t$ (d is the dimension). Subsections 5.2.1 and 5.2.2 show that sampling of the vorticity field at these particle locations indeed approximates the solution of the diffusion equation.

Equations (5.2.1) and (5.2.2) provide the recipe of the random walk in the absence of boundaries. As is evident, the method of random walk is simple and maintains the Lagrangian character of vortex methods. We will see in Chapter 6 how random walks of particles near boundaries account for no-slip boundary conditions.

5.2.1. Design

Although it is customary to analyze this method with tools from stochastic differential equations, following Ref. 36 we will adopt the viewpoint of quadrature rules based on random choice of quadrature points. This is consistent with the view that vortex methods operate on integrals discretized by use of the locations of the computational elements as quadrature points.

Let us first recall that the solution to the heat equation can be given the following integral representation:

$$\omega(\mathbf{x}, t) = \int \mathcal{G}(\mathbf{x} - \mathbf{y}, \nu t)\omega_0(\mathbf{y}) \, d\mathbf{y},$$

where $\mathcal{G}$ denotes the heat kernel. In order to derive a particle approximation of ω it is convenient to integrate the vorticity field with a test function ϕ. A particle representation of ω amounts to discretizing the integral by a quadrature on the particle locations. The accuracy of this approximation is discussed in Appendix A.

Integrating ω against ϕ gives

$$\langle \omega(\cdot, t), \phi \rangle = \int \mathcal{G}(\mathbf{x} - \mathbf{y}, \nu t)\omega_0(\mathbf{y})\phi(\mathbf{x}) \, d\mathbf{x} \, d\mathbf{y}.$$

We now approximate this integral by using quadrature points uniformly distributed according to the measure $\mathcal{G}(\mathbf{x} - \mathbf{y}, t) \, d\mathbf{x}$; more precisely, given the explicit form of $\mathcal{G}$ in Eq. (5.1.3) and using successively the changes of variables $(\mathbf{x}, \mathbf{y}) \to (\mathbf{x}, r) \to (\mathbf{x}, u)$ defined by

$$\mathbf{x} - \mathbf{y} = re^{i\theta}, \quad \text{then } u = \exp\left(-\frac{r^2}{4\nu t}\right),$$

we can rewrite

$$\langle \omega(\cdot, t), \phi \rangle = \frac{1}{2\pi} \int d\mathbf{y} \int_0^{2\pi} d\theta \int_0^1 \omega_0(\mathbf{y})\phi[\mathbf{y} + R(u, \nu t)e^{i\theta}] \, du,$$

where $R(u, t) = \sqrt{-4t \log u}$.

The integral is calculated numerically with N random quadrature points $(\mathbf{y}_p, u_p, \theta_p)$ uniformly distributed in $[0, 1]^2 \times [0, 1] \times [0, 2\pi]$. Formula (A.1.4) in Appendix A yields

$$\langle \omega(\cdot, t), \phi \rangle \approx \frac{1}{N} \sum_{p=1}^{N} \omega_0(\mathbf{y}_p) \phi\big[\mathbf{y}_p + R(u_p, \nu t)e^{i\theta_p}\big],$$

where the $\approx$ sign means an approximation in the sense of the expectation with respect to all possible drawings of random numbers. The approximation error is analyzed in Subsection 5.2.2.

This derivation shows that it is admissible to approximate the vorticity after one time step by a set of particles located at the points $\mathbf{y}_p - R(u_p, \nu t)e^{i\theta_p}$; in other words,

$$\omega(\mathbf{x}, t) \approx \frac{1}{N} \sum_{p=1}^{N} \omega_0(\mathbf{y}_p) \delta\big[\mathbf{x} - \mathbf{y}_p - R(u_p, \nu t)e^{i\theta_p}\big].$$

Note that, while the initial distribution had a compact support, the particles $\mathbf{y}_p + R(u_p, \nu t)e^{i\theta_p}$ may lie anywhere in $\mathbf{R}^2$. However, they are essentially located within a distance of $O(\sqrt{\nu t})$ from the support of ω_0. This reflects the fact that diffusion acts instantaneously to spread out the support of the vorticity field.

Simulating n steps of the diffusion equation is equivalent to convolving n times the diffusion kernel $\mathcal{G}$, for

$$\mathcal{G}(\cdot, n\nu\delta t) = \underbrace{\mathcal{G}(\cdot, \nu\delta t) \star \cdots \star \mathcal{G}(\cdot, \nu\delta t)}_{n \text{ times}},$$

and the solution at time $t_n = n\delta t$ can be expressed explicitly as

$$\langle \omega(\cdot, t_n), \phi \rangle = \int \omega_0(\mathbf{y}_1)\mathcal{G}(\mathbf{y}_2 - \mathbf{y}_1, \nu\delta t)\mathcal{G}(\mathbf{y}_3 - \mathbf{y}_2, \nu\delta t) \cdots$$
$$\mathcal{G}(\mathbf{y}_n - \mathbf{y}_{n-1}, \nu\delta t)\mathcal{G}(\mathbf{x} - \mathbf{y}_n, \nu\delta t)\phi(\mathbf{x}) \, d\mathbf{y}_1 \cdots d\mathbf{y}_n \, d\mathbf{x}.$$

Using now the change of variables,

$$\mathbf{z}_1 = \mathbf{y}_1, \mathbf{z}_2 = \mathbf{y}_2 - \mathbf{y}_1, \ldots, \mathbf{z}_n = \mathbf{y}_n - \mathbf{y}_{n-1}, \mathbf{z}_{n+1} = \mathbf{x} - \mathbf{y}_n,$$

we have $\mathbf{x} = \mathbf{z}_1 + \cdots + \mathbf{z}_{n+1}$; setting $\mathbf{z}_j = r^j e^{i\theta^j}$ and $u^j = \exp[-(r^j)^2/4\nu\delta t]$, we obtain the following approximation:

$$\omega(\mathbf{x}, t_n) \approx \frac{1}{N} \sum_p \omega_0(\mathbf{y}_p) \delta \Big[\mathbf{x} - \mathbf{y}_p - R\big(u_p^1, \nu\delta t\big) e^{i\theta_p^1}$$

$$+ R\big(u_p^2, \nu\delta t\big) e^{i\theta_p^2} + \cdots + R\big(u_p^n, \nu\delta t\big) e^{i\theta_p^n} \Big].$$

The N points $[\mathbf{y}_p, (\theta_p^i, u_p^i)_i]$ are randomly distributed with a uniform density in $[0, 1]^2 \times [0, 2\pi]^n \times [0, 1]^n$. It is important to note that, for a given number of particles N, the large dimension of the integration space, $2n + 1$ for n time steps, does not affect the accuracy of this approximation since it is based on the law of large numbers. This will be seen more clearly in the numerical analysis below.

We can now define the random-walk method for the original convection–diffusion problem (5.1.1) by inserting regular Euler advection steps between two successive random walks. The resulting approximate vorticity ω^h can be written as

$$\omega^h(\mathbf{x}, t) = \frac{1}{N} \sum_p \omega_0(\mathbf{y}_p) \delta \big[\mathbf{x} - \mathbf{X}_p^h(t) \big], \qquad (5.2.3)$$

with $\mathbf{X}_p^h$ defined by

$$\mathbf{X}_p^h(t) = \mathbf{X}^h\big(t; \mathbf{y}_p^n, t_n\big), \quad \text{for } t_n \le t \le t_{n+1}$$

and

$$\mathbf{y}_p^n = \mathbf{X}^h\big(t_n; \mathbf{y}_p^{n-1}, t_{n-1}\big) + R\big(u_p^n, \nu\delta t\big) e^{i\theta_p^n}.$$

As usual $\mathbf{X}^h$ denotes the characteristics associated with the velocity field $\mathbf{u}^h = \mathbf{K}_\varepsilon \star \omega^h$, where the mollification $\mathbf{K}_\varepsilon$ is defined from the Biot–Savart kernel $\mathbf{K}$, as discussed in Section 2.3.

5.2.2. *Numerical Analysis*

We present the convergence analysis of random-walk techniques for the solution of the linear convection–diffusion equation:

$$\frac{\partial \omega}{\partial t} + \text{div}\,(\mathbf{a}\omega) - \nu\Delta\omega = 0, \qquad (5.2.4)$$

$$\omega(\cdot, 0) = \omega_0. \qquad (5.2.5)$$

For the convergence analysis of random-walk techniques in viscous splitting algorithms for the solution of the full Navier–Stokes equations we refer to Refs. 94 and 139.

In our proof, we will follow the approach suggested by Brenier [36], which is based on quadrature estimates. This proof is quite simple and illustrates well the importance of the randomness involved in the simulation of the diffusion.

Since we have already proved the convergence of the splitting, we fix a time step δt and compare the random-walk solution with the sequence ω^n defined in Eq. (5.1.4) (in the present situation, the Euler step is replaced by a simpler linear advection step).

Let us now introduce some additional notation. We denote by Ψ the space of sequences in $[0, 1]^2$. A point in Ψ will be noted as

$$\mathbf{Y} = [\mathbf{Y}_0, \mathbf{Y}_1 = (u_1, \theta_1/2\pi), \ldots, \mathbf{Y}_n = (u_n, \theta_n/2\pi), \ldots],$$

where $\mathbf{Y}_0$ is the initial condition and (u_i, θ_i) are in $[0, 1] \times [0, 2\pi]$.

We next define the sequence of mappings G^k defined by induction for $k \in [0, n]$ in the following way:

$$G^0(\mathbf{Y}_0) = \mathbf{Y}_0; \; G^k(\mathbf{Y}_0, \ldots, \mathbf{Y}_k) = \mathbf{X}\big[t_k; G^{k-1}(\mathbf{Y}_0, \ldots, \mathbf{Y}_{k-1}) \\ + R(u_k, \nu\Delta t)e^{i\theta_k}, t_{k-1}\big],$$

where $t_k = k\delta t$ and $\mathbf{X}$ are the characteristics associated with the flow $\mathbf{a}$.

In other words, G^n consists of n successive random walks parameterized by the vectors $\mathbf{Y}_i, i \in [1, n]$, starting from the vector $\mathbf{Y}_0$ and alternating with n advection steps. An element $\mathbf{Y} \in \Psi$ defines through the mappings G^k the evolution of a single particle.

We then denote by $\Omega = \Psi^N$ the set of all possible random choices. These include the particle initializations and the successive drawings for the random walks. The random-walk method is thus parameterized on the choice of quadrature points in Ω. Given a sequence $(\mathbf{Y}^i) \in \Omega$, we will finally denote by $\omega^N[\cdot, t_n, (\mathbf{Y}^i)]$ random-walk approximation (5.2.3) using the N first elements of $(\mathbf{Y}^i)$, that is, particles $\mathbf{Y}_0^1, \ldots, \mathbf{Y}_0^N$ for the initialization and for the kth, $k \leq n$ random walk, the quadrature points $\mathbf{Y}_k^1 = (u_k^1, \theta_k^1/2\pi), \ldots, \mathbf{Y}_k^N = (u_k^N, \theta_k^N/2\pi)$.

In order to analyze the accuracy of the approximation we consider a mollification of the vorticity field discretized on the computed particle locations. This mollification is defined as usual by

$$\omega_\varepsilon^N[\mathbf{x}, t_n, (\mathbf{Y}^i)] = \int \omega^h[\mathbf{y}, t_n, (\mathbf{Y}^i)]\zeta_\varepsilon(\mathbf{x} - \mathbf{y}) \, d\mathbf{y},$$

where ζ_ε is a cutoff function. We can now state Theorem 5.2.1.

Theorem 5.2.1. *Assume that ω_0 is in $L^2(\mathbf{R}^2)$; then for almost all random choice $(\mathbf{Y}^i) \in \Omega$, we have*

$$\left\| \omega^n - \omega_\varepsilon^N[\cdot, t_n, (\mathbf{Y}^i)] \right\|_{L^2(\mathbf{R}^2)} \to 0,$$

when $N \to \infty$, $\varepsilon \to 0$ with $N\varepsilon^2 \to \infty$.

Proof. We compare ω_ε^N with $\omega_\varepsilon^n = \omega^n \star \zeta_\varepsilon$, as we know that ω_ε^n tends to ω^n at a rate depending on the moment conditions of the cutoff ζ_ε. On the one hand, in view of the definition of the mapping G^n and Eq. (5.1.4) we can write

$$\omega_\varepsilon^n(\mathbf{x}) = \int_\Psi \omega_0(\mathbf{Y}_0) \zeta_\varepsilon[\mathbf{x} - G^n(\mathbf{Y}_0, \ldots, \mathbf{Y}_n)] \, d\mathbf{Y}_0 \cdots d\mathbf{Y}_n.$$

On the other hand we also have

$$\omega_\varepsilon^N[\mathbf{x}, t_n, (\mathbf{Y}^i)] = \frac{1}{N} \sum_{i=1}^{N} \omega_0 \left(\mathbf{Y}_0^i \right) \zeta_\varepsilon \left[\mathbf{x} - G^n \left(\mathbf{Y}_0^i, \ldots, \mathbf{Y}_n^i \right) \right].$$

$\omega_\varepsilon^N[\cdot, t_n, (\mathbf{Y}^i)]$ is thus a quadrature estimate for ω_ε^n, with a random choice of quadrature points, as analyzed in Appendix A. More precisely, Lemma A.1.3 enables us to write

$$\int d\mathbf{x} \int_\Omega d(\mathbf{Y}^i)_i \left| \omega_\varepsilon^n(\mathbf{x}) - \omega_\varepsilon^N[\mathbf{x}, t_n, (\mathbf{Y}^i)] \right|^2$$

$$\leq \frac{1}{N} \int |\omega_0(\mathbf{Y}_0)\zeta_\varepsilon[\mathbf{x} - G^n(\mathbf{Y}_0, \ldots, \mathbf{Y}_n)]|^2 \, d\mathbf{x} \, d\mathbf{Y}_0 \cdots d\mathbf{Y}_n.$$

We then use the bound $|\zeta_\varepsilon| \leq C\varepsilon^{-2}$ to obtain

$$\int |\omega_0(\mathbf{Y}_0)\zeta_\varepsilon[\mathbf{x} - G^n(\mathbf{Y}_0, \ldots, \mathbf{Y}_n)]|^2 \, d\mathbf{x} \, d\mathbf{Y}_0 \cdots d\mathbf{Y}_n$$

$$\leq C\varepsilon^{-2} \int |\omega_0(\mathbf{Y}_0)|^2 |\zeta_\varepsilon[\mathbf{x} - G^n(\mathbf{Y}_0, \ldots, \mathbf{Y}_n)]| \, d\mathbf{x} \, d\mathbf{Y}_0 \cdots d\mathbf{Y}_n.$$

The function of $\mathbf{x}$ on the right-hand side above can be seen as the result of n advection–diffusion steps, acting on $|\omega_0|^2$, followed by a convolution with $|\zeta_\varepsilon|$. Since the advection velocity field has been assumed divergence free, none of these steps increase its integral, which is thus bounded by that of $|\omega_0|^2$.

This finally yields

$$\int d\mathbf{x} \int_\Omega d(\mathbf{Y}^i)_i \left| \omega_\varepsilon^n - \omega_\varepsilon^N[\mathbf{x}, t_n, (\mathbf{Y}^i)] \right|^2 \leq \frac{1}{N\varepsilon^2} \|\omega_0\|_{0,2}^2, \qquad (5.2.6)$$

from which our claim follows. $\qquad\qquad\square$

Combined with relation (5.1.5), this result shows the convergence of the random-walk method for almost all random drawings. In two dimensions, the number of particles N asymptotically scales as h^{-2}, where h is the mean spacing between nearby particles. The condition $N\varepsilon^2 \to \infty$ in the assumption of Theorem 5.2.1 thus relates to the usual overlapping condition $h \ll \varepsilon$. The low convergence rate h/ε^2 in relation (5.2.6), however, indicates that a large number of particles would be required for obtaining a reasonable pointwise accuracy for the computed vorticity.

5.2.3. Results for Random-Walk Simulations

To illustrate the method of random walk we first consider here some simple one-dimensional numerical experiments that confirm the above convergence analysis.

We consider the following problem:

$$\frac{\partial \omega}{\partial t} = \nu \Delta \omega, \tag{5.2.7}$$

$$\omega(x, 0) = x e^{-x^2}. \tag{5.2.8}$$

The solution to this problem may be expressed in integral form as

$$\omega(x, t) = \int_{-\infty}^{\infty} \mathcal{G}(x - y, \nu t)\, y e^{-y^2}\, d\mathbf{y},$$

where $\mathcal{G}$ is the one-dimensional heat kernel. The above integral may be calculated explicitly to give

$$\omega^{\mathrm{ex}}(x, t) = x e^{-x^2/(1+4\nu t)}/(1 + 4\nu t)^{3/2}.$$

In the context of the random walk, the integral solution may be interpreted as follows: Place N randomly spaced particles on the line $x \geq 0$, in the computational domain, and at positions x_i^0, $i = 1, \ldots, N$, and assign to each particle a strength of $f(x_i)/N$, where $f(x_i) = \omega(x_i, 0)$. Then let the particles undergo a random walk by changing the positions of the particles at each time step δt under the following rule:

$$x_i^{n+1} = x_i^n + \xi_i^n, \tag{5.2.9}$$

where ξ_i^n are Gaussian independent random variables having mean 0 and variance $2\nu\delta t$. Now as we let the number N of particles go to infinity we may observe that the expected distribution of the particle strength on the x axis would approximate the exact solution:

$$\lim_{N \to \infty} \frac{\sum_{i=1}^{N} f(x_i)(x \le x_i \le x + dx, t = T)}{N dx} = \omega^{\text{ex}}(x, t). \qquad (5.2.10)$$

In Figure 5.1 the results of the computation for several numbers of particles, with a time step $\delta t = 0.5$ and a sampling interval of $dx = 0.175$, are shown. One may observe the slow convergence of the method that, in accordance with relation (5.2.6), has been estimated to be of $O(1/\sqrt{N})$ in Ref. 153. In order to check this estimate a series of numerical experiments was performed for several numbers of particles, and the rms error of these experiments is plotted versus the number of particles in Figure 5.2. The characteristic law of $-1/2$ is observed.

It is evident from these results that to get a reasonable accuracy, large numbers of particles are necessary.

Note also that the sampling interval is important in the presentation of the solution. In Figure 5.3 we plot the results for sampling increments of $dx = 0.073$ and $dx = 0.292$ for 5000 particles spread over $0 \le x \le 3.5$.

In the field of incompressible flows, the random-walk method was the first numerical technique to incorporate in a consistent way viscous effects while not limiting the adaptive Lagrangian character of vortex methods. Since its inception in 1973 it has served the scientific community as a computational tool for the simulation/modeling of unsteady flows around complex configurations. These simulations have shown that a vortex method that uses the random-walk method can reproduce the global quantities of the flow in a consistent way. In that respect, a simple accuracy criterion is given by the evolution of the angular impulse $\int |\mathbf{x}|^2 \omega(\mathbf{x}) \, d\mathbf{x}$, which, for the Navier–Stokes equations, must increase proportionally to the viscosity and the total circulation. Random-walk methods can reproduce this law with a reasonable accuracy.

Note that, unlike in grid-based methods, the Reynolds number does not impose directly a limitation to the spacing of the computational elements to ensure stability. Another interesting feature of random-walk methods is that they are physically appealing. In particular they yield a natural expansion of the support of the vorticity.

The possibility of incorporating a consistent treatment of viscous effects while using a Lagrangian vortex method has led to research efforts in the use of random vortex methods as a computational tool for direct numerical simulations. Smith and Stansby [188] used the random-walk method in conjunction with a cloud in cell methodology for the simulation of the flow past an impulsively started cylinder. They demonstrated that the random-walk method can capture the main features of the vortices developed in the wake and provide

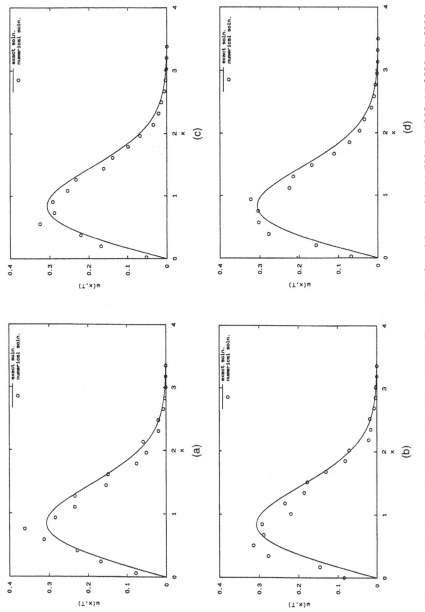

Figure 5.1. Random-walk solution of problem (5.2.8). The number of particles used is (a) 500, (b) 1000, (c) 2500, (d) 5000.

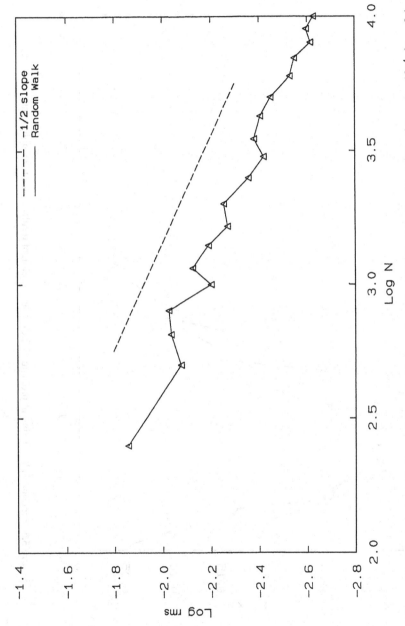

Figure 5.2. Convergence study for random-walk solution of problem (5.2.8). Sampling every 20 points. $\nu = 10^{-4}, \delta t = 0.1$, $T = 10$.

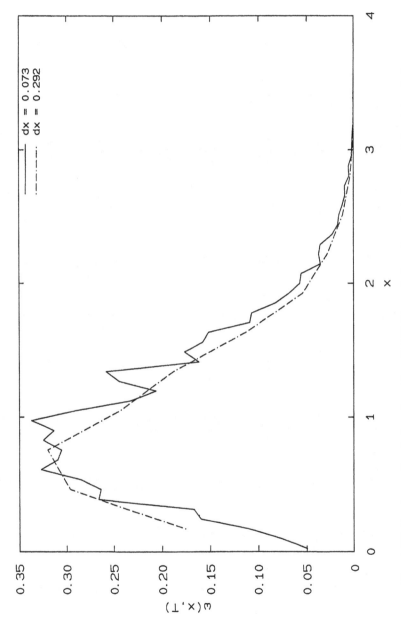

Figure 5.3. Random-walk solution of problem (5.2.8). The variation of results is due to sampling interval change.

reasonable estimates for quantities such as the forces experienced by the body and velocities in the wake. As perhaps could have been expected, the accuracy of their simulation increases in the later stages of the evolution of the flow as convective effects become more dominant. In the earlier stages, when viscous phenomena near the surface of the body induce large friction forces, the method seems inadequate to capture quantities such as the drag coefficient or instantaneous vorticity values.

Sethian and Ghoniem [185] conducted a systematic study of the random-walk method combined with the vortex sheet method (see Subsection 6.3.2) in simulations of the flow past a backward-facing step. Their two-dimensional simulations showed a convergence of the method in capturing quantities such as the eddy size and the average velocity profiles in close comparison with related experimental findings. These results demonstrated the ability of the method to model turbulent flows. The engineering community has extensively used vortex methods to obtain estimates of the flow field on a wide variety of applications such as estimating wind loads on buildings and bridges, hydrodynamic loads on offshore structures, and mixing in internal combustion engines.

However, because of their low convergence rate, random-walk methods that use a small number of elements should be viewed as providing a model of the flow, rather than an approximation of the Navier–Stokes equations. One of the drawbacks of the method is that it actually does not offer many degrees of freedom if one wishes to improve its pointwise accuracy. This lack of pointwise accuracy is probably responsible for the difficulties of this method in three dimensions. For three-dimensional flows, indeed, as we already mentioned, it is crucial to preserve a reasonable topology of the filaments, which can be guaranteed only if there is a good control of the accuracy everywhere in the flow and not only in the average. We will see how other Lagrangian methodologies can alleviate this difficulty. Finally the approximation of diffusion by random walk also implicitly affects the accuracy in the convection, as it makes regridding strategies difficult to implement in order to correct the effects of the flow strain on the particle distribution (this will be more apparent in the discussion in Chapter 7).

5.3. Resampling Methods

Resampling methods were originally proposed as an alternative to random-walk methods with possible higher accuracy. They were introduced in 1983 [66, 169] in the context of a viscous splitting algorithm and are based on the integral form of Eq. (5.1.2) of the explicit solution of the diffusion equation.

Simulations of viscous flows that use the resampling technique are performed with a fractional step algorithm. Each step consists of the following two substeps:

1. Move the particles to simulate the advection part of the vorticity equation.
2. Simulate the diffusion by resampling the vorticity field induced by one particle on its neighbors.

Unlike the random-walk method, in resampling schemes diffusion is simulated by a change in the strength of the particles and not their locations.

In this and the following section, we will consider particle approximations of the vorticity field while distinguishing between the values of circulation and vorticity at the particle locations. We recall that particle circulations are obtained by multiplying vorticity by the volume of the particle. In contrast with random-walk methods, which can be defined only in terms of circulations, the algorithms we describe can be stated in terms of local vorticity values as well.

Assume that at time step t_n the vorticity particle distribution is given by

$$\omega^h(\mathbf{x}, t_n) = \sum_p v_p \omega_p^n \delta\left[\mathbf{x} - \mathbf{x}_p^h(t_n)\right]. \tag{5.3.1}$$

In this formula, v_p are the volumes of the initial particles – or equivalently the weights of a quadrature formula for which these particles are used as integration points – that, in the case of a divergence-free flow field, are time independent and ω_p^n are the local values of the vorticity at time t_n at the particle locations $[\mathbf{x}_p^h(t_n)]$.

We solve for the diffusion equation with the vorticity field $\omega^h(\cdot, t_n)$ as the initial condition. We recall that the solution of the diffusion equation may be expressed as

$$\omega^h(\mathbf{x}, t_{n+1}) = \int \mathcal{G}(\mathbf{x} - \mathbf{y}, \nu\delta t)\, \omega^h(\mathbf{y}, t_n)\, d\mathbf{y},$$

where $\mathcal{G}$ is the diffusion kernel [given by Eq. (5.1.3) in two dimensions]. Substituting Eq. (5.3.1) into the above equation and evaluating it at the (unchanged) particle locations, we obtain that

$$\omega_p^{n+1} = \sum_q v_q \omega_q^n \mathcal{G}\left[\mathbf{x}_p^h(t_n) - \mathbf{x}_q^h(t_n), \nu\delta t\right]. \tag{5.3.2}$$

Formula (5.3.2) is the induction rule that allows us to update the particles weights from one time step to the next.

This method, although quite natural, has the drawback of not conserving the total circulation. In order to obtain a conservative scheme we use the property of the heat kernel

$$\int \mathcal{G}(\mathbf{x} - \mathbf{y}, \nu \delta t) \, d\mathbf{y} = 1,$$

so that we may also write

$$\omega^h(\mathbf{x}, t_{n+1}) = \omega^h(\mathbf{x}, t_n) + \int \mathcal{G}(\mathbf{x} - \mathbf{y}, \nu \delta t) \, [\omega^h(\mathbf{y}, t_n) - \omega^h(\mathbf{x}, t_n)] \, d\mathbf{y}.$$

Substituting now Eq. (5.3.1) into the above equation and evaluating it at the particle locations, we obtain a conservative scheme for updating of the particle strengths as

$$\omega_p^{n+1} = \omega_p^n + \sum_q \nu_q \left(\tilde{\omega}_q^n - \tilde{\omega}_p^n \right) \mathcal{G} \left[\mathbf{x}_p^h(t_n) - \mathbf{x}_q^h(t_n), \nu \delta t \right]. \qquad (5.3.3)$$

This last equation can be interpreted as the result of a balance of the vorticity received by and expelled from a given particle during one time step (see [112]). It can also be rewritten as

$$\omega_p^{n+1} = \sum_q \nu_q \tilde{\omega}_q^n \mathcal{G} \left[\mathbf{x}_p^h(t_n) - \mathbf{x}_q^h(t_n), \nu \delta t \right]$$

$$+ \tilde{\omega}_p^n \left\{ 1 - \sum_q \nu_q \mathcal{G} \left[\mathbf{x}_p^h(t_n) - \mathbf{x}_q^h(t_n), \nu \delta t \right] \right\}.$$

The term within braces on the right-hand side can be seen as a quadrature error based on the particle locations for the Gaussian function. It is therefore arbitrarily small, which shows that formula (5.3.3) is actually only a correction over formula (5.3.2).

The convergence of the method based on Eq. (5.3.2) has been proved in Refs. 66, 67, and 68. The modified scheme of Eq. (5.3.3) can be proved to be convergent by similar arguments. Not surprisingly, an essential ingredient in the convergence proof is that particles overlap enough to give a correct sampling of the Gaussian kernel. This condition may be written as

$$h \ll \sqrt{\nu \delta t}.$$

At first glance, this condition contradicts the usual stability condition ($\nu \delta t \leq Ch^2$) for explicit time discretization of diffusion equations. However, the parameter δt is the splitting time step, and not a diffusion time step, so that actually ν and δt should not be seen here as independent parameters. This point will be further discussed below.

5.3.1. Other Resampling Schemes

Other resampling methods have been more recently proposed. First it has been observed that the Gaussian function $\mathcal{G}$, which may be expensive to calculate, can be replaced in Eq. (5.3.2) or (5.3.3) with any cutoff function η_ε with radial invariance and sharing with $\mathcal{G}$ the following moment properties:

$$\int \eta(\mathbf{x})\, d\mathbf{x} = 1, \quad \int x_i^2 \eta(\mathbf{x})\, d\mathbf{x} = 2, \quad i = 1, 2, \qquad (5.3.4)$$

provided the parameters ε and δt are linked through the relation $\varepsilon^2 = \nu \delta t$.

In other terms one can use the numerical dissipation involved in the successive regularization of the particles into blobs of shape η_ε to simulate the physical diffusion. These regularizations, when properly scaled, do provide the right amount of diffusion. One can view Eq. (5.3.2), with $\mathcal{G}$ replaced with η as the discrete analog, through quadrature rules of

$$\omega^{n+1} = \eta_\varepsilon \star \omega^n.$$

A Taylor series expansion of ω^n by use of the symmetry properties of η readily shows that, under assumptions (5.3.4),

$$\eta_\varepsilon \star \omega^n = \omega^n + \varepsilon^2 \Delta \omega^n + O(\varepsilon^4).$$

Setting $\varepsilon^2 = \nu \delta t$, we thus obtain the right scaling to perform a time step of the diffusion equation. Note that by choosing η to be a Gaussian and the above scaling we recover Eq. (5.3.2).

In practice one keeps ε fixed (and more or less equal to the cutoff parameter used in the computation of the velocity for the Euler step) and varies δt as a function of ν. Note that if ν decreases, one can use only a few splitting steps (zero if $\nu = 0$!). It thus may become important to use a different time stepping for advecting the particles in the numerical integration of the Euler equations (in other words to perform several Euler substeps for one diffusion step).

We will not get into more details about this resampling technique. We will see in Section 5.4 that this class of methods can be viewed as resulting from a particular time discretization within a more general class of deterministic methods, and we will gather in Section 5.4 numerical illustrations corresponding to the various possible approaches.

Another resampling method, proposed in Ref. 157, consists of redistributing the vorticity on particles located on a uniform grid. This can be done to prevent numerical discrepancy resulting from an excessive distortion of the Lagrangian grid. However, this scheme is not conservative, and it is more advisable to decouple the viscous resampling and the regridding steps. Another possibility to avoid accuracy problems due to the distortion of the Lagrangian grid while keeping advantage of the localization feature of vortex methods is to homogenize the grid by moving randomly each particle within a square of size h before resampling (see Ref. 37 for the convergence proof of a method in this spirit in which resampling is done with a completely random choice).

All these methods, except the last one based on random rezoning, are second order in space. Their time accuracy is limited only by the splitting of the Navier–Stokes equations. This means that a high-order splitting method, like the one analyzed in Ref. 24, which alternates half steps of advection and diffusion, virtually yields second-order methods.

5.4. The Method of Particle Strength Exchange

In this section we derive a general framework for the construction of high-order viscous algorithms for vortex methods introduced by Mas-Gallic in 1987 [147].

The starting point is the Lagrangian description of the Navier–Stokes equations as approximated on the particle locations:

$$\frac{d\omega_p}{dt} = \nu \Delta \omega_p. \tag{5.4.1}$$

The goal of the method is to derive formulas for consistent evaluations of $\Delta \omega_p$ on the particle locations.

5.4.1. Design

The key idea is to replace the diffusion operator by an integral one. The links between integral and diffusion operators have long been exploited in the field of kinetic equations, but in general in the other direction, namely, to derive a diffusion approximation of integral operators that model collisions of particles. The motivation here is that integrals are much better suited for particle methods than second-order differential operators: As already observed on many occasions, inserting particles inside an integral is indeed equivalent to writing a numerical quadrature of this integral.

Let us now describe in detail the resulting methods. Let η be a kernel satisfying the following moment properties:

$$\int x_i x_j \eta(\mathbf{x}) \, d\mathbf{x} = 2 \delta_{ij} \quad \text{for } i, j = 1, 2,$$

$$\int x_1^{i_1} x_2^{i_2} \eta(\mathbf{x}) \, d\mathbf{x} = 0 \quad \text{if } i_1 + i_2 = 1 \text{ or } 3 \leq i_1 + i_2 \leq r + 1,$$

$$\int |\mathbf{x}|^{r+2} |\eta(\mathbf{x})| \, d\mathbf{x} < \infty. \tag{5.4.2}$$

For $\varepsilon > 0$ we set $\eta_\varepsilon(\mathbf{x}) = \varepsilon^{-2}\eta(\mathbf{x}/\varepsilon)$ (recall that we consider the two-dimensional case; here and subsequently appropriate modifications need to be done for three dimensions). We then consider the integral operator Δ_ε defined by

$$\Delta_\varepsilon \omega(\mathbf{x}) = \varepsilon^{-2} \int [\omega(\mathbf{y}) - \omega(\mathbf{x})] \, \eta_\varepsilon(\mathbf{y} - \mathbf{x}) \, d\mathbf{y}. \tag{5.4.3}$$

This provides an integral approximation of the Laplace operator according to the following result.

Proposition 5.4.1. *Under assumptions (5.4.2), if $\omega \in W^{r+2,p}(\mathbf{R}^2)$ for some $p \in [1, +\infty]$, then the following estimate holds (with the notation defined in Appendix A):*

$$\|\Delta_\varepsilon \omega - \Delta\omega\|_{0,p} \le C\varepsilon^r \|\omega\|_{r+2,p}. \tag{5.4.4}$$

Proof. Let us check this assertion only for $p = +\infty$. The Taylor expansion of ω at the point $\mathbf{x}$ to the order $r + 2$ yields

$$\omega(\mathbf{y}) = \omega(\mathbf{x}) + (\mathbf{y} - \mathbf{x}, \nabla)\omega(\mathbf{x}) + \cdots + \frac{1}{(r+1)!}(\mathbf{y} - \mathbf{x}, \nabla)^{(r+1)}\omega(\mathbf{x})$$
$$+ O(|\mathbf{y} - \mathbf{x}|^{r+2} \|u\|_{r+2,\infty}).$$

When this expansion is used on the right-hand side of Eq. (5.4.3), by assumptions (5.4.2), the contributions of all the terms of the form $\int (\mathbf{y} - \mathbf{x})^\alpha \eta_\varepsilon(\mathbf{y} - \mathbf{x}) \, d\mathbf{y}$ vanish except for the terms

$$\frac{\partial^2 \omega}{\partial x_1^2}(\mathbf{x}) \int (y_1 - x_1)^2 \eta_\varepsilon(\mathbf{y} - \mathbf{x}) \, d\mathbf{y} + \frac{\partial^2 \omega}{\partial x_2^2}(\mathbf{x}) \int (y_2 - x_2)^2 \eta_\varepsilon(\mathbf{y} - \mathbf{x}) \, d\mathbf{y}$$
$$+ \|\omega\|_{r+2,\infty} O\left[\int |\mathbf{y} - \mathbf{x}|^{r+2} \eta_\varepsilon(\mathbf{y} - \mathbf{x}) \, d\mathbf{y}\right],$$

which, after the change of variables $\mathbf{z} = [(\mathbf{y} - \mathbf{x})/\varepsilon]$ in the integrals, yields

$$\Delta_\varepsilon \omega(\mathbf{x}) = \Delta\omega(\mathbf{x}) + O(\varepsilon^r). \tag*{$\square$}$$

Particle approximation of the diffusion can then be defined from numerical integrations of Δ_ε. The resulting scheme is written as

$$\omega^h(\mathbf{x}, t) = \sum_p v_p \omega_p^h(t) \, \delta(\mathbf{x} - \mathbf{x}_p^h), \tag{5.4.5}$$

where the trajectories of the particles $\mathbf{x}_p^h$ are computed with the velocity field

$\mathbf{u}^h = \mathbf{K}_\varepsilon \star \omega^h$ and $\omega_p^h(t)$ obey the following system of ODEs

$$\frac{d\omega_p^h}{dt} = \nu\varepsilon^{-2} \sum_q \left(v_q\omega_q^h - v_q\omega_p^h\right)\eta_\varepsilon\left(\mathbf{x}_q^h - \mathbf{x}_p^h\right). \tag{5.4.6}$$

Concerning moment conditions (5.4.2), let us first observe that, for symmetry reasons, any function η with radial invariance would satisfy

$$\int x_i\eta(\mathbf{x})\,d\mathbf{x} = \int x_1x_2\eta(\mathbf{x})\,d\mathbf{x} = 0,$$

$$\int x_1^2\eta(\mathbf{x})\,d\mathbf{x} = \int x_2^2\eta(\mathbf{x})\,d\mathbf{x}$$

and thus would provide, on proper rescaling, a kernel satisfying conditions (5.4.2) for $r = 2$. Several constructions of the kernel η are proposed in Ref. 73. They in particular relate moment conditions (5.4.2) to the cutoff conditions determining the accuracy of vortex blobs in inviscid calculations (see Section 2.4). One possibility is to use

$$\eta = \Delta\zeta.$$

If ζ is a cutoff of the order of r, then it is straightforward to check that η satisfies conditions (5.4.2) with the same value of r. This choice was later rediscovered by Fishelov. However, Fishelov's scheme, which is derived in Ref. 83 from the approximation $\Delta\omega \simeq \omega \star \Delta\zeta_\varepsilon$, is given by

$$\frac{d\omega_p^h}{dt} = \nu \sum_q \left(v_q\omega_q^h\right)\Delta\zeta_\varepsilon\left(\mathbf{x}_q^h - \mathbf{x}_p^h\right).$$

Therefore it does not incorporate the correction term $-\omega_p \sum_q v_q\Delta\zeta_\varepsilon(\mathbf{x}_p^h - \mathbf{x}_q^h)$, which in Eq. (5.4.5) implicitly results from the integral approximation approach. Like the resampling scheme of Eq. (5.3.2), it therefore fails to be conservative.

Another systematic construction of arbitrarily high-order kernels consists of using the formula

$$\eta(\mathbf{x}) = -2\frac{\nabla\zeta(\mathbf{x}) \cdot \mathbf{x}}{|\mathbf{x}|^2}. \tag{5.4.7}$$

If ζ satisfies the cutoff moment properties at order r (in the sense of Section 2.3), then it is readily seen that η satisfies Eq. (5.4.2) at the same order. For a cutoff with radial symmetry, as is most often the case, this formula gives

$$\eta(\mathbf{x}) = -2\frac{\zeta'(|\mathbf{x}|)}{|\mathbf{x}|}.$$

It is noteworthy that, if one starts from the Gaussian kernel $\mathcal{G}(\mathbf{x}) = (4\pi)^{-1}$ $\exp(-|\mathbf{x}|^2/4)$, the above construction gives $\eta = \mathcal{G}$. For a more general presentation of these methods and their extension to anisotropic diffusion, we refer to Ref. 73.

One advantage of the PSE schemes just described over the resampling methods introduced in Section 5.3 is that their definition does not require any viscous splitting of the Navier–Stokes equations. The time discretization of Eq. (5.4.6) can be done through classical time-stepping schemes, leading to a method that can be virtually of any order in space and time, provided the proper choice of η is made. It is, however, interesting to note that, if system (5.4.6) is discretized by the explicit forward Euler scheme with a time step δt, one advances from time $t_n = n\delta t$ to time t_{n+1} by

$$\omega_p^{n+1} = \omega_p^n + \frac{\nu\delta t}{\varepsilon^2} \sum_q v_q \left(\omega_q^n - \omega_p^n\right)\eta_\varepsilon\left(\mathbf{x}_q^n - \mathbf{x}_p^n\right), \qquad (5.4.8)$$

which is the resampling scheme of Eq. (5.3.3), with η instead of $\mathcal{G}$ and $\varepsilon^2 = \nu\delta t$. As stated in Section 5.3, resampling schemes can thus be seen as a particular low-order time discretization of PSE schemes. Finally, as we will see in Subsection 5.4.3, the PSE schemes can also be formulated for spatially varying ε.

5.4.2. Numerical Analysis

Before sketching the analysis of these methods in the linear case, let us point out the particular role played by nonnegative even kernels η. For this class of kernels, one can easily check that the algorithm is unconditionally enstrophy decreasing, which is what can be expected for the Navier–Stokes equations in the absence of boundaries.

If we indeed define by $\mathcal{E}^h(t) = \sum_p v_p |\omega_p^h(t)|^2$ a discrete version of the enstrophy $\mathcal{E}(t) = \int |\omega(\mathbf{x}, t)|^2 \, d\mathbf{x}$, we get from Eq. (5.4.6)

$$\frac{1}{2}\dot{\mathcal{E}}^h = \sum_p v_p \dot{\omega}_p^h \omega_p^h = \nu\varepsilon^{-2} \sum_{p,q} v_p v_q \left(\omega_q^h - \omega_p^h\right)\omega_p^h \eta_\varepsilon\left(\mathbf{x}_q^h - \mathbf{x}_p^h\right).$$

Writing then $\omega_p = \frac{1}{2}(\omega_p + \omega_q) + \frac{1}{2}(\omega_p - \omega_q)$ and observing that, by symmetry, since η is even, the contribution of the first term in the sum is zero, we obtain

$$\dot{\mathcal{E}}^h = -\nu\varepsilon^{-2} \sum_{p,q} v_p v_q \left(\omega_q^h - \omega_p^h\right)^2 \eta_\varepsilon\left(\mathbf{x}_q^h - \mathbf{x}_p^h\right) \le 0,$$

provided that $\eta > 0$. Furthermore it can be shown (see Ref. 62) that, on reinitialization of the particle distribution to guarantee the overlapping of particles for all time, the discrete enstrophy has the same decay property as for the continuous equation, that is,

$$\mathcal{E}^h(t) \le \frac{C}{vt},$$

where C depends on only the initial vorticity.

The drawback of nonnegative kernels is that they are second order only, as the moments of order 4 cannot vanish.

We give now the main steps of the convergence analysis in the linear case for general diffusion approximations (we refer to Ref. 73 for details). We focus again on the linear convection–diffusion equations (5.2.4) and (5.2.5), and we assume that the initial particle distribution consists of particles located at the nodes of a uniform grid of mesh size h.

We first give some notation: $S^v(t)$, and $S^v_\varepsilon(t)$ will respectively denote the linear operators that give the solutions, at time t, of Eq. (5.2.4) and of the same equation with Δ replaced by Δ_ε (susequently referred to as the Δ_ε equation). We first observe that

$$\Delta_\varepsilon \omega = \varepsilon^{-2}(\omega \star \eta_\varepsilon - \lambda \omega),$$

where $\lambda = \int \eta(\mathbf{x})\, d\mathbf{x}$. This implies that, for all $p \in [1, \infty]$,

$$\|\Delta_\varepsilon \omega\|_{L^p} \le C\varepsilon^{-2}\|\omega\|_{L^p}, \tag{5.4.9}$$

and the Δ_ε equation immediately yields

$$\frac{d}{dt}\left\|S^v_\varepsilon(t)\omega_0\right\|_{L^2} \le Cv\varepsilon^{-2}\left\|S^v_\varepsilon(t)\omega_0\right\|_{L^2}.$$

We thus obtain a finite bound for the enstrophy (for finite times) only if $v\varepsilon^{-2} = O(1)$. As a result, the Δ_ε model and the particle method that has been derived from it, must be seen as a vanishing viscosity model (however, if $\eta \ge 0$, it does provide a valid approximation for fixed v and $\varepsilon \to 0$).

From now on we assume that the parameter ε is linked with v by

$$v/\varepsilon^2 \approx O(1). \tag{5.4.10}$$

If we subtract the Δ_ε equation from Eq. (5.2.4) and use relation (5.4.4), we obtain, under the assumptions of Eq. (5.4.2) and provided that $S^v \omega_0$ is smooth enough ($\omega_0 \in W^{r+2,\infty}$ is actually a sufficient condition),

$$\frac{\partial (S_\varepsilon^\nu - S^\nu)\omega_0}{\partial t} + \mathrm{div}\left[\mathbf{a}(S_\varepsilon^\nu - S^\nu)\omega_0\right] - \nu\Delta_\varepsilon\left[(S_\varepsilon^\nu - S^\nu)\omega_0\right] = O(\nu^{r/2}).$$

We now use relation (5.4.9) and approximation (5.4.10) together with the stability properties of the advection equation to deduce, for $t \leq T$,

$$\left\|S_\varepsilon^\nu(t)\omega_0 - S^\nu(t)\omega_0\right\|_{L^\infty} \leq C(T)\nu^{r/2}. \tag{5.4.11}$$

The Δ_ε equation can thus be viewed as giving the $[r/2]$ first terms in an asymptotic expansion, with respect to ν, of the solution to the convection–diffusion equation with small viscosity. This result easily extends to the nonlinear Navier–Stokes equations (one must then use the elliptic regularity properties stated in Section A.3 to deal with the nonlinear terms in the advection equation).

Let us now continue our analysis and in particular investigate the effect of the particle approximation itself. To denote the particle approximation of Δ_ε on the moving particles, let us set

$$\Delta_{\varepsilon,h,t}\omega = \varepsilon^{-2}\sum_p v_p\{\omega[\mathbf{x}_p(t)] - \omega(\mathbf{x})\}\eta_\varepsilon[\mathbf{x}_p(t) - \mathbf{x}],$$

where the subscript h refers to the particle discretization and $\omega = S^\nu\omega_0$, $\omega_\varepsilon = S_\varepsilon^\nu\omega_0$. We also define

$$e_p(t) = \omega_p(t) - \omega_\varepsilon[\mathbf{x}_p(t), t].$$

By subtracting Eq. (5.4.6) from Eq. (5.2.4), we obtain

$$\frac{de_p}{dt} = (\Delta_\varepsilon - \Delta_{\varepsilon,h,t})\omega_\varepsilon[\mathbf{x}_p(t), t] + \nu\varepsilon^{-2}\sum_q v_q(e_q - e_p)\eta_\varepsilon(\mathbf{x}_p - \mathbf{x}_q). \tag{5.4.12}$$

We next observe that the definition of the particle paths $\mathbf{X}$ together with a straightforward application of the quadrature error estimate (A.1.3) allows us to write

$$\Delta_{\varepsilon,h,t}\omega_\varepsilon(\mathbf{x}) = \varepsilon^{-2}\int\{\omega_\varepsilon[\mathbf{X}(t; \mathbf{y}, 0)] - \omega_\varepsilon(\mathbf{x})\}\eta_\varepsilon[\mathbf{X}(t; \mathbf{y}, 0) - \mathbf{x}]\,d\mathbf{y}$$

$$+ O(h^m\varepsilon^{-m-2}) = \Delta_\varepsilon\omega_\varepsilon(\mathbf{x}) + O(h^m\varepsilon^{-m-2}). \tag{5.4.13}$$

This estimate is actually valid on suitable smoothness assumptions on ω_ε and on the particle paths $\mathbf{X}$, which in turn respectively follow from the smoothness of ω_0 and the velocity field $\mathbf{a}$.

We also note that

$$\left| \sum_q v_q (e_q - e_p) \eta_\varepsilon (\mathbf{x}_p - \mathbf{x}_q) \right|$$

$$\leq 2 \max_q |e_q| \sum_q v_q |\eta_\varepsilon (\mathbf{x}_q - \mathbf{x}_p)| \leq \max_q |e_q|$$

$$\left[\int |\eta(\mathbf{x})| \, d\mathbf{x} + O(h^m \varepsilon^{-m}) \right]. \tag{5.4.14}$$

Combining relations (5.4.12), (5.4.13), and (5.4.14) thus leads to

$$\max_p |e_p(t)| \leq C \left[h^m \varepsilon^{-m} + (1 + h^m \varepsilon^{-m}) \int_0^t \max_p |e_p(s)| \, ds \right].$$

If the usual overlapping condition $h \leq \varepsilon$ is met, this shows, by Gronwall's theorem, that

$$\max_p |e_p(t)| = O(h^m \varepsilon^{-m}).$$

It remains to recall relation (5.4.11) to obtain the error estimate (stated in terms of ν for convenience):

$$\max_p |\omega_p(t) - \omega[\mathbf{x}_p(t), t]| = O(\nu^{r/2} + h^m \nu^{-m/2}).$$

This result is the expected convergence result for the semidiscrete particle method. Let us now briefly comment on the stability requirements for the fully discrete algorithm, and in particular the already mentioned explicit first-order time discretization (5.4.8) of Eq. (5.4.6). Under the scaling $\nu = C\varepsilon^2$ one easily deduces from Eq. (5.4.8) that

$$\max_p |\omega_p^{n+1}| \leq \max_p |\omega_p^n| \left\{ 1 + C\delta t \left[\sum_q v_q |\eta_\varepsilon (\mathbf{x}_q - \mathbf{x}_p)| \right] \right\}$$

$$\leq \max_p |\omega_p^n| \{ 1 + C\delta t [\lambda + O(h^m \varepsilon^{-m})] \}.$$

This shows that the algorithm is stable, provided that $h = O(\varepsilon)$. Now if one uses nonnegative cutoff, which allows for smaller values of ε, one gets

$$\max_p \left| \omega_p^{n+1} \right| \le \max_p \left| \omega_p^n \right|,$$

if

$$\nu \delta t \varepsilon^{-2} \sum_q v_q \eta_\varepsilon (\mathbf{x}_q - \mathbf{x}_p) \le 1. \tag{5.4.15}$$

Since

$$\sum_q v_q \eta_\varepsilon (\mathbf{x}_q - \mathbf{x}_p) \simeq \lambda = \int \eta(\mathbf{x}) \, d\mathbf{x}, \tag{5.4.16}$$

relation (5.4.15) is a stability condition similar to that found in finite-difference methods. However, if one uses relation (5.4.15) to choose the maximal time step allowed, one has to be careful that, in the absence of remeshing, $\sum_q v_q |\eta_\varepsilon (\mathbf{x}_q - \mathbf{x}_p)|$ can significantly differ from λ in parts of the flow where particle spacings get highly distorted, e.g., around stagnation points. Of course at these points the accuracy of the algorithm is questionable (the apparent conflict with the error estimates comes from the fact that, as in the inviscid case, these estimates involve constants that depend, in particular, on the smoothness of the flow map; these constants can grow exponentially in time), and local regridding is certainly necessary. However, it may be desirable to have a method that at least does not blow up in these situations and does not require choosing a time step much smaller than needed in the rest of the flow. One way suggested in Ref. 62 is to modify PSE scheme (5.4.8) into

$$\omega_p^{n+1} = \omega_p^n + \nu \delta t \varepsilon^{-2} \lambda \frac{\sum_q v_q (\omega_q - \omega_p) \eta_\varepsilon (\mathbf{x}_q - \mathbf{x}_p)}{\sum_q v_q |\eta_\varepsilon (\mathbf{x}_q - \mathbf{x}_p)|}. \tag{5.4.17}$$

In this formulation the algorithm is stable under the condition

$$\nu \delta t \varepsilon^{-2} \lambda \le 1.$$

5.4.3. PSE Schemes with Variable-Size Blobs

The PSE methodology has enough flexibility in dealing with viscous effects to allow the treatment of variable viscous scales in the vorticity redistribution scheme. The use of variable-blob size has already been considered in the inviscid case (see Subsection 2.6.3). The underlying idea is that one may wish to use fewer particles and thus reduce the computational cost in zones of the flow where vorticity gradients are small, for example in the far wake behind obstacles. This can be done through merging of nearby vortices or, more generally, by periodically remeshing the particle distribution on a variable size mesh.

As we have seen above, a consistent treatment of the diffusion requires the overlapping of the particles on a scale given by the kernel used in the PSE formula. As a result, a locally coarser particle resolution must go with an increasing diffusion range ε. The simple recipe used to incorporate variable blobs in the Biot–Savart law [Eq. (2.2.13)] would consist of replacing ε with $\varepsilon(y)$ in definition (5.4.3) of Δ_ε. However, this technique would not be consistent. The reason is that, if one follows the argument developed in Subsection 2.6.3, the Taylor expansion of ε would bring new terms in the factor of the second-order moments of the cutoff η.

The correct way to implement a variable-blob size in a PSE scheme is through a change of variables mapping the variable grid to a uniform one. For simplicity let us focus on the one-dimensional case. We denote by x, y locations in the physical space, with variable grid size, and by $\hat{x}$, $\hat{y}$ locations on the mapped coordinates, where the grid size is uniform. We assume that the mapping is given by the formulas

$$x = f(\hat{x}), \hat{x} = g(x), \omega(x) = \hat{\omega}(\hat{x}).$$

Writing derivatives in the mapped coordinates yields

$$\frac{d^2\omega}{dx^2} = h(\hat{x})\frac{d}{d\hat{x}}\left[h(\hat{x})\frac{d\hat{\omega}}{d\hat{x}}\right],$$

where $h(\hat{x}) = g'(x)$. Next, we use the following integral approximation, the proof of which relies on Taylor expansions at second order similar to those in the proof of relation (5.4.4):

$$\frac{d}{d\hat{x}}\left[h(\hat{x})\frac{d\hat{\omega}}{d\hat{x}}\right] \simeq \varepsilon^{-3}\int\frac{h(\hat{x})+h(\hat{y})}{2}[\hat{\omega}(\hat{x})-\hat{\omega}(\hat{y})]\eta\left(\frac{\hat{x}-\hat{y}}{\varepsilon}\right)d\hat{y}.$$

In the above formula, the kernel η satisfies the moment properties of Eqs. (5.4.2) and ε is a constant blob size. This leads to the following PSE scheme for the heat equation in one dimension with $\nu = 1$:

$$\frac{d\omega_p}{dt} = \varepsilon^{-3}h(\hat{x}_p)\sum_q \hat{v}_q\frac{h(\hat{x}_p)+h(\hat{x}_q)}{2}(\omega_q - \omega_p)\eta\left(\frac{\hat{x}_p-\hat{x}_q}{\varepsilon}\right), \qquad (5.4.18)$$

where $\hat{v}_q$ denote the volumes of the mapped particles. Note that the volumes of the physical and the mapped particles are related through the Jacobian of the mapping:

$$\hat{v}_q = v_q h(\hat{x}_q),$$

which establishes that scheme (5.4.18) is indeed conservative. To extend this approach to two- or three-dimensional flows, one must handle integral approximations of differential terms involving diffusion tensors. We refer to Ref. 73 for general formulas to deal with this case.

In complete convection–diffusion problems, the use of variable-blob sizes must be combined with regridding techniques, such as those designed in Section 7.2, to make sure that the particle discretization is everywhere consistent with the local blob sizes. The overlapping condition needed to derive Eq. (5.4.18) in particular requires that the frequency of regridding is such as to deter particles in the low-resolution regions from traveling too far in the high-resolution zones.

5.4.4. Results

In order to illustrate the particular aspects of the method and to compare its advantages and drawbacks with the random-walk method, we consider first the one-dimensional initial boundary-value problem for the diffusion equation described by Eqs. (5.2.7) and (5.2.8). In Figure 5.5 we present the rms error by the PSE scheme for which a Gaussian kernel is used. Comparing with the respective results obtained by using the random-walk method, we can clearly observe a 2-order-of-magnitude improvement in the results of the present calculations. Note that in this case the particles remain uniformly distributed. For convection–diffusion problems, the accuracy of the PSE scheme to handle diffusion is based on the ability of the method to approximate the Laplacian operator on the irregular particle locations, while providing a natural way for the diffusive expansion of the support of the vorticity field. Unlike the random-walk method, resampling schemes and the method of PSE require that particles overlap at all times and that ghost particles always surround the vorticity-carrying elements to provide a natural expansion of the vorticity field. We will see in Section 7.1 how regridding techniques allow us to accommodate these constraints.

Our next illustration deals with the merging of two like-signed Gaussians subject to an axisymmetric strain field. Figure 5.5 is a comparison of a PSE scheme and a spectral finite-difference calculation of Buntine and Pullin [38]. The good agreement between these results confirm that PSE methods give an

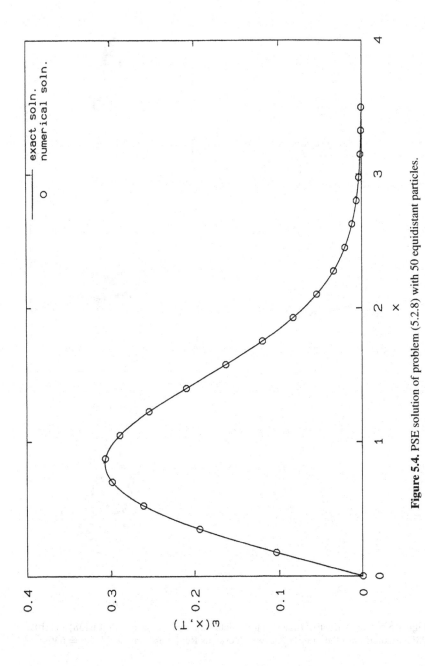

Figure 5.4. PSE solution of problem (5.2.8) with 50 equidistant particles.

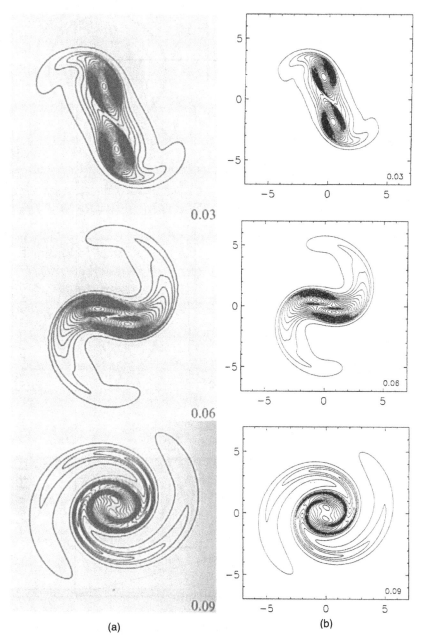

Figure 5.5. Comparison of (a) a spectral calculation by Buntine and Pullin [38] and (b) a PSE calculation of the merging of two Gaussians Reynolds number $\Gamma/2\pi\nu = 640$.

Figure 5.6. Vorticity contours of vortex shedding behind a circular cylinder at Re = 200, using a variable blob vortex method.

accurate treatment of the diffusion. Illustrations of PSE schemes for three-dimensional flows will be given in Chapters 6 and 8.

Our final example illustrates the flexibility of PSE schemes to handle variable diffusion scales. Vortex methods are very well suited for external flows because they allow an exact treatment of the far-field conditions and their computational effort is restricted to vortical zones of the flow. They can be made even more efficient if one can take advantage of the decay of vorticity gradients in the wake to save computational elements. This can be done by using the methodology explained in Subsection 5.4.3. In the case of a wake behind a cylinder, shown in Figures 5.6 and 5.7, the blob size is proportional to the distance to the center of the obstacle. Particles are mapped to a uniform mesh through the mapping $(\mathbf{x}, \mathbf{y}) \rightarrow (r' = \log r, \theta)$ where r, θ are the polar coordinates. In the mapped coordinates the diffusion is given as $\Delta \omega = r^{-2} \Delta_{r', \theta} \hat{\omega}$ which allows us to use simple PSE formulas. Figure 5.6 shows the classical development of a Karman street in the wake for a Reynolds number of 200. Figure 5.7 shows the corresponding particle locations. The total number of particles in the last picture is approximately 6,000.

The efficiency and the simplicity of the variable-blob approach in this flow results from the fact that there is a natural global mapping that accounts for the scales that are to be resolved. For more complex geometries, it is possible to combine several local mappings, corresponding to different parts of the flow. Figure 5.8 gives an example in which particles around each cylinder have volumes given by $r^2 dr d\theta$, whereas in the intermediate zone they have uniform volumes. The grids shown in the figure illustrate the resolution of the particle distribution (for clarity it corresponds to $1/4$ of the actual resolution used in this computation), but of course particles are present only in the support of the vorticity.

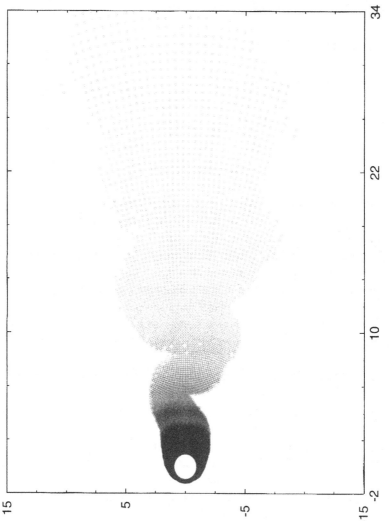

Figure 5.7. Particle locations for vortex shedding behind a circular cylinder at Re $= 200$.

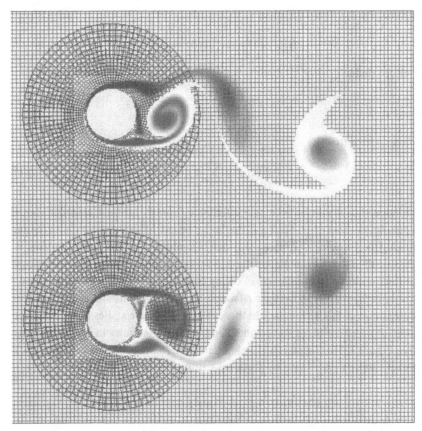

Figure 5.8. Simulation of flow past a pair of circular cylinders (Re $= 400$) through the combination of several local mappings.

5.5. Other Redistribution Schemes

We describe here two additional schemes in which the diffusion is approximated by redistributing the vorticity among neighboring particles.

5.5.1. Russo's Triangulated Viscous Vortex Method

The method we now describe is in the spirit of the triangulated vortex method that will be described later in Chapter 7. It has been introduced by Russo [174, 175] as a deterministic alternative to the random-walk method and is based on the observation that a particle method can deal with viscous terms as soon as one is able to project second derivatives of Dirac masses onto particles. This can be done in an elegant way by use of the weak formulation of derivatives.

More specifically, let us denote by ϕ a test function vanishing at the boundary of the computational domain and consider a single particle, located at $\mathbf{x}_p$ and denoted by $\delta_{\mathbf{x}_p}$. From the identity

$$\langle \Delta \delta_{\mathbf{x}_p}, \phi \rangle = \langle \delta_{\mathbf{x}_p}, \Delta \phi \rangle = \Delta \phi(\mathbf{x}_p)$$

one can deduce

$$\Delta \delta_{\mathbf{x}_p} \simeq \sum_q \alpha_{pq} \delta_{\mathbf{x}_q}$$

if one is able to write

$$\Delta \phi(\mathbf{x}_p) \simeq \sum_q \alpha_{pq} \phi(\mathbf{x}_q). \tag{5.5.1}$$

Relation (5.5.1) can be seen as a difference formula for the Laplacian on the Lagrangian mesh $(\mathbf{x}_q)$. Such a formula can be established once a triangulation linking the particles is built, in a way that is reminiscent of the procedure used in the so-called free Lagrangian method [86] (the method also bears some analogy to finite-volume techniques). Following Ref. 174, we focus on triangulations resulting from a Voronoi diagram (we postpone to Section 7.1 a more complete description of this technique). In this case, each particle $\mathbf{x}_p$ lies in the center of a cell P_p, and the triangles are obtained by linking particles in connected cells (see Figure 7.2 in Chapter 7). To obtain relation (5.5.1), the idea is to introduce fluxes through integration by parts, by writing

$$|P_p| \Delta \phi(\mathbf{x}_p) \simeq \int_{P_p} \Delta \phi(\mathbf{y}) \, d\mathbf{y} = \int_{\partial P_p} \frac{\partial \phi}{\partial n}(\mathbf{y}) \, d\mathbf{y},$$

where $|P_p|$ denotes the cell area (or volume in three dimensions). The normal derivative on the right-hand side above is then approximated by finite differences across the corresponding cell edge:

$$\int_{\partial P_p} \frac{\partial \phi}{\partial n}(\mathbf{y}) \, d\mathbf{y} \simeq \sum_{q \sim p} \frac{\phi(\mathbf{x}_q) - \phi(\mathbf{x}_p)}{|\mathbf{x}_q - \mathbf{x}_p|} l_{pq},$$

where the notation $\sim$ means that the sum is restricted to the particles in adjacent cells and l_{pq} denotes the length of the common edge (or, in three dimensions, the area of the common face) to the cells centered in $\mathbf{x}_p$ and $\mathbf{x}_q$.

Formula (5.5.1) follows then with

$$\alpha_{pq} = \begin{cases} \frac{1}{|P_p|} \frac{l_{pq}}{|\mathbf{x}_p - \mathbf{x}_q|} & p \sim q \\[2ex] \frac{1}{|P_p|} \sum_{r \sim q} \frac{l_{pr}}{|\mathbf{x}_p - \mathbf{x}_r|} & p = q \end{cases}. \tag{5.5.2}$$

The diffusion scheme that arises from formulas (5.5.1) and (5.5.2) now is given by

$$\omega(\mathbf{x}, t) \simeq \sum_p \Gamma_p \delta(\mathbf{x} - \mathbf{x}_p),$$

where particles are advected by the local velocity and circulations are updated with

$$\frac{d\Gamma_p}{dt} = \nu \sum_q \alpha_{pq} \Gamma_q. \tag{5.5.3}$$

From Eq. (5.5.2) it is readily seen that

$$\sum_q \alpha_{pq} = 0,$$

from which results the conservativity of scheme (5.5.3). The scheme is also enstrophy decreasing: if one defines the enstrophy with the formula

$$\mathcal{E}^2 = \sum_p \frac{\Gamma_p^2}{v_p},$$

where v_p are the time-independent volumes of the particles (not to be confused with the varying cell volumes), then

$$\mathcal{E}\frac{d\mathcal{E}}{dt} = \nu \sum_{p,q} \Gamma_p \Gamma_q \alpha_{pq}.$$

By using elementary tools from linear algebra (in particular Gershgorin theorem) and formula (5.5.2), we readily see that the matrix $[\alpha_{pq}]$ has only negative eigenvalues, implying that the right-hand side above is negative.

Compared with resampling methods devised earlier, this method avoids the introduction of blobs for the diffusion equation, thus bypassing the quadrature error and the associated overlapping condition. However, its applicability has been hampered so far by difficulties with stability. Indeed, an explicit Euler time discretization of Eq. (5.5.3) with time step δt would yield

$$\Gamma_p^{n+1} = \Gamma_p^n(1 - \nu\delta t\alpha_{pp}) + \sum_{q \sim p, q \neq p} \alpha_{pq} \Gamma_q^n. \tag{5.5.4}$$

Since $\alpha_{pq} > 0$ for $q \neq p$, this scheme is stable if one can check that, for all p,

$$\nu\delta t\alpha_{pp} < 2,$$

or, equivalently,

$$\delta t < \frac{2}{\nu \max_p \sum_{q \sim p, q \neq p} \frac{l_{pq}}{|\mathbf{x}_p - \mathbf{x}_q|}}.$$

In view of the incompressibility of the flow, one can assume that the cell volumes remain of the order of the particle volumes v_p, say h^2 (in two dimensions). If the particle distribution is uniform, then l_{pq} and $|\mathbf{x}_p - \mathbf{x}_q|$ are of the same order and the stability condition is similar to that obtained for explicit time discretization of finite-difference schemes. However, if the flow strain elongates the cells, making l_{pq} large compared with $|\mathbf{x}_p - \mathbf{x}_q|$, the condition may lead to drastic time-step restrictions. Note that the same difficulty has been mentioned for PSE schemes. The remedies suggested at this time [namely correction algorithm (5.4.17) and/or remeshing strategies] would be difficult to translate into the present scheme. Remeshing in particular would affect the cost of the construction of Voronoi diagram, which can be limited to $O(N)$ operations insofar as it is only locally updated at each time step.

Russo [174] presents an implicit alternative to Eq. (5.5.4). The resulting scheme is unconditionally stable but requires the inversion of a matrix that, as a consequence of particle distortion, can become highly irregular. The experiments reported in Ref. 174 show that a conjugate gradient approach to the solution of the linear system that must be solved at each iteration leads to a considerable computational overhead over the repeated cost of the construction of the Voronoi diagram.

5.5.2. The Redistribution Scheme of Shankar–van Dommelen

This scheme [186] consists of updating at each time step the circulations of the particles with the formula

$$\Gamma_p^{n+1} = \sum_q \Gamma_q^n f_{pq}^n, \tag{5.5.5}$$

where f_{pq}^n represents the amount of vorticity to transfer through diffusion between particles during a time step δt. The influence range, which determines the number of particles involved in the above sum, is taken of the order of $\sqrt{\nu \delta t}$. The parameters f_{pq}^n are computed at each time step in order to satisfy the following constraints:

$$\sum_q f_{pq}^n = 1, \tag{5.5.6}$$

$$\sum_q f_{pq}^n (\mathbf{x}_p - \mathbf{x}_q) = 0, \tag{5.5.7}$$

$$\sum_q f_{pq}^n \left(x_p^i - x_q^i\right)\left(x_p^j - x_q^j\right) = 2\nu \delta t \delta_{ij}. \tag{5.5.8}$$

The superscripts i and j indicate the component indices. Because of Eq. (5.5.6), scheme (5.5.5) can be rewritten in a PSE setting as

$$\Gamma_p^{n+1} = \Gamma_p^n + \sum_q \left(\Gamma_q^n - \Gamma_p^n \right) f_{pq}^n \tag{5.5.9}$$

and conditions (5.5.7) and (5.5.8) can be seen as a discrete version of the first- and the second-order PSE conditions in Eqs. (5.4.2). The numerical analysis follows along the same lines and reveals that the method is second order.

It is interesting to remark that the method is reminiscent of an earlier version of the adaptive method of Strain that we will outline in Chapter 7. In this method [190], proposed in the context of inviscid flows, the volumes of the particles are processed at each time step in such a way as to decrease the quadrature error in the calculations of the particle velocities. This is achieved by constraining the quadrature rule to be exact for the Biot–Savart kernel multiplied by polynomials up to a given degree. Transposing this idea in the context of PSE schemes, with a second-order PSE kernel instead of the Biot–Savart kernel, would precisely lead to discrete moment properties (5.5.7) and (5.5.8).

The present method also bears some similarity with the triangulated method discussed earlier in this section in that it relies on only particle quantities without resorting to any *a priori* cutoff functions and the associated overlapping constraints; as far as viscous effects are concerned, it is thus potentially free of any remeshing need. Reference 186 reports results for flows around an impulsively cylinder that compare well with the results obtained by the PSE method (see Chapter 6).

This method, however, carries some limitations and uncertainties. For stability reasons the amount of transfered vorticity has to be positive, which precludes the higher-order momentum properties that would be necessary to obtain higher-order schemes. Moreover system (5.5.6)–(5.5.8) is most likely either underdetermined or overdetermined, depending on the number of particles in the diffusion range of a given particle. In the former case, the authors use a linear programming technique to determine a solution, which results in a significant computational overhead. The results given in Ref. 186 indicate that the computation of the f_{pq} more than doubles that of the velocity evaluations, whereas, in PSE schemes, the diffusion evaluation, even when combined with a remeshing procedure, has a relatively minimal ($\sim 2\%$ of the convection step) cost. In the proposed redistribution scheme (second case) it is necessary to insert new particles, thus reintroducing remeshing but in a less clear-cut way than through the interpolation techniques devised in Chapter 7.

5.6. Subgrid-scale Modeling in Vortex Methods

Subgrid-scale modeling is commonly associated with the concept of Large Eddy Simulations (LES). Its need stems from the fact that, while one is generally interested in only large-scale structures of the flow, because of the nonlinearity of the Navier–Stokes equations, one cannot ignore the effect of the nonresolved small scales.

In grid-based methods, small scales must be explicitly parameterized in order to allow energy transfers to and from large scales. In contrast, vortex methods do not involve any minimal scale in their dynamics and have the ability to allow small scales to proliferate. This is particularly clear in the example of the evolution of a single filament subject to stretching and folding. The creation of small loops and hairpins is the signature of small scales appearing in the course of the calculation.

One may wonder whether these microstructures are physical or merely numerical artifacts. On the one hand, the fact that vortex methods solve exactly the inviscid equations pleads in favor of the former interpretation. On the other hand, one may argue that round-off error can at least partially trigger these small scales, suggesting a numerical artifact. The important practical issue is actually that, no matter how relevant the small scales produced in a vortex calculation are, their accurate resolution would require an increasing number of points, eventually driving the computation to overwhelming computational costs.

A natural action to overcome this difficulty is to remove, or at least control, the developing microstructures, hopefully without affecting the significant large scales. We describe in this section two approaches to address this issue.

The first one consists, in the context of inviscid three-dimensional filament calculations, of modeling dissipation effects by locally modifying the topology of filaments. This covers the filament surgery techniques introduced by Leonard [134] and the hairpin removal algorithms of Chorin [52, 53]. In the second approach, dissipation is explicitly introduced to remove small scales. In this class of methods we describe the core expansion technique and a more recent turbulent viscosity model derived within the framework of PSE schemes [64].

5.6.1. Filament Reconnection and Hairpin Removal

In a vortex filament calculation, nearby filaments may attract each other until the corresponding numerical vortex tubes overlap, eventually canceling the overall vorticity. Filament reconnection is a way to mimic viscous effects, on the scale of the filament separations, to reconnect the filaments on the basis of their combined vorticity field. Reconnection can also take place within a single filament whenever it creates entangled loops. Similarly, a single filament submitted to stretching can fold on itself, producing small hairpins. Rather than refining the filament discretization to resolve these hairpins, which presumably do not contribute to the large-scale features of the flow, it is advisable to remove them. It is interesting to observe here that the concept of hairpin removal was first discussed by Feynman in 1957 [82] in the context of turbulent flows in superfluids. In Figure 5.9 we sketch Feynman's ideal of the breakup of an inviscid vortex ring into "rotons."

The procedure suggested by Chorin [52] for hairpin removal consists of considering the angles made by successive segments along each filament. Hairpin removal takes place whenever this angle goes beyond a certain predetermined threshold. The two segments are then replaced by their sum, thus reducing the length of the filament. Conversely, splitting of a filament into two filaments is done by consideration of the segments along the filaments that are distant from each other by less than a given threshold (which must be set smaller than the length of individual segments) and application of the same angle condition. The filament is split into two new filaments by the introduction of two counterrotating segments. The same strategy can be used for the reconnection of approaching filaments (in this case distances and angles segments belonging to two filaments are monitored). As noted in Ref. 53, the methods bear, at least formally, important similarities to the contour surgery techniques devised for contour dynamics methods (we refer to Subsection 3.2.3 for the analogy between vortex filament and contour dynamics methods).

Numerical experiments are reported in Ref. 53 concerning the simulation of vortex rings. They indicate that these procedures allow considerable savings through the reduction of computational elements while preserving reasonably well the invariants of the flow. The limitations of the method are in the number of ad hoc parameters that require tuning (minimum distance, maximum angle, as well as the frequency of the operation). Also it has difficulties, inherent in vortex filament methods, in handling Reynolds-number-dependent effects, if viscous, rather than inviscid, solutions are sought.

5.6.2. The Core-Spreading Technique

This method was originally proposed as an alternative to the random-walk method to solve the Navier–Stokes equations. It was based on the observation that Gaussians are explicit solutions to the heat equations and consisted of using spreading Gaussian blobs around the particles while keeping constant the circulation carried by the particles.

However, this method cannot be used if consistent approximations to the Navier–Stokes equations are sought, and the core-spreading method should be

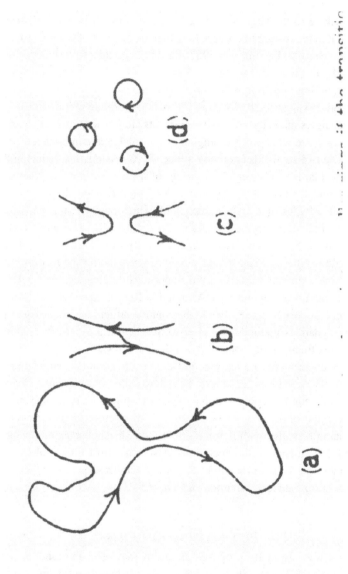

Figure 5.9. From Ref. 82: "*A vortex ring (a) can break up into smaller rings if the transition between states (b) and (c) is allowed when the separation of vortex lines becomes of atomic dimensions. The eventual small rings (d) may be identical to rotons.*"

seen merely as a diffusion model that is valid only in the limit of vanishing viscosity. The reason for this inconsistency is that, for a fixed viscosity value, the size of the blobs can no longer be considered as a small smoothing parameter. Because of the strain in the flow, the spreading blobs cease therefore to yield a consistent approximation of the flow in the advection phase. This was clearly demonstrated by Greengard [90].

A plausible cure for this inconsistency would be to break the Gaussian cores after each diffusion step into new particles. This may be viewed as another version of the resampling schemes described earlier. The prospect of injecting physical considerations in the evolution of the particle core size has prompted several research efforts in this topic. These are works presently in progress that we will not discuss further in this book.

5.6.3. A Turbulent Viscosity Model

We outline here a subgrid-scale model based on a rigorous analysis of the truncation error involved in vortex calculations.

In the following we denote by $\overline{(\ )}$ the filtering operation resulting from the mollification by a given cutoff. As we have seen in Chapter 3, all three-dimensional vortex schemes (based on either point particles or filaments) share the property that, in the absence of time discretization, they satisfy the equation

$$\frac{\partial \omega}{\partial t} + \operatorname{div}(\overline{\mathbf{u}}\omega) - (\omega \cdot \nabla)\overline{\mathbf{u}} = 0 \tag{5.6.1}$$

where $\overline{\mathbf{u}}$ is obtained from the particle vorticity field ω through a mollified Biot–Savart law. On mollifying Eq. (5.6.1), we obtain the following equation for the vortex blobs (or vortex tubes, if a vortex filament method is used):

$$\frac{\partial \overline{\omega}}{\partial t} + \operatorname{div}\overline{\overline{\mathbf{u}}\omega} - \overline{(\omega \cdot \nabla)\overline{\mathbf{u}}} = 0. \tag{5.6.2}$$

When compared with the Euler equation for the fields $\overline{\mathbf{u}}$ and $\overline{\omega}$, this equation involves a truncation error $\mathbf{e}$ that we can split into two terms, $\mathbf{e}_1$ and $\mathbf{e}_2$, corresponding to the convection and the stretching terms, respectively:

$$\mathbf{e}_1 = \operatorname{div}\overline{\overline{\mathbf{u}}\omega} - \overline{\overline{\mathbf{u}}\omega}, \tag{5.6.3}$$

$$\mathbf{e}_2 = \overline{(\omega \cdot \nabla)\overline{\mathbf{u}}} - (\overline{\omega} \cdot \nabla)\overline{\mathbf{u}}. \tag{5.6.4}$$

One may observe that these error terms are reminiscent of the comutation error resulting from filtering nonlinear terms in the Navier–Stokes equations. These are the terms that are commonly modeled to obtain the governing equations of LES. They account for the transfer of enstrophy (in the case of a vorticity formulation) between large and small scales.

The fact that vortex methods solve filtered equation (5.6.2) indicates that, by themselves, these methods realize some kind of turbulence modeling. This confirms the already mentioned fact that, unlike that of grid-based methods, their dynamics is not constrained to any minimal scale (beyond the initialization stage, of course). From Taylor expansions that are closely related to those involved in the derivation of PSE schemes, one can show that, if the cutoff is radially symmetric, the error term $\mathbf{e}_1$ can be approximated by

$$\mathbf{e}_1 \simeq \operatorname{div}(\mathbf{u} - \bar{\mathbf{u}})\omega + m_2\varepsilon^2 \operatorname{div}([D\bar{\mathbf{u}}]\nabla\omega), \qquad (5.6.5)$$

where ε is the cutoff length and $m_2 = 1/2 \int x_i^2 \zeta(\mathbf{x})\,d\mathbf{x}$. The first term on the right-hand side above is an advective term that does not contribute to enstrophy transfers between large and small scales and thus can be disregarded. In contrast, the second one involves dissipation as well as antidiffusion, as is seen by consideration of the eigenvalues of the tensor $D\bar{\mathbf{u}}$: Since div $\mathbf{u} = 0$, the trace of the tensor vanishes, implying that dissipation in one direction comes with antidiffusion in a transverse direction (note that in this discussion we have implicitly assumed that the cutoff has a nonvanishing second-order momentum; if this were not the case, the error term $\mathbf{e}_1$ would involve a hyperviscosity tensor).

It is now natural to assert that the antidiffusion (or backscatter, to use turbulence modeling terminology) is responsible for the spontaneous appearance of microstructures in vortex calculations. In three-dimensional calculations, hairpins are one particular manifestation of these microstructures, and hairpin removal can be viewed as a backscatter control device. In two-dimensional calculations, the amplification of round-off errors in vortex sheet calculations [130] or in periodic turbulence experiments [64, 100] can be traced back to the same origin. In vortex sheet calculations, filtering and/or mesh refinement has been used by Krasny to prevent the appearance of spurious microstructures that overwhelm the large-scale features of the flow.

A more clear-cut way to control the backscatter has been proposed in Ref. 64, based on the truncation error analysis just outlined. It consists of identifying the antidiffusive part of $\mathbf{e}_1$ through an enstrophy balance calculation. Recalling that filtering is the result of the convolution by a kernel ζ_ε and multiplying Eq. (5.6.3) by ω, we obtain

$$\int \mathbf{e}_1(\mathbf{x})\omega(\mathbf{x})\,d\mathbf{x} = \int \omega(\mathbf{y})\omega(\mathbf{x})[\mathbf{u}_\varepsilon(\mathbf{x}) - \mathbf{u}_\varepsilon(\mathbf{y})] \cdot \nabla\zeta_\varepsilon(\mathbf{x} - \mathbf{y})\,d\mathbf{x}\,d\mathbf{y},$$

where $\mathbf{u}_\varepsilon = \mathbf{u} \star \zeta_\varepsilon$. We then write $\omega(\mathbf{x}) = \omega(\mathbf{y}) + [\omega(\mathbf{x}) - \omega(\mathbf{y})]$. Because of the incompressibility of the flow, it is readily seen through integration by parts that the contribution of $\omega(\mathbf{y})$ vanishes. This contribution indeed reflects the advective part of the truncation error already mentioned about on the right-hand side of relation (5.6.5). We are thus left with

$$\int \mathbf{e}_1(\mathbf{x})\omega(\mathbf{x})\,d\mathbf{x} = \int [\omega(\mathbf{x}) - \omega(\mathbf{y})]\omega(\mathbf{y})[\mathbf{u}_\varepsilon(\mathbf{x}) - \mathbf{u}_\varepsilon(\mathbf{y})] \cdot \nabla\zeta_\varepsilon(\mathbf{x} - \mathbf{y})\,d\mathbf{x}\,d\mathbf{y}.$$

We now write $\omega(\mathbf{y}) = 1/2\{[\omega(\mathbf{y}) + \omega(\mathbf{x})] + [\omega(\mathbf{y}) - \omega(\mathbf{x})]\}$. By symmetry the first term does not contribute to the integral and, finally,

$$\int \mathbf{e}_1(\mathbf{x})\omega(\mathbf{x})\,d\mathbf{x} = -\frac{1}{2}\int [\omega(\mathbf{y}) - \omega(\mathbf{x})]^2[\mathbf{u}_\varepsilon(\mathbf{x}) - \mathbf{u}_\varepsilon(\mathbf{y})] \cdot \nabla\zeta_\varepsilon(\mathbf{x} - \mathbf{y})\,d\mathbf{x}\,d\mathbf{y}.$$
$$(5.6.6)$$

Backscatter and dissipation produced by vortex methods at a given location $\mathbf{x}$ as the results of nearby particles located at $\mathbf{y}$ are thus distinguished by the sign of the quantity

$$[\mathbf{u}_\varepsilon(\mathbf{x}) - \mathbf{u}_\varepsilon(\mathbf{y})] \cdot \nabla\zeta_\varepsilon(\mathbf{x} - \mathbf{y}).$$

In the case of a cutoff that is a positive decaying function of the radius, approaching particles contribute to dissipation, whereas diverging particles yield backscatter.

To remove the backscatter a natural scheme is to modify the weights of the particle in such a way as to cancel the positive contribution in relation (5.6.6). This leads to the scheme

$$\frac{d\omega_p}{dt} = \sum_q (\omega_p - \omega_q)v_q\{[\mathbf{u}(\mathbf{x}_p) - \mathbf{u}(\mathbf{x}_q)] \cdot \nabla\zeta_\varepsilon(\mathbf{x}_p - \mathbf{x}_q)\}_-, \qquad (5.6.7)$$

where $\{a\}_- = \min\{0, a\}$.

This scheme can be seen as a PSE method for the diffusion model $-\mathrm{div}([D\mathbf{u}]_+ \nabla\omega)$. In the above scheme, the choice was to remove all the backscatter produced by the flow strain. This has the advantage of providing us with a parameter-free method. The numerical experiments reported in Ref. 64 for two-dimensional turbulence show that this choice seems appropriate for inviscid flows. Figure 5.10 shows the evolution of a random initial vorticity field with a periodic vortex-in-cell method (see Chapter 8). The bottom pictures correspond to a purely inviscid scheme and show an overwhelming proliferation of small scales, whereas the LES model of Eq. (5.6.7) allows these small scales to organize into filaments (which later on will merge into a unique dipole). In Figure 5.11, we have represented the intensity of the subgrid-scale disssipation at all particle locations

Figure 5.10. Successive stages, from left to right, in the evolution of a random vorticity field in an inviscid calculation (bottom pictures) and with the LES model of Eq. (5.6.7) (top pictures).

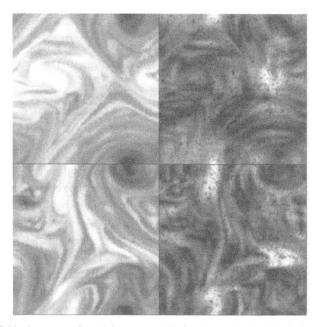

Figure 5.11. Contours of vorticity strength (left) & magnitude of turbulent viscosity (right) for two successive times (high values of viscosity correspond to light gray levels).

for two selected times. The vorticity is shown on the top pictures and the gray levels of the dissipation on the bottom pictures. These pictures show the correlations between hyperbolic zones in the flow and regions of high dissipation, whereas in the core of the eddies the flow remains essentially inviscid.

It is natural to wonder whether a less strict backscatter control would be more efficient in other situations. The additional coefficient to introduce in Eq. (5.6.7) would somehow play the role of the minimum distance/angle used in hairpin removal algorithms. The added flexibility could be also seen as reflecting the possibility of using a different cutoff (or a different scale) in the averaged equation (5.6.2) and in the mollification of the Biot–Savart law. In viscous calculations, one way to choose this coefficient would be by ensuring that the backscatter does not exceed the molecular dissipation.

6

Vorticity Boundary Conditions for the Navier–Stokes Equations

In this chapter we present boundary conditions for the vorticity–velocity formulation of the Navier–Stokes equations and we describe their implementation in the context of vortex methods. We restrict our discussion to flows bounded by impermeable, solid walls, although several of the ideas can be extended to other cases such as free-surface flows.

The direct numerical simulation of wall-bounded flows requires accurately resolving the unsteady physical processes of vorticity creation and evolution in small regions near the boundary. Vortex methods directly resolve the vorticity field, and they automatically adapt to resolve strong vorticity gradients in regions near the wall, but they are faced with the algorithmic complication of dealing with the no-slip boundary condition. The no-slip boundary condition is expressed in terms of the velocity field at the wall and does not involve explicitly the vorticity.

Mathematically we may understand this difficulty by considering the kinematic and dynamic description of the flow motion and observing that there is an inconsistency between the number of equations and the number of boundary conditions. The kinematic description of the flow, relating the velocity to the vorticity, is an overdetermined set of equations if we prescribe all the components of the velocity at the boundary. On the other hand, no vorticity boundary condition is readily available for the Navier–Stokes equations that govern the dynamic description of the flow.

Physically, the no-slip boundary condition expresses the requirement that the flow field must adhere to the boundary. This condition imposes a torque onto the fluid elements adjacent to the wall, which in turn, may impart a rotational motion to the fluid. Hence the no-slip boundary condition is physically manifested by the creation of vorticity at the boundary. In a vorticity–velocity formulation, vorticity boundary conditions are required for formulating and quantifying this vorticity creation process. Indeed, vorticity boundary conditions are often presented as models of vorticity creation rather than rigorous mathematical constraints.

In general we may distinguish between schemes that involve the boundary value of the vorticity (Dirichlet-type boundary condition) or the wall-normal vorticity flux (Neumann-type boundary condition). The Neumann form of the boundary condition is in general preferred, because, as we will see, it is directly related to the local production of vorticity at a no-slip wall.

The first sucessful attempt to introduce boundary conditions in calculations of viscous flows by use of vortex methods is due to Chorin [49]. He introduced a fractional step algorithm in which vortex blobs are created on the surface of the body in order to enforce the no-slip boundary condition. However, the existence of such blobs on the boundary introduces an artificial smoothing region for the vorticity field near the surface of the body that significantly increases the overall numerical dissipation of the algorithm. To alleviate this difficulty Chorin [50] proposed the vortex sheet method based on a coupling of the solution of the Prandtl boundary-layer equations near the body and the full Navier–Stokes equations away from it. However, this approach seems to encounter difficulties with fully separated flows [12] in which a well-defined boundary-layer region fails to exist and requires tuning of several critical numerical parameters, such as the extent of the region of validity of the boundary-layer equations. Convergence analysis of this algorithm is hindered by the fact that it does not result from a well-defined vorticity boundary condition, independent of the geometry of the domain. As a matter of fact, the only convergence proof so far for this algorithm [29] is restricted to the two-dimensional half-plane case, and it relies on the extension of the vorticity and the use of integral representations over the entire plane.

In an effort to derive exact vorticity boundary conditions, free of any ad hoc parameters, several methods have been proposed. Anderson [6] computes the time derivative of the tangential component of the velocity in terms of the time derivative of the vorticity in the domain. Using the vorticity–velocity form of the Navier–Stokes equations leads to an integrodifferential equation for the wall-normal vorticity flux. Another way to construct exact boundary conditions was proposed by Cottet [59]. In order to alleviate the mathematical inconsistency mentioned above, he proposed to solve for the Poisson's equation that relates the vorticity and the velocity by using velocity boundary conditions. A vorticity boundary condition is then derived by enforcing the identity $\omega = \nabla \times \mathbf{u}$ at the boundary.

These formulations hint at a direct coupling of the kinematic and the dynamic descriptions of the flow. The resulting set of equations is equivalent to the original set of the Navier–Stokes equations in primitive variables. However, energy estimates are not always available for these models and it is not clearly understood under which conditions natural discretizations of the equations will lead to stable numerical algorithms.

Returning to the realm of fractional step algorithms, Cottet [62] and Koumoutsakos et al. [124] independently proposed a splitting algorithm to replace the no-slip boundary condition with an equivalent vorticity flux condition. In Ref. 62 this algorithm is presented as an extension to general geometries of Chorin's vorticity creation algorithm. In Ref. 124, the scheme is viewed as a quantification and extension of the vorticity creation process as envisioned by Lighthill [138] and formulated by Kinney and his co-workers [115, 116]. The vorticity creation process is considered as a fractional step algorithm bridging the kinematic and dynamic description of the flow. Given a certain vorticity field the presence of boundaries is accounted for by the distribution on their surface of a vortex sheet whose role is to enforce the kinematic boundary condition of no-through flow. This vortex sheet is then distributed diffusively into the flow by translating its strength into a Neumann boundary condition for the vorticity equation, which can be handled in an integral representation of the vorticity field. In the context of vortex methods, this consists of modifying the strength of the particles in the vicinity of the boundary.

In Section 6.1 we present the no-slip boundary condition and its formulation in terms of a vorticity boundary condition from the physical and mathematical point of view. Section 6.2 is devoted to the derivation of exact vorticity boundary conditions in the absence of any numerical approximation of the Navier–Stokes equations. We then discuss in Section 6.3 several fractional step algorithms for the Navier–Stokes equations and we emphasize the underlying vorticity boundary conditions and their implementation in the context of vortex methods in two and three dimensions.

6.1. The No-Slip Boundary Condition

We formulate the no-slip boundary condition for incompressible flow in a domain V where a body enclosed by a surface S is traveling with a speed $\mathbf{U}_b(t)$ and rotating with angular velocity $\Omega_b(t)$ around its center of mass located at $\mathbf{x}_b$ (Fig. 6.1).

At an impermeable boundary, kinematics dictate that the normal component of the flow velocity ($\mathbf{u}$) at any location on the body surface ($\mathbf{x}_s$) must be equal to the normal velocity of the body. Hence, at the boundary S of V,

$$\mathbf{u}(\mathbf{x}_s) \cdot \mathbf{n} = \mathbf{U}_s \cdot \mathbf{n}, \tag{6.1.1}$$

where $\mathbf{U}_s = \mathbf{U}_b + \Omega_b \times (\mathbf{x}_s - \mathbf{x}_b)$.

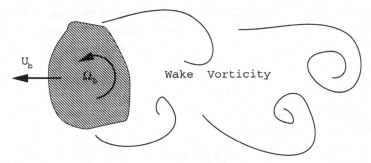

Figure 6.1. Sketch showing the different contributions to the flow field of a body in translation and rotation immersed in a viscous incompressible flow field.

The boundary condition for the tangential component of the velocity results from an experimental fact. Molecular viscosity acts so that, at their interface, the tangential components of the velocity of the fluid and the velocity of the body are the same:

$$\mathbf{u}(\mathbf{x}_s) \cdot \mathbf{s} = \mathbf{U}_s \cdot \mathbf{s}. \tag{6.1.2}$$

This experimental fact is valid in situations in which the fluid is to a good approximation a continuum. Equations (6.1.1) and (6.1.2) are the constituents of the no-slip boundary condition:

$$\mathbf{u}(\mathbf{x}_s) = \mathbf{U}_s. \tag{6.1.3}$$

Fluid elements in contact with the wall are subjected to the flow velocity and the motion of the wall. This may result in a net torque onto the fluid elements that may in turn impart a rotational motion to the fluid. The no-slip boundary condition is then physically manifested by the creation of vorticity at the boundary.

In viscous incompressible flows, boundaries are the only source of vorticity. As the vorticity field determines the evolution of the flow it is of paramount importance that numerical simulations resolve accurately this physical vorticity creation process by accurately enforcing the no-slip boundary condition.

6.1.1. Vorticity and its Production at a Solid Wall

In the vorticity–velocity formulation of the Navier–Stokes equations we seek boundary conditions for the vorticity field, equivalent to the no-slip boundary condition. These boundary conditions may appear in a Dirichlet (wall vorticity) or a Neumann (wall-normal vorticity flux) form.

Before the numerical implementation of vorticity boundary conditions is discussed, it is helpful to outline the physical character of the wall vorticity and the wall-normal vorticity flux.

The wall-normal vorticity flux is a measure of the vorticity that enters the flow at the boundaries. The equation for the evolution of the vorticity field at the wall degenerates into a diffusion type equation that may be expressed as:

$$\frac{D\omega}{Dt}\Big|_{\text{wall}} = \text{div}\,(\nu\nabla\omega)|_{\text{wall}}. \tag{6.1.4}$$

The fluid elements adjacent to the wall acquire vorticity according to the source term defined by the tensor:

$$J_f = \nu\nabla\omega.$$

The tensor J_f is defined as the wall vorticity flux tensor [105]. We are interested in the vorticity acquired by the fluid elements near the wall, and hence in the wall normal component of this source tensor, defined as the wall vorticity flux vector:

$$\sigma = \mathbf{n}\cdot J_f = \nu\frac{\partial\omega}{\partial\mathbf{n}}.$$

This flux is responsible for the overall production of vorticity: integrating the vorticity equation over the fluid domain gives the following form of Kelvin's theorem of conservation of circulation.

$$\frac{d\Gamma}{dt} = \nu\int_S \frac{\partial\omega}{\partial\mathbf{n}}(s)\,ds.$$

Another important feature of the vorticity fluxes is their relationship with pressure gradients on the walls. To clarify this point, let us assume that the velocities are evaluated in a frame moving with the body, so that we have to deal with a velocity vanishing at the wall ($\mathbf{U}_s = 0$). We consider a local curvilinear orthogonal coordinate system around the body, with axes $(\mathbf{e}_1, \mathbf{e}_2, \mathbf{e}_3)$, where $\mathbf{e}_1, \mathbf{e}_2$ are parallel to the surface S. Let us denote by ξ_1, ξ_2, ξ_3 the coordinates in these axes and by h_1, h_2, h_3 the corresponding scale factors, such that $dx = h_1 d\xi_1\mathbf{e}_1 + h_2 d\xi_2\mathbf{e}_2 + h_3 d\xi_3\mathbf{e}_3$.

In these coordinates the wall is defined by $\xi_3 = 0$. We may write:

$$\omega = \nabla\times\mathbf{u} = \frac{\mathbf{e}_1}{h_2 h_3}\left[\frac{\partial(h_3 u_3)}{\partial\xi_2} - \frac{\partial(h_2 u_2)}{\partial\xi_3}\right]$$

$$- \frac{\mathbf{e}_2}{h_3 h_1}\left[\frac{\partial(h_1 u_1)}{\partial\xi_3} - \frac{\partial(h_3 u_3)}{\partial\xi_1}\right] + \frac{\mathbf{e}_3}{h_1 h_2}\left[\frac{\partial(h_2 u_2)}{\partial\xi_1} - \frac{\partial(h_1 u_1)}{\partial\xi_2}\right].$$

Since $u_1 = u_2 = 0$ on $\xi_3 = 0$, we already obtain

$$\omega_3 = 0 \quad \text{on } \xi_3 = 0.$$

In other words, the wall-normal component of the vorticity is zero at the wall. If we now write the velocity–pressure formulation of the Navier–Stokes equation and take its tangential components on the wall, we get, on S,

$$(\nabla \times \omega) \cdot \mathbf{s} = -\Delta \mathbf{u} \cdot \mathbf{s} = \frac{1}{\nu} \frac{\partial p}{\partial \mathbf{s}}$$

for all tangent vectors $\mathbf{s}$. Since $\omega_3 = 0$, taking successively $\mathbf{s} = \mathbf{e}_1, \mathbf{s} = \mathbf{e}_2$ gives

$$-\frac{\nu}{h_2 h_3} \frac{\partial(h_2 \omega_2)}{\partial \xi_3} = \frac{1}{h_1} \frac{\partial p}{\partial \xi_1},$$

$$-\frac{\nu}{h_3 h_1} \frac{\partial(h_1 \omega_1)}{\partial \xi_3} = \frac{1}{h_2} \frac{\partial p}{\partial \xi_2}.$$

These relations link the wall-normal fluxes of the tangential components of the vorticity to the pressure gradients in the corresponding directions. In particular they imply that vorticity fluxes can be deduced from measurements of the pressure at the walls. Reciprocally, if one wishes to recover the pressure from a velocity–vorticity calculation, identifying the proper vorticity boundary condition allows one to specify the pressure boundary condition needed to complement the Poisson equation $\Delta p = \text{div}\,[(\mathbf{u} \cdot \nabla)\mathbf{u}]$.

Finally, let us mention that the vorticity at the boundary is related to the values of the wall shear stresses: at a solid wall

$$\mathbf{n} \cdot \tau = -\mu \mathbf{n} \times \omega,$$

where τ is the shear-stress tensor, which is defined by

$$\tau_{ij} = 1/2 \left(\frac{\partial u_i}{\partial x_j} + \frac{\partial u_j}{\partial x_i} \right)$$

and μ, the dynamic viscosity, is the molecular viscosity ν divided by the density. Thus, prescribing the vorticity field at the boundary is physically equivalent to prescribing the wall shear stress.

6.1.2. Mathematical Formulation

For simplicity, we present the mathematical formulation for two-dimensional flows past stationary solid walls or a single closed body. The case of three-dimensional flows will be discussed in Section 6.3.

The set of governing equations and boundary conditions may be expressed as follows.

- The governing Navier–Stokes equations in velocity–vorticity formulation:

$$\frac{\partial \omega}{\partial t} + (\mathbf{u} \cdot \nabla \omega) = \nu \Delta \omega \quad \text{in } V. \tag{6.1.5}$$

- Initial conditions:

$$\omega(\cdot, 0) = \omega_0 \quad \text{in } V.$$

- For an incompressible flow the velocity field may be expressed in terms of the streamfunction Ψ as

$$\mathbf{u} = \nabla \times \Psi \quad \text{in } V,$$

and the streamfunction is related to the vorticity by

$$-\Delta \Psi = \omega \quad \text{in } V. \tag{6.1.6}$$

- The boundary condition of no-through flow ($\mathbf{u} \cdot \mathbf{n} = 0$) can be expresssed as

$$\Psi = 0 \quad \text{on } S. \tag{6.1.7}$$

- The no-slip boundary condition, which requires additionally that the tangential component of the velocity vanishes at the wall:

$$\frac{\partial \Psi}{\partial \mathbf{n}} = 0 \quad \text{on } S. \tag{6.1.8}$$

We have discussed in Chapter 4 that, for a given motion of the body, the vorticity field determines the flow field satisfying the condition of no-through flow. In fact, the set of equations (6.1.6)–(6.1.8) is overdetermined, as only the normal component of the velocity at the wall is necessary to fix the flow uniquely. At the same time, there is not a vorticity boundary condition available for convection–diffusion equation (6.1.5).

There is a restricted class of vorticity distributions that satisfy the no-slip condition on the tangential component of the velocity. Proper vorticity boundary conditions for the Navier–Stokes equations must result in a vorticity field in the interior of the domain that satisfies the set of equations

$$-\Delta \Psi = \omega \quad \text{in } V,$$
$$\Psi = 0 \quad \text{on } \partial S,$$

which automatically yields

$$\frac{\partial \Psi}{\partial \mathbf{n}} = 0 \quad \text{on } S.$$

We call such a vorticity field admissible. As noted in Ref. 168, a characterization of admissible vorticity fields is that they are orthogonal to any harmonic function in Ω. This results from straightforward integration by parts.

It is clear that it is possible to derive proper vorticity boundary conditions for the Navier–Stokes equations. In a primitive variable $(\mathbf{u} - p)$ formulation of the Navier–Stokes equations the no-slip boundary condition is explicitly enforced. At the same time, one may envision the vorticity field as evolving with an implicit boundary condition, the vorticity or the vorticity flux that is established at the boundary.

The task, of course, is to derive these vorticity boundary conditions without *a priori* knowledge of the flow field. We distinguish between two types of methodologies to achieve this goal. In the first type, one attempts to derive vorticity boundary conditions that can directly complement the governing equations in the absence of any discretization. We call this approach the continuous algorithm, as it allows the set of the governing Navier–Stokes equations and the boundary conditions to be solved simultaneously. The second type, which is more widely implemented, comprises the so-called viscous splitting algorithms. In this case the boundary conditions are implemented in a fractional step algorithm similar to that seen in Chapter 5 that formulates the problem as a succession of inviscid and viscous substeps.

6.2. Vorticity Boundary Conditions for the Continuous Problem

We have chosen to outline here two particular formulations because of their apparent simplicity (for a thorough discussion of vorticity boundary conditions we refer to the book of Quartapelle [168]). The first one, due to Anderson [6], results in a Neumann boundary condition, whereas the second, suggested by Cottet [59], may result in either a Dirichlet or a Neumann boundary condition.

The idea behind the method suggested in Ref. 6 is that, if one assumes that the initial vorticity field is consistent with no-slip boundary conditions, one needs only to enforce that the time derivative of the vorticity is admissible, i.e.,

$$\frac{\partial}{\partial t} \frac{\partial}{\partial \mathbf{n}} \int_V G_d(\mathbf{x} - \mathbf{y}) \omega(\mathbf{y}, t) \, d\mathbf{y} = 0; \quad \mathbf{x} \in S,$$

where G_d is the Green's function associated with the Laplace equation in V with a homogeneous Dirichlet boundary condition on S. After the integral and derivatives are exchanged, the Navier–Stokes equation yields

$$\frac{\partial}{\partial \mathbf{n}} \int_V G_d(\mathbf{x} - \mathbf{y})[-(\mathbf{u} \cdot \nabla)\omega + \nu \Delta \omega](\mathbf{y}, t) \, d\mathbf{y} = 0; \quad \mathbf{x} \in S,$$

which, after integration by parts of the term $G_d \Delta \omega$, leads to

$$
\nu \frac{\partial \omega}{\partial \mathbf{n}} + \frac{\partial}{\partial \mathbf{n}} \int_V \frac{\partial G_d}{\partial \mathbf{n}} (\mathbf{x} - \mathbf{y}) \omega(\mathbf{y}, t) \, d\mathbf{y}
$$

$$
= \frac{\partial}{\partial \mathbf{n}} \int_V G_d(\mathbf{x} - \mathbf{y})[(\mathbf{u} \cdot \nabla)\omega](\mathbf{y}, t) \, d\mathbf{y}. \tag{6.2.1}
$$

This is an integrodifferential equation for the wall-normal derivative of the vorticity field. Observe that Eq. (6.2.1) couples values of the vorticity on the boundary with values of both the velocity and the vorticity in the fluid domain. In turn the velocity and vorticity in the fluid domain are conditioned by the boundary values of the vorticity.

The approach followed in Ref. 59 (see also [72]) is based on the the representation of the velocity in terms of the vorticity in the flow and the boundary values of the velocity through the Poincaré formula (see Chapter 4). This formula consists of viewing the velocity as the solution of the following system:

$$
\Delta \mathbf{u} = -\nabla \times \omega.
$$

This system can now be supplemented with boundary conditions on both components of the velocity at the wall. The expected pitfall of this approach is that, for a given vorticity field ω, the velocity satisfying

$$
-\Delta \mathbf{u} = \nabla \times \omega \quad \text{in } V,
$$

$$
\mathbf{u} = 0 \quad \text{on } S
$$

is not necessarily divergent free, thus violating the assumption of incompressibility. Therefore this velocity does not necessarily satisfy $\nabla \times \mathbf{u} = \omega$. However, it turns out that enforcing $\nabla \times \mathbf{u}$ and ω to coincide on the boundary is enough to make them coincide everywhere. This requirement supplements the set of the governing equations, giving us the following system of equations:

$$
\frac{\partial \omega}{\partial t} + (\mathbf{u} \cdot \nabla)\omega - \nu \Delta \omega = 0 \quad \text{in } V,
$$

$$
\omega(\cdot, 0) = \omega_0 \quad \text{in } V,
$$

$$
-\Delta \mathbf{u} = \nabla \times \omega \quad \text{in } V,
$$

$$
\mathbf{u} = 0 \quad \text{on } S,
$$

$$
\omega = \nabla \times \mathbf{u} \quad \text{on } S.
$$

From this formulation, it is also possible to derive a Neumann-type boundary condition. For a nonrotating body, the correct boundary condition is

$$\frac{\partial \omega}{\partial \mathbf{n}} = \frac{\partial}{\partial \mathbf{n}}(\nabla \times \mathbf{u}) - \frac{1}{|S|}\int_S \frac{\partial}{\partial \mathbf{n}}[\nabla \times \mathbf{u}(\mathbf{y})]\,d\mathbf{y}. \qquad (6.2.2)$$

The role of the constant substracted to $\partial/\partial \mathbf{n}(\nabla \times \mathbf{u})$ on the right-hand side above is to ensure that at all times Kelvin's theorem, for the conservation of circulation, is enforced (we refer to Ref. 63 for a proof that the resulting system is equivalent to the Navier–Stokes equations with the no-slip boundary condition).

In closing this section, we emphasize that, from a numerical point of view, the equivalence with the original problem is not enough to ensure the stability of the numerical methods based on these formulations. To ensure this stability, one has to derive the so-called energy estimates. These estimates indicate how the errors in the computation of the velocity will interfere, through the vorticity boundary condition, with the numerical scheme used for the discretization of the vorticity equation in the fluid domain. To our knowledge such estimates are not available for either one of the above two methodologies. However, the computations performed by these authors (in general with finite-difference discretizations) do indicate that these boundary conditions are stable.

6.3. Viscous Splitting Algorithms

We have already seen in Chapter 5 that the viscous splitting is a natural approach to design diffusion algorithms for vortex methods. It is also a popular methodology for enforcing the no-slip boundary condition in the context of vortex methods. It is a fractional step algorithm, handling successively the inviscid and the viscous physical processes as described by the different terms of the Navier–Stokes equations. In 1963, Lighthill [138] used the concept of a fractional step algorithm to describe the generation of vorticity at a solid boundary. However, he never explicitly formulated this process. Ten years later a number of numerical methodologies appeared, formulating the concept of vorticity generation at the boundary and deriving vorticity boundary conditions for the vorticity–velocity formulation of the Navier–Stokes equations.

Starting from the formulation of Lighthill's algorithm, we outline the different steps of viscous splitting. This helps elucidate the differences and the similarities of the various methodologies that have been proposed.

6.3.1. Vorticity Creation at a Solid Wall: Lighthill's Model

Let us assume that at the nth time step (corresponding to time $t - \delta t$) an admissible vorticity field has been computed and we seek to advance the solution to the next time step (time t). This is accomplished in the following two sub-steps:

Substep 1: Using as initial conditions $f(\mathbf{x}) = \omega^n(\mathbf{x}^n, n\delta t)$, we solve the system

$$\frac{\partial \omega}{\partial t} + \mathbf{u} \cdot \nabla \omega = 0 \quad \text{in } V, \tag{6.3.1}$$

$$\omega(\cdot, t - \delta t) = f \quad \text{in } V, \tag{6.3.2}$$

$$-\Delta \Psi = \omega \quad \text{in } V, \tag{6.3.3}$$

$$\Psi = 0 \quad \text{on } S. \tag{6.3.4}$$

The solution of the Poisson equation gives

$$\Psi = \Psi_\omega + \Psi_\gamma + \Psi_{\text{ext}},$$

where (see Chapter 4):

$$\Psi_\omega = \int_V \omega(\mathbf{x}', t) G|\mathbf{x} - \mathbf{x}'| \, d\mathbf{x}',$$

Ψ_γ represents the solution to the homogeneous equation satisfying the boundary conditions, and Ψ_{ext} results from the external field (Fig. 6.1). As was discussed in Chapter 4, Ψ_γ may be represented by the flow field induced by a vortex sheet γ distributed on the surface of the body:

$$\Psi_\gamma = \int_S \gamma(\mathbf{x}', t) G|\mathbf{x} - \mathbf{x}'| \, d\mathbf{x}',$$

and the solution of this problem is reduced to identifying the strength of the vortex sheet in order to satisfy the no-through-flow boundary condition. It was shown in Chapter 4 that a convenient method of solving for the vortex sheet is to solve the integral equation:

$$\gamma(\mathbf{x}) + 2 \int_S \frac{\partial}{\partial \mathbf{n}} (G|\mathbf{x} - \mathbf{x}'|) \, \gamma(\mathbf{x}) \, d\mathbf{x}' = -2h(\mathbf{x}), \tag{6.3.5}$$

where

$$h(\mathbf{x}) = \frac{\partial \Psi_{\text{ext}}}{\partial \mathbf{n}}(\mathbf{x}) + \frac{\partial \Psi_\omega}{\partial \mathbf{n}}(\mathbf{x}) - \mathbf{U}_s \cdot \mathbf{s}. \tag{6.3.6}$$

We obtain a unique solution for this equation by adding a constraint on the strength of the vortex sheet. This constraint is usually provided by Kelvin's circulation conservation theorem:

$$\int_S \gamma(\mathbf{x}) \, d\mathbf{x} = -2A[\Omega_b(t + \delta t) - \Omega_b(t)], \tag{6.3.7}$$

where A denotes the area of the body.

At the end of this step a vorticity field has been established in the fluid and a vortex sheet has been established on the surface of the body. We may think of the vortex sheet γ as part of the interior vorticity, so that the resulting vorticity is

$$\omega' = \omega - \gamma \otimes \delta_S.$$

Assume further that the vortex sheet $\gamma \otimes \delta_S$ is slightly detached inside V. Then one can view the velocity field associated with ω' as satisfying the no-slip condition at the boundary and undergoing a sudden jump inside the domain.

Substep 2: Lighthill concludes the description of his model by stating that the vorticity per unit area has been created and is equal to the negative of this vortex sheet strength. What remains incomplete in this model is how this vorticity enters the fluid adjacent to the wall or how the vortex sheet strength may be incorporated in a vorticity-type boundary condition.

We will try to clarify this point in the following subsections. One may already observe that the strength of the vortex sheet has dimensions of velocity (or length over time). To obtain an appropriate (dimensionally correct) vorticity boundary condition, this vortex sheet strength can be manipulated so that a Dirichlet- or a Neumann-type boundary condition may be obtained. This is basically the point of diversion of the various formulations involving vorticity boundary conditions. In order to get a Dirichlet-type boundary condition, Chorin [49] divides the strength of the vortex sheet by a length equal to the elementary discretization length on the body surface, whereas Wu [202] multiplies the strength by the mesh size in the normal direction to obtain the vorticity on the body. Alternatively, Kinney and his co-workers [115, 116] envision this vortex sheet as equivalent to a vorticity flux over a small time interval (thus dividing the sheet strength by time to obtain the proper units of acceleration). An integral constraint is imposed on all formulations on the vorticity created at the wall so as to satisfy Kelvin's theorem of production of circulation.

6.3.2. Chorin's Algorithm and the Vortex Sheet/Vortex Blob Method

The goal of Chorin's algorithm is to reformulate the vorticity production mechanisms described above in terms of a rigorous splitting algorithm.

As in the case of an unbounded domain (see Section 5.1), it is natural to split the Navier–Stokes equation into an inviscid and a viscous part. Compared with the full space, the situation is complicated by the fact that the inviscid part can only handle the no-through-flow boundary condition. Thus each viscous substep in the algorithm is singular in the sense that it starts with a velocity field violating its boundary condition. As a result, the convergence proof [21] for the splitting algorithm is far more involved than the one given in Section 5.1.

If one considers the velocity–pressure formulation, one way to enforce the no-slip boundary condition in the viscous step is to provide an odd extension of the velocity field across the boundary and to solve the Stokes equation in the full space. The algorithm described in Ref. 54 proceeds as follows.

- Solve the Euler equation in Ω with $\mathbf{u} \cdot \mathbf{n} = 0$ on S.
- Extend the velocity across $\partial\Omega$ and solve the Stokes equation in $\mathbf{R}^2$.

In a vorticity–velocity formulation, this odd extension is equivalent to the creation of a singular vorticity distribution at the boundary – a vortex sheet – with strength equal to the jump of the tangential component of the velocity that is twice the slip at the boundary. The Stokes step then preserves the symmetry of the velocity across the boundary, while regularizing the fields. Thus the vorticity is even across the boundary and smooth so it satisfies $\partial\omega/\partial\mathbf{n} = 0$ on S.

If one wishes now to reformulate this algorithm in V and in terms of the vorticity, a natural candidate is the following algorithm, where $\Delta t > 0$, $t_n = \Delta t$, and ω_n approximates the vorticity at time t_n

Substep 1: Solve over one time step the Euler equation with a no-through-flow boundary condition and initial vorticity field ω_n; let $\omega_{n+1/2}$ be the resulting vorticity and $\mathbf{u}_{n+1/2}$ be the related velocity, with the normal component vanishing at the boundary.

Substep 2: Cancel the slip by incorporating a vortex sheet in the interior vorticity: this gives a new vorticity field

$$\tilde{\omega}_{n+1/2} = \omega_{n+1/2} - (\mathbf{u}_{n+1/2} \cdot \mathbf{s}) \otimes \delta_S. \tag{6.3.8}$$

Substep 3: Solve over one time step the heat equation with the homogeneous Neumann boundary condition,

$$\frac{\partial\omega}{\partial t} - \nu\Delta\omega = 0 \quad \text{in } V, \tag{6.3.9}$$

$$\omega(\cdot, 0) = \tilde{\omega}_{n+1/2} \quad \text{in } V, \tag{6.3.10}$$

$$\frac{\partial\omega}{\partial\mathbf{n}} = 0 \quad \text{on } S \tag{6.3.11}$$

to finally obtain ω_{n+1}.

One recognizes in Eq. (6.3.8) the vorticity field already described in Lighthill's formulation. As it is clear in the formulation that uses the extension across the

Figure 6.2. Evolution of the velocity profile in the initial stage of a Stokes step.

boundary, the vortex sheet has to be considered as part of the interior vorticity, and the effect of the diffusion step that follows is in particular immediate regularization of the discontinuous velocity profile, establishing a boundary layer (see Fig. 6.2). Note that the factor of 2 in the strength of the vortex sheet when the Stokes equation is solved in the full space has disappeared here. The reason is that in the full-space version only half of the boundary layer spreads into V. The circulation conservation principle can be again called for as a check that the coefficients are correct in both cases. Subsection 6.3.3 will further elucidate these points.

The Vortex Sheet/Vortex Blob Algorithm

This is a vortex method [50] based on the vorticity creation algorithm just outlined, with the addition of two main ingredients. The first one is the use of the Prandtl boundary-layer approximation for the flow field near the surface of the body. This approximation in particular yields a simple expression of the slip at the end of the first substep. The second one is the use of flat vortex sheet elements, rather than vortex blobs, in the boundary layer. This prevents vortex elements from leaking outside the fluid domain and allows for a more natural discretization of the vortex sheet involved in the second step of the algorithm.

The Prandtl equations are based on the assumption that, at the boundary, velocity gradients are essentially in the component parallel to the wall, and diffusion of vorticity is essentially normal to the wall. To exploit these assumptions, one needs to work in local coordinates adapted to the boundary geometry. This is in general done through a mapping of the physical domain onto a half-space geometry. For simplicity we thus focus on the half-space case: If we denote by x and y the tangential and the normal coordinates, respectively (V is defined by $y > 0$) and by (u, v) the velocity vector in this coordinate system, the Prandtl equations then are given by

$$\frac{\partial \omega}{\partial t} + u \frac{\partial \omega}{\partial x} + v \frac{\partial \omega}{\partial y} - \nu \frac{\partial^2 \omega}{\partial y^2} = 0, \tag{6.3.12}$$

$$\omega = -\frac{\partial u}{\partial y}, \tag{6.3.13}$$

supplemented by the continuity equation, no-slip boundary conditions at $y = 0$, and a far-field condition in general written for the u component at a distance r_∞ of the wall:

$$u(x, r_\infty, t) = U_\infty(x, t). \tag{6.3.14}$$

A vortex sheet approximation of ω consists of writing

$$\omega \simeq \sum_j \alpha_j b_\varepsilon(x - x_j)\delta(y - y_j). \tag{6.3.15}$$

In the above equation δ is the one-dimensional Dirac mass in the y direction and b_ε is a one-dimensional cutoff function in the x direction, given by $b_\varepsilon(x) = \varepsilon^{-1}b(x/\varepsilon)$. The weight α_j is the result of the multiplication of dl, a grid size in the direction x, and a quantity of the size of a velocity.

A direct integration of Eq. (6.3.13), with the right-hand side given by Eq. (6.3.15), yields

$$u(x, y) = U_\infty(x) + \sum_j \alpha_j b_\varepsilon(x - x_j)H(y_j - y), \tag{6.3.16}$$

where H denotes the Heaviside function. The continuity equation in turn allows us to deduce the value of v:

$$v(x, y) = -\frac{\partial U_\infty(x)}{\partial x} y - \sum_j \alpha_j b'_\varepsilon(x - x_j)\min(y, y_j). \tag{6.3.17}$$

If one starts from a given collection of sheets, one step of the overall algorithm includes the following substeps.

Substep 1: Compute u and v with Formulas (6.3.16) and (6.3.17); move the sheets with this velocity.

Substep 2: Compute u at $y = 0$; create new sheets at the boundary with strengths $\alpha_j = -u(x_j)\,dl$.

Substep 3: Random walk the sheets in the y direction, using the ideas of Section 5.2, to solve for the diffusive part of Eq. (6.3.12). If a sheet crosses the boundary, reflect it on the other side. If a sheet moves beyond r_∞, transform it into a vortex blob.

For substep 2, instead of using one vortex sheet with varying strength at every wall point, one in general prefers to introduce several elementary vortex sheets that have all the same strengths. These originally collocated vortex sheets scatter in the first subsequent random walk.

Note that the reflection of particles at the boundary amounts to extending in an even fashion the vorticity outside the domain, as prescribed in Ref. 54. It thus translates homogeneous Neumann boundary condition (6.3.11). There is no factor of 2 in the strengths of the vortex sheets because, because of this reflection, all sheets created at the boundary end up in the fluid domain V.

One may observe that the use of vortex sheets near the wall allows avoiding the inconsistency mentioned in Section 4.5 resulting from the vortex blobs that overlap the boundary. However, this is at the expense of using the Prandtl approximation, which requires relying on a mapping of the domain into a half-space geometry. Moreover, it is well known that this approximation ceases to be valid as soon as the flow detaches from the wall. Another difficulty associated with the method is related to the interface condition, in which vortex sheets turn into vortex blobs. The value of r_∞ is in general taken of the order of the boundary-layer width. As for U_∞, it is computed on the basis of the Biot–Savart law, with a vorticity field consisting of both the vortex sheets and blobs. Formula (6.3.16) certainly ensures a continuous value of the tangential component of the velocity, but, as pointed out by Anderson and Reider [12], not necessarily of the normal component, creating oscillations in the interface zone. This problem can be fixed in the context of a finite-difference method if an appropriate linear system is solved to enforce the continuity of both components of the velocity, but, to our knowledge, not in the context of random-walk methods.

The vortex sheet/vortex blob method has been extensively used in conjunction with the random-walk method for the simulation of slightly viscous flows around obstacles. We refer for example to Ref. 188 for flows around cylinders and to Ref. 185 for a validation study of the method for flow over a backward-facing step at high and moderate Reynolds numbers. These studies show that the method is appropriate to obtain the qualitative features of the flows.

Pucket [166] gives an extensive numerical study of the convergence properties of the vortex sheet/vortex blob method. The relevant numerical parameters that govern the accuracy of the method are the circulations of the individual vortex sheet introduced at the wall, the length of the sheets, and the time step. It appears that reducing the first two parameters does not allow us to obtain converged results if the time step is not reduced at the same time. We present in Figures 6.3 and 6.4 two illustrations of a convergence study by Mortazavi, of the vortex sheet method in which the circulations of the sheets and the time step are reduced. Figure 6.3 shows the averaged velocity profiles, and Figure 6.4 shows the averaged streamlines.

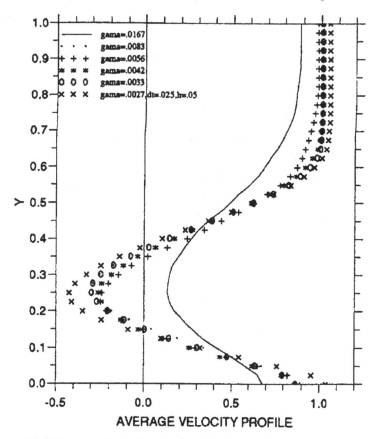

Figure 6.3. Velocity profiles behind a backward-facing step for decreasing values of the sheet circulations (Courtesy of I. Mortazavi).

6.3.3. A Vorticity Creation Algorithm with Vorticity Flux Boundary Conditions

Our goal in this section is to clarify further the links among the no-slip condition, vorticity creation, and vorticity flux boundary condition in the context of splitting-based vortex algorithms.

We first focus on the continuous problem in two dimensions, that is, in the absence of spatial discretization, and we give a reformulation of vorticity creation algorithms that is intrinsic in the sense that it carries on to general geometries. We then show how integral formulations allow us to handle vorticity flux boundary conditions in a vortex method. We finally discuss the three-dimensional case.

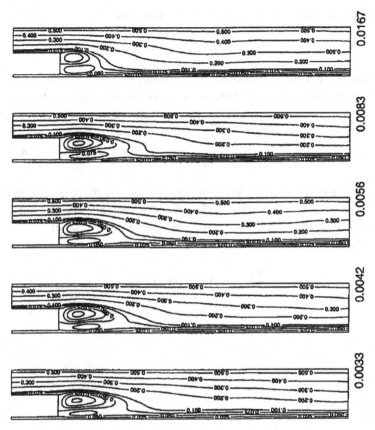

Figure 6.4. Convergence study for the vortex sheet/vortex blob method: averaged streamlines for the flow behind a backward-facing step for different numerical parameters values (Courtesy of I. Mortazavi).

The Continuous Problem

The formulation of the no-slip condition, in terms of a splitting algorithm with a vorticity flux boundary condition, has been proposed independently by Cottet [63], and Koumoutsakos et al. [124]. The latter approach consists of an enhanced implementation, in the context of vortex methods, of the algorithm originally proposed by Kinney and his co-workers [115, 116]. To describe this algorithm in more detail, we follow the approach described in Ref. 63.

To understand the links between vorticity creation and vorticity flux boundary conditions, let us go back to the form of Eqs. (6.3.8)–(6.3.11) of the vorticity creation algorithm, and first observe that, if we assume some kind of time continuity of the heat equation, it is natural to expect that (recall that, for simplicity, we deal with vanishing velocity at the boundary)

$$\mathbf{u}^h \cdot \mathbf{s}(\cdot, t) \to 0 \qquad \text{when} \quad t \to t_n, t > t_n.$$

This claim can actually be proved in an appropriate functional framework, at least for the half-space case, through the use of integral representations. As a result, one can rewrite step 2 of the vorticity creation algorithm [Eq. (6.3.8)] as

$$\tilde{\omega}_{n+1/2} = \omega_{n+1/2} - \int_{t_{n-1}}^{t_n} \frac{\partial}{\partial t} (\mathbf{u}^h \cdot \mathbf{s})(\cdot, t) \otimes \delta_S \, dt. \tag{6.3.18}$$

This leads us to reformulate the algorithm in a slightly different way; in this version we consider two vorticity fields ω_1^h and ω_2^h that are respective solutions of the following two problems, in the same time interval $[t_n, t_{n+1}]$:

Problem 1:

$$\frac{\partial \omega_1^h}{\partial t} - \nu \Delta \omega_1^h = 0 \quad \text{in } V, \tag{6.3.19}$$

$$\omega_1^h(\cdot, t_n) = \omega_n \quad \text{in } V, \tag{6.3.20}$$

$$\frac{\partial \omega_1^h}{\partial \mathbf{n}} = 0 \quad \text{on } S. \tag{6.3.21}$$

Problem 2:

$$\frac{\partial \omega_2^h}{\partial t} - \nu \Delta \omega_2^h = 0 \quad \text{in } V, \tag{6.3.22}$$

$$\omega_2^h(\cdot, t_n) = \omega_n \quad \text{in } V, \tag{6.3.23}$$

$$\nu \frac{\partial \omega_2^h}{\partial \mathbf{n}} = -\frac{\partial}{\partial t} (\mathbf{u}_1^h \cdot \mathbf{s}) \quad \text{on } S, \tag{6.3.24}$$

where $\mathbf{u}_1^h$ refers to the velocity field associated with ω_1^h.

We then set $\omega_{n+1} = \omega_2^h(\cdot, t_{n+1})$. This is the result of one step of the algorithm. Note immediately that system (6.3.19)–(6.3.21) does not produce any net vorticity in the flow, and thus the circulation of $\mathbf{u}_1$ is constant. Therefore boundary condition (6.3.24) prescribes the right amount of total circulation to enter the flow (zero in the present case).

It is important to note that, in the half-plane case, the only difference between the original algorithm and formulation (6.3.19)–(6.14.24) is that the former scheme uses one additional time discretization when it incorporates the vortex sheet in the flow (while the modified scheme does it in a continuous way). We can readily check this by extending the solution of Eq. (6.3.22) to the whole plane: If $V = \{(x_1, x_2), x_2 > 0\}$ and we set

$$\bar{\omega}_2^h(x_1, x_2) = \begin{cases} \bar{\omega}_2^h(x_1, x_2) & \text{if } x_2 > 0 \\ \bar{\omega}_2^h(x_1, -x_2) & \text{if } x_2 < 0 \end{cases},$$

then $\bar{\omega}_2^h$ satisfies

$$\frac{\partial \bar{\omega}_2^h}{\partial t} - \nu \Delta \bar{\omega}_2^h = 2\nu \frac{\partial \omega_2^h}{\partial \mathbf{n}}(x_1) \otimes \delta(x_2) \text{ in } \mathbf{R}^2.$$

Using Eq. (6.3.24), we can split this last equation into two substeps: first a diffusion equation with zero right-hand side, which, translated as a boundary-value problem, would be equivalent to Eqs. (6.3.9)–(6.3.11), then the equation

$$\frac{\partial \omega}{\partial t} = -2 \frac{\partial}{\partial t} \left(\mathbf{u}_1^h \cdot \mathbf{s} \right) \otimes \delta_S.$$

The result of this last step is precisely Eq. (6.3.8) (the apparent discrepancy caused by the factor of 2 comes from the fact that if the vortex sheet is diffused, only half of it will fall in V).

The advantage of formulation (6.3.19)–(6.3.24) is that it is free of any assumption on the geometry. Another interesting feature of this algorithm is that its convergence can be rigorously proved. This scheme is actually enstrophy decreasing, which is what should be expected for the solution to the Stokes problem. This stability property is a key ingredient in proving its convergence (we refer to Ref. 63 for this proof).

If we now turn to the full Navier–Stokes equations, a natural generalization of the method summarized by system (6.3.19)–(6.3.24) would consist of restoring advection terms and replacing Eq. (6.3.24) with

$$\nu \frac{\partial \omega_2^h}{\partial \mathbf{n}} = -\left[\frac{\partial u_1^h}{\partial t} + (\mathbf{u}_1^h \cdot \nabla) u_1^h \right] \cdot \mathbf{s}. \tag{6.3.25}$$

If we recognize on the right-hand side a particle derivative along the flow determined by $\mathbf{u}_1$ and note that this flow is parallel to the wall, the same arguments as those given above show that this formulation is closely related to Chorin's algorithm in the half-space case. The convergence of the algorithm based on Eq. (6.3.25) is an open question.

Integral Techniques

In this subsection we describe the algorithm proposed in Ref. 124. This algorithm is implemented through integral technique vorticity flux boundary conditions in a vortex code based on the PSE method described in Section 5.4.

The method is again based on a viscous splitting of the Navier–Stokes equations, and we will focus on the approximation of a diffusion step. We thus have successively to solve for a diffusion equation, compute the slip, and turn it into a vorticity flux boundary condition, along the lines of system (6.3.19)–(6.3.24). We first observe, that if we write $\omega_2 = \omega_1 + \omega$, ω solves

$$\frac{\partial \omega}{\partial t} - \nu \Delta \omega = 0, \quad \text{in } V \times [0, t],$$

$$\omega(\mathbf{x}, 0) = 0 \quad \text{in } V,$$

$$\nu \frac{\partial \omega}{\partial \mathbf{n}} = F(\mathbf{x}, t) = -\frac{\partial}{\partial t}(\mathbf{u} \cdot \mathbf{s}) \quad \text{on } S \times [0, t]. \qquad (6.3.26)$$

The solution of the above equation may be expressed in integral form [85] as

$$\omega(\mathbf{x}, t) = \int_0^t \int_S \mathcal{G}[\mathbf{x} - \boldsymbol{\xi}, \nu(t - \tau)]\, \mu(\boldsymbol{\xi}, \tau)\, d\boldsymbol{\xi}\, d\tau, \qquad (6.3.27)$$

where the diffusion potential $\mu(\mathbf{x}, t)$ is determined as the solution of

$$-\frac{1}{2}\mu(\mathbf{x}, t) + \nu \int_0^t \int_S \frac{\partial \mathcal{G}}{\partial \mathbf{n}}[\mathbf{x} - \boldsymbol{\xi}, \nu(t - \tau)]\, \mu(\boldsymbol{\xi}, \tau)\, d\boldsymbol{\xi}\, d\tau = F(\mathbf{x}, t)$$

$$(6.3.28)$$

and $\mathcal{G}$ is the Gaussian kernel:

$$\mathcal{G}(\mathbf{x}, \tau) = \frac{1}{4\pi \tau}\, \exp\left(-\frac{|\mathbf{x}|^2}{4\tau}\right).$$

The resulting expressions for the vorticity field involve integrals over only the surface of the body. We may discretize those integrals with a boundary integral method by assuming that the surface of the body comprises a set of discrete panels (straight or curved) and assuming a certain variation (constant, linear, etc.) of the unknown function $\mu(\mathbf{x}, t)$ in space (over the panels) and time.

For small values of ν the kernel $\mathcal{G}$ is stiff, so in order to evaluate accurately the surface integrals, it is advisable to use explicit integration formulas. Hence the vorticity field induced by a panel with index i and length d, centered at $\mathbf{x}_i$, may be expressed as an integral over time:

$$\omega_i(\mathbf{x}, t) = \frac{1}{2} \int_0^t \mu_i(\tau)\, \phi(\mathbf{x} - \mathbf{x}_i, t - \tau)\, d\tau, \qquad (6.3.29)$$

where, in a coordinate system $\mathbf{x} = (x, y)$ with the x axis parallel to the wall,

$$\phi(\mathbf{x}, t - \tau) = \frac{e^{\frac{-y^2}{4\nu(t-\tau)}}}{\sqrt{4\pi \nu(t - \tau)}} \left\{ \mathrm{erf}\left[\frac{d + x}{\sqrt{4\nu(t - \tau)}}\right] + \mathrm{erf}\left[\frac{d - x}{\sqrt{4\nu(t - \tau)}}\right] \right\}.$$

The time integral is evaluated with the midpoint rule. Finally, the strength of the particles $\Gamma_j = \omega(\mathbf{x}_j)v_j$ are updated by

$$\Gamma_j(t) = \Gamma_j(0) + \frac{\delta t v_j}{2} \sum_{i=1}^{M} \mu_i \, \phi\left(\mathbf{x}_j - \mathbf{x}_i, \frac{\delta t}{2}\right), \tag{6.3.30}$$

where M is the number of panels on the wall.

To complete the evaluation of the vorticity field in the domain it remains to determine the surface density μ. Following the derivation of Ref. 93, we solve Eq. (6.3.28) explicitly by exploiting the local character of the Green's function $\mathcal{G}$ and its normal derivative on the body.

Consider the double-layer heat potential, which is defined as

$$\mathcal{H}\mu(\mathbf{x}, t) = \int_S \int_0^t \frac{\partial \mathcal{G}}{\partial \mathbf{n}} [\mathbf{x} - \boldsymbol{\xi}, \nu(t - \tau)] \, \mu(\boldsymbol{\xi}) \, d\boldsymbol{\xi} \, d\tau.$$

If we describe the shape of the body around the panel location $\mathbf{x}_w$ by using a Taylor's series expansion, we have that

$$\mathcal{H}\mu(\mathbf{x}_w, \delta t) \approx \frac{\mu(\mathbf{x}_w)\kappa}{2} \sqrt{\pi \nu \, \delta t/2} + O\left[(\nu \, \delta t)^{3/2}\right].$$

Substituting the above result into the equation for the heat potential (6.3.28), we find for points (s) along the surface of the body that

$$\mu(s) \approx -2F(s) \left[1 - \kappa(s)\sqrt{\pi \nu \, \delta t/2}\right]^{-1}. \tag{6.3.31}$$

Note that for the case of a cylinder of radius R, the curvature is constant ($\kappa = 1/R$), and for the case of a flat plate the curvature is zero so that the surface potential is a function of only the vorticity flux.

Combining now Eq. (6.3.30) and approximation (6.3.31) we obtain an algorithm for updating the particle strengths in the domain so that the no-slip boundary condition is enforced:

$$\Gamma_j^{n+1} = \Gamma_j^n + v_j \, \delta t \sum_{i=1}^{M} \frac{q_i}{\left(1 - \kappa_i \sqrt{\pi \nu \frac{\delta t}{2}}\right)} \, \phi(\mathbf{x}_j - \mathbf{x}_i, \delta t/2), \tag{6.3.32}$$

where the i index refers to the panels, the j index to the particles, and the n index to time; the vortex sheet strength is derived from Eq. (6.3.26) by the first-order formula

$$q_i = \frac{1}{\delta t}(\mathbf{u}_1 \cdot \mathbf{s})(s_i).$$

In computations, ϕ can be calculated efficiently with tabulated values, thus avoiding the costly evaluation of the error functions that are involved. Also the local character of ϕ requires the interaction of each panel only with its nearby particles, resulting in a computational cost that scales as $O(M)$. To summarize, the algorithm implemented in Ref. 124 proceeds as follows:

- Advect particles and solve for the diffusion in V with a PSE scheme.
- Compute the slip on S.
- Use formula (6.3.32) to distribute the resulting boundary vorticity flux onto the particles in V.

Two observations should be made here concerning this algorithm. First, the curvature term in formula (6.3.31) clearly indicates the discrepancy that would result from the direct implementation of the half-space formula of vorticity creation, as used in vortex sheet algorithms. This error is probably hidden by the other source of errors involved in vortex sheet methods (random walk and Prantl approximation in particular). Second, in the algorithm just described, unlike the one described in the previous paragraph [Eqs. (6.3.19)–(6.3.24)], the second diffusion step is not implemented with an homogeneous Neumann boundary condition. However, it conserves the total circulation, and the complete algorithm is therefore consistent with Kelvin's theorem; the results that we present now seem to indicate that this version of the vorticity flux has the same convergence features as the splitting method of Eqs. (6.3.19)–(6.3.24).

Figure 6.5 shows the results of the algorithm of Ref. 124 for the viscous flow around a purely rotating cylinder. We may construct an analytic solution to this problem by assuming the streamfunction of the flow to be of the form [96]

$$\Psi = \Psi(r)e^{i\sigma t}.$$

If the angular velocity of the cylinder is $Q \sin \sigma t$, under the assumption of symmetry, the vorticity equation reduces to a diffusion-type equation whose analytic solution is given by

$$\omega(r, t) = A\cos(\sigma t)\,[kei_1(c)K^-(cr) - ker_1(c)K^+(cr)]$$
$$- A\sin(\sigma t)\,[kei_1(c)K^+(cr) - ker_1(c)K^-(cr)],$$

with the definitions

$$K^{\pm}(cr) = ker(cr) \pm kei(cr)$$

$$c = \sqrt{\frac{\sigma}{\nu}}, \quad A = \frac{Qc}{\sqrt{2}}\,\frac{1}{ker_1^2(c) + kei_1^2(c)},$$

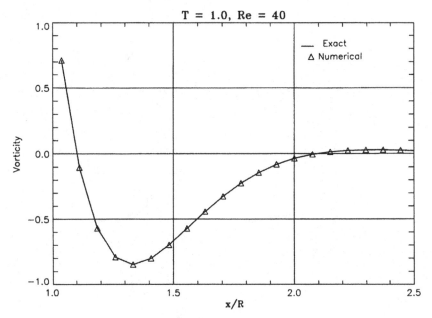

Figure 6.5. Vorticity field around a purely rotating cylinder with $Q = 1$, $\sigma = 1$ at Re $=$ UD$/v = 40$, $M = 180$, $\delta t = 0.025$.

where $ker_1(x)$, $kei_1(x)$ and $ker(x)$, $kei(x)$ are the Kelvin's functions of order 1 and 0, respectively. In Figure 6.5 we show the results of the computed and analytical vorticity field for $T = 1$ and $v = 0.5$. In order to avoid the computation of the transient solution in this computation, the vorticity field was initialized with the analytical solution at the end of a period.

Another illustration of the accuracy of the vorticity flux boundary conditions is the computation of the drag of an impulsively started cylinder. In the early stages of the flow development, the drag coefficient is inversely proportional to the diffusion length $\sqrt{(vt)}$, thus exhibiting a rather singular behavior. For small times analytical solutions exist [16, 57] for various quantities of the flow, including the drag coefficient. In Figure 6.6, we present the drag coefficient obtained using the vorticity flux boundary conditions (a PSE scheme accounts for diffusion) and we compare with the analytic results for various Reynolds numbers.

At later times the analytic solutions are no longer valid, and we compare the time history of the drag coefficient (see Figure 6.7) at a Reynolds number of 3000 with the drag coefficient obtained by using a spectral-element calculation (Henderson, (private communication)). The plateau at approximately $t = 2$

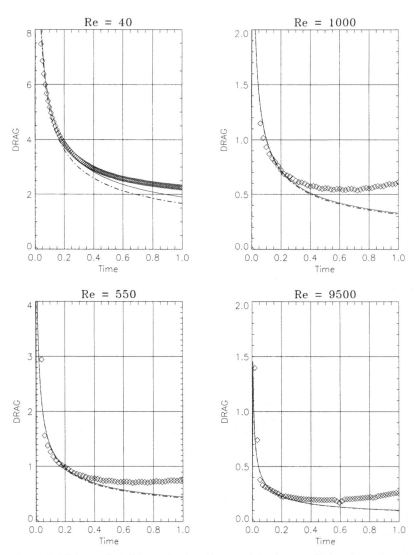

Figure 6.6. Linear plot of the early time history of the drag coefficient for an impulsively started circular cylinder. Solid curves [57], dashed curves [16], symbols, present computations.

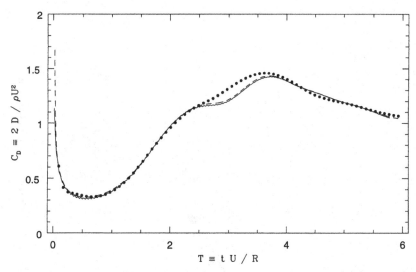

Figure 6.7. Drag coefficient for an impulsively started cylinder at Re = 3,000. Solid line (Vortex methods), Bullets 14th order spectral element method, Dashed line, 12th order spectral element method, with refinement near the separation points.

corresponds to the formation and the eruption of secondary vorticity (see Figure 6.8 for the flow evolution) and cannot be captured if the near-wall vorticity field is not accurately resolved. This is one of the features of this flow that makes this problem particularly challenging for various numerical schemes. As is shown in Figure 6.7, a spectral-element method produces the same results as a vortex method when the grid is properly refined in the region of secondary separation.

In vortex methods the computational elements automatically adapt to resolve these sensitive regions of the flow, resulting in robust calculations. Simulations with finite-difference [12] and spectral-element methods [84] exemplify these difficulties and use large numbers of grid points and relatively small time steps to produce results that are in good quantitative agreement with the results presented in Figure 6.8.

Finally, Figure 6.9 shows the evolution of the vorticity field in a dipole wall interaction at a Reynolds number of 3200. The method uses a vortex-in-cell technique to evaluate the particle velocities and a PSE scheme for the diffusion. The vorticity boundary conditions are handled by the integral just described, but the diffusion step preceding the computation of the slip is done with homogeneous Neumann boundary conditions. In other words, the method that is implemented corresponds exactly to system (6.3.19)–(6.3.24). Because of the simplicity of the geometry, homogeneous Neumann boundary condition (6.3.21) is easily

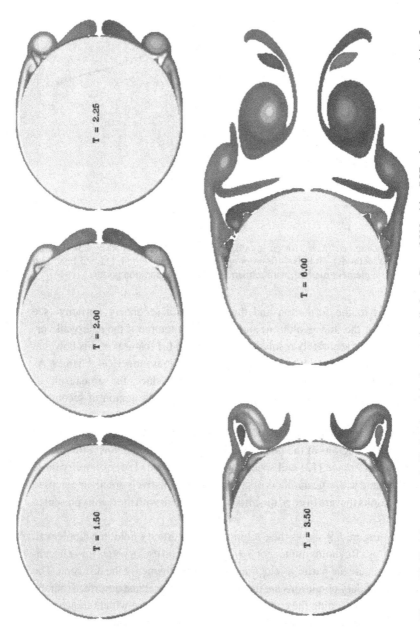

Figure 6.8. Vorticity contours behind an impulsively started cylinder at Re = 9500 with a PSE scheme that uses vorticity flux boundary conditions [125].

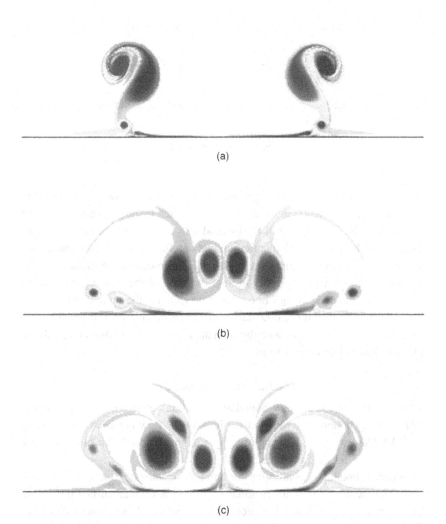

(a)

(b)

(c)

Figure 6.9. Successive stages for the rebound of a dipole impinging at a wall at Re = 3200 with a PSE algorithm and vorticity flux boundary conditions.

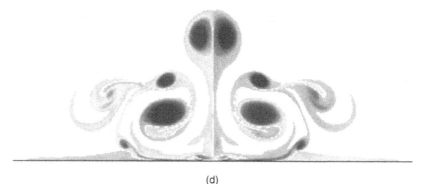

(d)

Figure 6.9. (*Continued*)

enforced by use of particle images with identical circulations. It is interesting to mention that comparisons of the results obtained by this method with those obtained in Ref. 128 by use of the method of Ref. 124 showed no discernible differences. It is also worth mentioning that careful comparisons of these simulations with results obtained by high-order finite-difference methods [159] showed that, although first order in time (because of the splitting involved in the method), the present algorithms compare well, for the same grid size, with high-order finite-difference boundary vorticity formulas (see Ref. 133 for more discussions on these tests), with the advantage (as usual for vortex methods) of allowing much bigger time steps.

The Three-Dimensional Case

The formulation of vorticity boundary conditions in terms of integral equations linking boundary terms and vorticity in the flow has been addressed in Ref. 40. Here we focus on a splitting-type algorithm in the spirit of what we just presented for the two-dimensional case.

Vorticity boundary conditions for three-dimensional viscous flows have, compared with the two-dimensional case, two additional difficulties. First, since vorticity is a vector, one needs three instead of one boundary condition. Secondly, vorticity created at the boundary must be divergence free, and this constraint must somehow enter the boundary conditions. To simplify the exposition, we assume that the boundary is a flat plate located at $x_3 = 0$. The general case follows by use of a local coordinate axis parallel and orthogonal to the wall, along the same lines as given in Subsection 6.1.1. We also assume a velocity vanishing at the wall.

As we already mentioned in Subsection 6.1.1, one vorticity boundary condition immediately follows from the no-slip condition: The normal component of the vorticity vanishes at the wall:

$$\omega_3 = 0 \quad \text{on } \partial\Omega. \tag{6.3.33}$$

It is worthwhile to note that a consequence of this condition is that at the wall $(\mathbf{u} \cdot \nabla)\omega_3 = (\omega \cdot \nabla)u_3 = 0$. Hence the normal component of the vorticity equation written at the wall yields

$$\frac{\partial^2 \omega_3}{\partial x_3^2} = 0 \quad \text{on } S. \tag{6.3.34}$$

Let us now turn to the tangential components of the vorticity. A natural extension of the two-dimensional vorticity flux boundary conditions is to enforce no-slip for the components u_1 and u_2 of the velocity by creation of vorticity for the components ω_2 and ω_1, respectively.

This leads to the following three-dimensional version of algorithm (6.3.19)–(6.3.24):

Substep 1: Solve the convection diffusion for the three components of the vorticity with the homogenuous Dirichlet boundary condition for the normal component and homogeneous Neumann boundary conditions for the tangential components:

$$\omega_3 = \frac{\partial \omega_1}{\partial x_3} = \frac{\partial \omega_2}{\partial x_3} = 0. \tag{6.3.35}$$

Substep 2: Compute the slip (u_1, u_2) at the boundary.

Substep 3: Repeat substep 1 for the tangential components, with the new Neumann boundary conditions

$$\frac{\partial \omega_1}{\partial x_3} = -\frac{u_2}{\Delta t}, \quad \frac{\partial \omega_2}{\partial x_3} = \frac{u_1}{\Delta t}. \tag{6.3.36}$$

Let us now check that this procedure guarantees that the vorticity remains for all time divergence free: Since (u_1, u_2) is associated with a vorticity field with a vanishing normal component, it satisfies

$$\frac{\partial u_2}{\partial x_1} - \frac{\partial u_1}{\partial x_2} = 0. \tag{6.3.37}$$

In addition we observe that writing the Navier–Stokes equation as solved in step 1 at the wall yields

$$\frac{\partial^2 \omega_3}{\partial x_3^2} = 0. \tag{6.3.38}$$

Now if we compute the normal derivative of the vorticity divergence at the wall, we find

$$\frac{\partial}{\partial x_3}(\text{div}\,\boldsymbol{\omega}) = \frac{\partial}{\partial x_1}\left(\frac{\partial \omega_1}{\partial x_3}\right) + \frac{\partial}{\partial x_2}\left(\frac{\partial \omega_2}{\partial x_3}\right) + \frac{\partial^2 \omega_3}{\partial x_3^2}. \tag{6.3.39}$$

In view of approximation (6.3.36) and Eq. (6.3.38), this gives

$$\frac{\partial}{\partial \mathbf{n}}(\text{div}\,\boldsymbol{\omega}) = 0 \qquad \text{for} \quad x_3 = 0. \tag{6.3.40}$$

On the other hand, taking the divergence of the vorticity equation solved in substep 3 gives

$$\frac{\partial}{\partial t}(\text{div}\,\boldsymbol{\omega}) + (\mathbf{u} \cdot \nabla)\text{div}\,\boldsymbol{\omega} - \nu\Delta(\text{div}\,\boldsymbol{\omega}) = 0 \quad \text{in } V. \tag{6.3.41}$$

Supplemented with boundary condition (6.3.40), Eq. (6.3.36) implies that $\boldsymbol{\omega}$ remains divergence free for all times.

Let us now briefly describe some of the existing techniques for vorticity creation used in vortex methods. Three-dimensional versions of the vortex sheet/vortex blob method described in Section 6.3 have been implemented, for example, in Ref. 88. They consist of creating vortex tiles parallel to the wall at the boundary, which then undergo random walk and reflection at the boundary. As in two dimensions, they are based on the Prandtl equations, and vortex tiles are transformed into vortex blobs as they leave the boundary layer. Besides the limitations of this technique already mentioned for the two-dimensional case related to the validity of the Prandtl model and curvature effects, the three-dimensional case raises additional difficulties in this scheme. The boundary condition on the normal component is implicitly dealt with by the fact that vortex blobs entering the boundary layer lose their normal component. This is, however, at the price of having no normal vorticity at all present in the whole boundary layer. Moreover the algorithm does not give a clear-cut way to control the divergence of the vorticity created at the wall, which probably makes it inappropriate for simulating strong vortex–wall interactions.

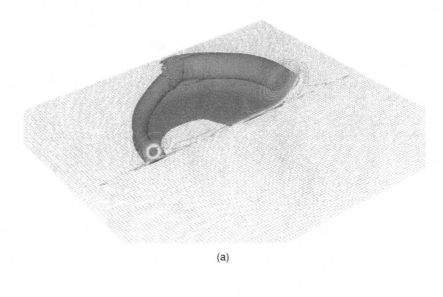

(a)

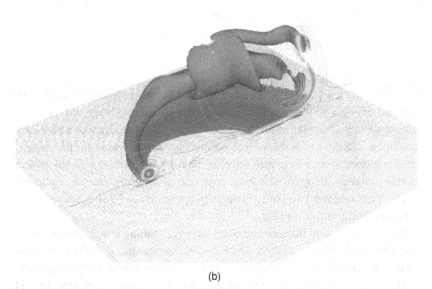

(b)

Figure 6.10. Vorticity magnitude isosurface and contours in a cross section of the orthogonal component of the vorticity for a vortex ring hitting a wall at an angle (for clarity only half of the isosurfaces are shown). (a): $t = 24$; (b): $t = 40$; (c): $t = 64$.

(c)

Figure 6.10. (*Continued*)

In a PSE scheme, the boundary conditions derived above [namely Eqs. (6.3.35) and (6.3.36)] can be implemented in a rigorous way, through integral representations formulas similar to those derived above. The homogeneous boundary condition – Dirichlet for the normal component, and Neumann for the tangential components – can be treated either by image techniques (which means that vortex sources with opposite normal components and equal tangential components in a ε neighborhood of the boundary are used in the PSE formula) or by a heat potential at the boundary.

Figure 6.10 shows the rebound of a vortex ring hitting a wall at an angle of 30° simulated with this method. The Reynolds number based on the circulation of the ring and the viscosity is 1400. The cross sections indicate the magnitude of the component of the vorticity orthogonal to this cross-section. Velocity and stretching were computed on a Eulerian 128^3 grid and then interpolated back to the particles through a vortex-in-cell approach, as described in Section 8.2. At this time of the calculation, the number of particles has increased from

600,000, to cover the initial ring, to approximately 1.2 million, in response to stretching, vorticity flux from the wall, and diffusion (regridding based on the techniques designed in Section 7.2 was done in this simulation at every time step).

7

Lagrangian Grid Distortions: Problems and Solutions

In vortex methods the flow field is recovered at every location of the domain when one considers the collective behavior of all computational elements. The length scales of the flow quantities that are been resolved are characterized by the particle core rather than the interparticle distance. These observations, which stem from the definition itself of vortex methods and are confirmed by its numerical analysis, differentiate particle methods from schemes such as finite differences.

The essense of the method is the "communication" of information between the particles, that requires a particle overlap. As a result, a computation is bound to become inaccurate once the particles cease to overlap. Computations involving nonoverlapping finite core particles should be regarded then as modeling and not as direct numerical simulations. Excluding case-specific initial particle distributions (e.g., particles placed on concentric rings to represent an azimuthally invariant vorticity distribution) the loss of overlap (and excessive overlap) is an inherent problem of purely Lagrangian methods.

The cause of the problem is the flow strain that may cluster particles in one direction and spread them in another in the neighborhood of hyperbolic points of the flow map, resulting in nonuniform distributions. At the onset of such particle distributions no error is usually manifested in the global quantities of the flow such as the linear and the angular impulse. However, locally the vorticity field becomes distorted and spreading of the particles results in loss of naturally present vortical structures, whereas particle clustering results in the appearance of unphysical ones on the scale of the interparticle separation.

A mathematical explanation of this can be found in the numerical analysis of Section 2.6. The equation governing the error involves a right-hand side that is related to the truncation error of the method and that is amplified exponentially in time at a rate given by the first-order derivatives of the flow. These derivatives are precisely related to the amount of strain contained in the flow. Ultimately, although the method can in principle be of high order, the strain-related large constants in the error estimate prevent this order from being achieved in practice.

As we have already seen in Section 2.4, a characteristic example of this pathology is the simulation of the evolution of an axisymmetric vortex patch by use of particles initially distributed on a rectangular grid. Although the linear and quadratic diagnostics of the flow do not exhibit any unphysical behavior, an examination of the vorticity field reveals unphysical nonsmooth vortical structures that eventually destroy the whole calculation and the error between the computed and the exact (steady-state) velocity fields dramatically increases. It is worth noting that, at least for this particular case, the accuracy deterioration does not go beyond a certain level. One reason is that it is reasonable to think that, whatever strain is developed by the flow, the first-order accuracy predicted in the case of a random initial choice will always hold, as this error does not involve any derivatives of the flow, but only the fact that the flow is divergence free. Moreover, in this specific case, because particles are moving on concentric circles, a minimal overlapping is always ensured.

In a numerical simulation, the clustering and the spreading of the particles have various consequences, depending on the particular numerical schemes that are implemented with the vortices. For two-dimensional inviscid flows they result in a loss of accuracy on the computed velocity, which in severe cases can result in the appearance of undesirable small scales; this has been observed, for example, for long-time vortex sheet calculations [130], in the evolution of elliptical vortices, or for periodic turbulence experiments starting for random initial data [100]. For three-dimensional flows, in regions of high strain the depletion of particles gets more dramatic as this flow geometry is in general associated with vorticity intensification. For viscous flows, when the diffusion is calculated through the PSE scheme defined in Section 5.4 or other related schemes, constraints on the overlapping of particles are even more severe: If a given particle does nor overlap with its neighbors, it simply cannot diffuse its vorticity at all. When, on the other hand, particles accumulate, the local Reynolds number diminishes, which can lead to numerical instabilities related to the explicit time discretization of the diffusion term used in this method [64].

To overcome these difficulties there are two possible strategies, which can be used either independently or combined: The first one consists of restarting the particles every few time steps at fresh locations where the overlapping is well controlled. The second one consists of processing the circulation carried by the particles in order to correct the effect of the distortion of the flow and allow particles to still give an accurate description of the vorticity. We will successively investigate these two strategies.

7.1. Circulation Processing Schemes

We discuss here several methods of remeshing that may seem different in nature. However, they all have as a common strategy the modification of the weights of the particles, either explicitly for the first method or through quadrature rules for the triangulated and adaptive quadrature methods.

One may wonder about the justification of modifying the weights of the particles, since the method is essentially based on the conservation of the circulation of each particle. One argument that can be used here is that the incompressibility of the flow is not satisfied at the discrete level. The discrete notion of incompressibility requires that the particle density remain constant, i.e., a constant number of particles is contained in each box for a uniform subdivision of the domain. In practice, however, this is satisfied only in the limit of infinite-size boxes or an infinite number of points. One may view all the methods that we will describe as a way to correct the volumes of the particles in order to improve this aspect of the calculation.

7.1.1. Beale's Method

In this method [20] corrected circulation values are computed at each time step in order to recover the vorticity field at the particle locations.

Assume that particles, located at $(\mathbf{x}_p)$, carry vorticity values ω_p. The idea is to compute circulations β_p such that

$$\sum_q \beta_q \zeta_\varepsilon(\mathbf{x}_p - \mathbf{x}_q) = \omega_p \tag{7.1.1}$$

for all particle of index p. The function ζ_ε above is the mollifier to be used, together with the processed values (β_q), for the computation of the velocity.

It is important to realize that Eq. (7.1.1) is actually a discrete deconvolution problem. One can thus anticipate difficulties in its solution, in particular when the right-hand side is not smooth. We will come back to this issue later in this section.

Let us denote by v_q the volumes of the particles and by $\alpha_q = v_q \omega_q$ their circulations. If we use the matrix notation $[A_{pq}] = [v_p \zeta_\varepsilon(\mathbf{x}_p - \mathbf{x}_q)]$, A is a sparse $N \times N$ matrix, if N is the number of particles, and Eq. (7.1.1) is given by

$$A\beta = \alpha. \tag{7.1.2}$$

For more than a few hundred particles, the size of the matrix does not allow for a direct inversion and one has to resort to iterative methods for the solution of the equations. The iterative procedure suggested by Beale to solve Eq. (7.1.1)

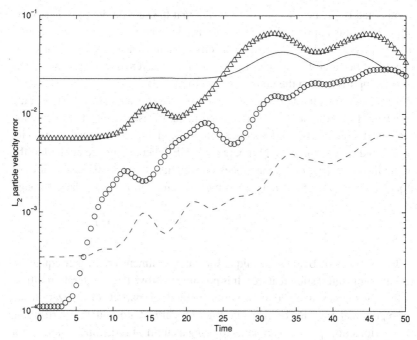

Figure 7.1. Improved accuracy through iterations (7.1.3) for the vorticity $\omega(\mathbf{x}) = (1 - |\mathbf{x}|^2)^3$. Solid curve, original scheme, $\varepsilon = 3h$; $\triangle$, original scheme, $\varepsilon = 2h$; $\bigcirc$, Beale's method, $\varepsilon = 2h$; dashed curve, Beale's method, $\varepsilon = 3h$.

is based on the observation that the circulations α_q are natural guesses for β_q; in other words $A \simeq I$, where I denotes the identity. Rewriting Eq. (7.1.2) as

$$(A - I)\beta + \beta = \alpha$$

leads thus to the following natural iterations:

$$\beta_p^{n+1} = \alpha_p + \beta_p^n - \sum_q v_q \beta_q^n \zeta_\varepsilon(\mathbf{x}_p - \mathbf{x}_q). \qquad (7.1.3)$$

Two-dimensional calculations based on circular-patch explicit solutions show a dramatic improvement in the accuracy of the velocity when only two or three iterations (7.1.3) are performed. Figure 7.1 illustrates that this improvement appears even at the initial stage. Iterations (7.1.3) are therefore sometimes used to prepare the particles to better fit the initial vorticity field. The main feature of iterations (7.1.3) is, that the gain in accuracy remains all along the calculation despite the distortions developed in the particle distribution. This procedure is, however, rather sensitive to the value of the ratio ε/h (it is clear in Figure 7.1 that the performance of the iterations is better for $\varepsilon/h = 3$ than for $\varepsilon/h = 2$).

As noted by Beale, one has to be aware that the sum on the right-hand side of Eq. (7.1.3) is in principle infinite, as particles having initially no circulation may receive some. Even when the vorticity has compact support, in some cases one has to be careful to lay ghost particles in an ε layer (if ε is the width of the cutoff support) around the nonzero-strength particles.

The iterative method of Eq. (7.1.3) has been also successfully tested in the context of PSE schemes by Choquin and Lucquin-Desreux [47]. In this case the processed weights are used inside the integral, giving the amount of vorticity exchanged by the particles. Of course the PSE scheme must be applied to the original weights and not to the processed weights; if n processing iterations of Eq. (7.1.3) are performed at a given time step, circulations are updated through

$$\frac{d\alpha_p}{dt} = \nu\varepsilon^{-2} \sum_q \left(\beta_q^n v_p - \beta_p^n v_q\right) \eta_\varepsilon(\mathbf{x}_p - \mathbf{x}_q).$$

The analysis of Beale's technique for the two-dimensional Euler equations reveals some interesting features. It is possible to prove that the iterations affect both error sources involved in the vortex method: the particle discretization and the regularization. If the original cutoff is of order r, it turns out that performing m iterations of Eq. (7.1.3) amounts to using a cutoff of order mr, and the error estimate proved in [20] is

$$\mathbf{u} - \mathbf{u}_h = O(\varepsilon^{mr}). \tag{7.1.4}$$

An alternative point of view on the convergence of iterations (7.1.3) is given by a Fourier analysis, assuming that the particles lie on a uniform (one dimensional for simplicity) mesh, with mesh size h. If we denote by $\hat{\alpha}$ and $\hat{\beta}$ the discrete Fourier transforms of α and β, Eq. (7.1.2) is equivalent to

$$\hat{\beta} = \frac{\hat{\alpha}}{h\hat{\zeta}_\varepsilon}. \tag{7.1.5}$$

On the other hand the iterative scheme of Eq. (7.1.3) can be rewritten as

$$\hat{\beta}^{n+1} = \hat{\alpha} + \hat{\beta}^n(1 - h\hat{\zeta}_\varepsilon), \tag{7.1.6}$$

which gives

$$\hat{\beta}^n = \hat{\alpha}\frac{1 - (1 - h\hat{\zeta}_\varepsilon)^n}{h\hat{\zeta}_\varepsilon}. \tag{7.1.7}$$

We thus recover Eq. (7.1.2) for n tending to infinity, provided that $\hat{\zeta}$ has a positive real part.

Formula (7.1.7) further allows us to predict the speed of convergence of iterations (7.1.3). Note first that, because of the aliasing errors (which will be analyzed later), the smoothness of ζ does not imply that $\hat{\zeta}_\varepsilon(k)$ tends to 0 as k tends to infinity. This means that the discrete deconvolution problem (7.1.1), or its Fourier version (7.1.5), is better behaved than the continuous one. However, the iterations can be very slowly convergent, depending on the parameter values. More precisely, assume that ζ has a bandlimited Fourier transform $\mathcal{F}\zeta$ with, say, a unit bandwidth. Further assume that $\varepsilon \gg h$, which is the usual overlapping condition that has to be satisfied in vortex methods. In one dimension, the discrete Fourier transform of ζ_ε can be written as

$$\hat{\zeta}_\varepsilon(k) = \sum_n \mathcal{F}\zeta \left(\varepsilon k + 2\pi n\right).$$

This shows that the denominator of Eq. (7.1.5) can vanish for wave numbers of the order of $1/\varepsilon$. The discrete deconvolution problem then becomes ill posed, unless α itself has vanishing Fourier modes for $k > 1/\varepsilon$ (which would mean that the vorticity is analytic).

In more realistic situations in which the Fourier transform of the cutoff is not bandlimited, but decays fast at infinity – because of its smoothness – if its Fourier transform is positive (as for a Gaussian), one never strictly runs into the above ill-conditioning problems. However, the right-hand side of Eq. (7.1.5) still gets small denominators for wave numbers of the order of $1/\varepsilon$, implying slow convergence for Eq. (7.1.3), except if the corresponding Fourier coefficients of the vorticity are small. Another observation is that increasing the overlapping of particles has the unpleasant effect of making the solution of Eq. (7.1.2) more difficult, except if one can make sure that the vorticity is smooth (which is the case in the radial vorticity test we have shown), in which case there is a compensating effect in the decay of the Fourier coefficients of the cutoff and of the vorticity.

To summarize, lack of regularity of the vorticity and strong overlapping are two factors that can substantially slow down the iterations, which makes this method difficult to use in many practical situations. This confirms observations of Winckelmans and co-workers in the context of three-dimensional flows (see Refs. 200 and 201 and Section 3.4). Efficient procedures to invert system (7.1.1) are still a current research topic.

7.1.2. Triangulated Vortex Methods

Although this method has been originally proposed in a more general context [41], we chose to present it here as we believe that its main interest is to give an elegant way to deal with distorted particle distributions.

The method is actually reminiscent of the Free Lagrange methods. It consists of constructing at each time step a triangulation whose nodes are the particle locations.

We are concerned here with two-dimensional flows. The circulations of the particles in this case remain unchanged, and the first task is to derive on this triangulation a piecewise linear interpolation to these circulations.

If τ is a given triangle, one has to deal there with a vorticity field of the form

$$\omega_h(\mathbf{x}) = a + bx_1 + cx_2.$$

The coefficients a, b, and c can be easily obtained from the coordinates of the vertices of the triangle and the vorticity values at these points. In view of the Biot–Savart law, the computation of the velocity will then be obtained by addition of the contributions of each over a triangle integral:

$$\int_\tau \mathbf{K}(\mathbf{x} - \mathbf{y})\omega_h(\mathbf{y})\,d\mathbf{y}.$$

The calculation of these integrals breaks down to evaluating the quantities

$$\mathbf{K}^{ij} = \int_\tau \mathbf{K}(\mathbf{x} - \mathbf{y})y_1^i y_2^j\,d\mathbf{y}, \qquad \text{for } 0 \le i + j \le 1. \tag{7.1.8}$$

These can be computed analytically in polar coordinates. The efficiency of the overall procedure is then conditioned by

- the accuracy of the piecewise linear interpolation,
- the computational cost of the repeated triangulations,
- the possibility of using a fast code to compute the integrals of Eq. (7.1.8) for all triangles τ and all particles $\mathbf{x}$.

Let us address these issues successively. Concerning the first point, we remark that this is a classical problem in finite-element analysis. We derive the interpolation error by mapping each triangle onto a reference triangle. The error estimates are then driven by the derivatives of the mapping, which themselves can be evaluated in terms of the size of the triangle.

For a piecewise linear interpolation, one typically gets

$$\|\omega - \omega_h\|_{0,2} = O\left(h_{\max}^2\right)\|\omega\|_{2,2}$$

(we refer to Appendix A for the norm notation), where $h_{\max}$ denotes the maximum size of the triangles. Note that the desingularization of the particles is implicitly done in the linear interpolation step, and there is no further need to introduce any cutoff for the velocity evaluations. The $O(h^2)$ accuracy is a definite advantage over vortex blob methods that use second-order cutoff, which would yield $O(\varepsilon^2)$ accuracy with $\varepsilon \gg h$. However, we have seen in Section 2.6 that the control of the flow derivatives appears in an essential way in the convergence proof [see estimate (2.6.4), for example]. In the context of the triangulated method, these estimates are classically obtained through so-called inverse inequalities, which involve the internal diameters of the triangles. Stability of the method thus requires that all sides of the triangles be approximately equal lengths so that the triangulation is regular. For the simplest choice, which would be to connect the initially neighboring particles, the stretching of the flow map would result in stretching the triangles (or increasing their aspect ratio) in such a way as to deteriorate the theoretical second-order accuracy.

As a remedy, Russo and Strain [176] suggested a strategy based on Delaunay triangulations. Given a cloud of nodes to connect, the principle of these triangulations is precisely to optimize the aspect ratio of the triangles. There are several possible ways to construct fast Delaunay triangulations. One popular way is through the construction of so-called dual Voronoi polygons, which are defined as

$$P_i = \left\{ \mathbf{x}; \ |\mathbf{x} - \mathbf{x}_i| \leq |\mathbf{x} - \mathbf{x}_j| \text{ for all } j \neq i \right\},$$

where $\mathbf{x}_j$ are the particles. Delaunay triangles are then obtained by connecting points whenever they belong to adjacent polygons (see Figure 7.2).

Figure 7.3 shows the Delaunay triangulation built on particles rotating on circles of different speeds, subject to a vorticity $\omega(r) = (1 - r^2)^7$, together with the triangulations obtained by following the initial triangulation along the flow map. The gain in regularity is evident. Except for a few triangles at late times, the Delaunay triangulation essentially remedies the distortion due to the stretching of the flow map and effectively leads to a triangulated vortex scheme that is second-order accurate. This is confirmed by Figure 7.4, which shows the velocity errors corresponding to the two types of triangulation. While the effect of the distortion in the original triangulation is a deterioration in the accuracy, the Delaunay triangulation allows us to stabilize the error.

Let us now discuss the computational complexity of triangulated methods. The construction of Delaunay triangulations always involves a distance minimization step. This minimization can be efficiently done in $O(N)$ operations if particles are presorted in cells, so that the search is limited to neighbors. The cell division goes together with a linked list similar to the one needed by all fast solvers (see Appendix B). If the particle distribution is highly nonuniform it may, however, happen that the optimization cost degenerates to the $O(N^2)$ worst case. To balance the calculation load among all cells optimally and further speed up the calculation, Russo and Strain use an adaptive strategy to make sure that

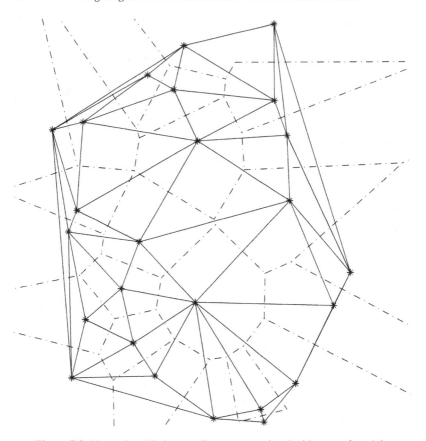

Figure 7.2. Voronoi and Delaunay diagrams associated with a set of particles.

all cells contain approximately the same number of points. Although this adds
to the complexity of the algorithm, this strategy allows us to gain a considerable
speed up for $N > 200$. Timing results reported by Russo and Strain indicate
that the triangulation time is less than one tenth of the velocity evaluation time.

This leads us to the third issue raised above concerning the fast evaluation of
velocities. Fast solvers of the kind described in Appendix B have not been implemented, as the triangulations hinder the transfer of the multipole expansions,
therefore making difficult the construction of tree data structures.

However, the computational cost of the velocity evaluations can be reduced
by use of a single level of refinement in these calculations. First, on dividing the
computational domain into N_C cells, one can split the velocity into a local- and
a far-field component. When necessary, cells are refined so that their borders
are not crossed by triangle edges. The local component resulting from triangles

Figure 7.3. Lagrangian (top row) and Delaunay (middle row) triangulations for the evolution of circular patches at times 0, 2π, and 4π. The bottom figures correspond to the Delaunay triangulations at later times ($t = 8\pi$, 16π, and 32π)(courtesy of J. Strain).

in a given cell is computed directly. For the far-field component,

$$\mathbf{u}_F(\mathbf{x}) = \sum_{C \neq \mathbf{x}} \sum_{T \subset C} \int_T \mathbf{K}(\mathbf{x} - \mathbf{x}')\omega(\mathbf{x}') \, d\mathbf{x}',$$

one expands $\mathbf{K}(\mathbf{x} - \mathbf{x}')$ in a Laurent series about the center of the cell containing $\mathbf{x}$; in complex notations, up to a 2π factor,

$$\mathbf{K}(z - z') = \frac{1}{z - z'} = \frac{1}{z - c} \sum_{n=0}^{\infty} \left(\frac{z' - c}{z - c} \right)^n.$$

If p terms are retained in the expansion [where p is related to the desired

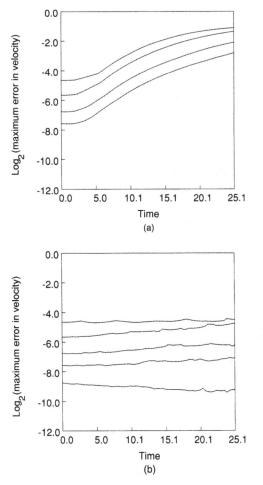

Figure 7.4. L^∞ relative error for the velocity in the triangulated vortex method for various mesh sizes by use of (a) a Lagrangian triangulation or (b) a Delaunay triangulation for a circular patch where $\omega(\mathbf{x}) = (1 - |\mathbf{x}|^2)^7$ (courtesy of J. Strain).

accuracy ε by $p = O(\log \varepsilon)$] the far field is thus given by

$$\mathbf{u}_F(z) = \sum_{C \not\ni z} \sum_{T \subset C} \sum_{n=0}^{p} d_n^C (z - c)^{-n-1}, \qquad (7.1.9)$$

where the coefficients d_n^C have the form

$$\int_C (z' - c)^n \omega(z') \, dz'.$$

Since the vorticity is known as a polynomial of degree 1 on each triangle, one has to evaluate only the terms

$$\int_T (\mathbf{x}' - c)^n x_1'^{\alpha} x_2'^{\beta} \, d\mathbf{x}'$$

for $0 \leq \alpha, \beta \leq 1$. If N is the number of triangles, the coefficients d_n^C are evaluated in $O(pN)$ operations and expansions (7.1.9) for all vertices require $O(pNN_C)$ operations. For a relatively uniform particle distribution the optimal value of N_C is $N^{1/2}$, yielding $O(N^{3/2} \log \varepsilon)$ operations.

A further speed up of the algorithm can be obtained if the above Laurent series is translated into Taylor series. We refer to Ref. 176 for details. The resulting algorithm has an $O(N^{4/3})$ computational cost.

In closing this section, let us mention that the triangulated method can be adapted to allow local refinements in the spirit of finite-element methods (see Ref. 41 for details).

7.1.3. Adaptive Quadrature

Besides the technical difficulties that one would face in implementing the method in three dimensions, one drawback of the triangulated vortex method just described is that it is limited to second order. From this point of view, the method proposed by Strain [191] that we now describe can be seen as a generalization of the triangulated vortex method.

Since particles give access to only pointwise values of the vorticity (although one can imagine particles carrying informations about derivatives as well), it is hard to propose higher-order triangulated vortex methods, unless one goes to some kind of macroelements that would combine several triangles together. Adaptive quadratures are in this spirit while avoiding as much as possible technicalities that in general go with macroelements.

The second order of the triangulated vortex method was related to the fact that vorticity values yield exact quadrature for linear functions on triangles. The idea is therefore to construct quadrature rules that are exact for polynomials of arbitrary degrees. To get order p accuracy for the quadrature rule requires exactness for polynomials of a degree less than or equal to $p - 1$ and thus $m = p(p + 1)/2$ degrees of freedom (assuming that we are in two dimensions). In principle one thus should divide the computational domain into cells containing m points $\mathbf{a}^j$ and find quadrature weights w_j in each cell B so that

$$\int_B x_1^{\alpha} x_2^{\beta} \, d\mathbf{x} = \sum_j w_j \left(a_1^j\right)^{\alpha} \left(a_2^j\right)^{\beta} \tag{7.1.10}$$

for all α, β such that $\alpha + \beta \leq p$.

Even so, a solution of this system does not necessarily exist. For example, if all points are on the same line, say the x axis, they will not give any information concerning the dependence on the y direction. The method suggested by Strain is to design a tree data structure across all particles dividing the computational domain into cells containing m or $m+1$ points. System (7.1.10) is then solved in the least-squares sense. If the solution yields large values for the weights, the cell is merged with its neighbor, thus increasing the number of quadrature points, and the procedure is repeated. The accuracy of the quadrature rule is then $O(H^p)$, where H is the side of the largest cell resulting from this merging procedure. Of course this technique does not require any particular treatment, compared with classical fast solvers (see Appendix B) for the fast evaluation of the velocities. Figure 7.5 shows the relative L^1 velocity errors obtained for the circular patch

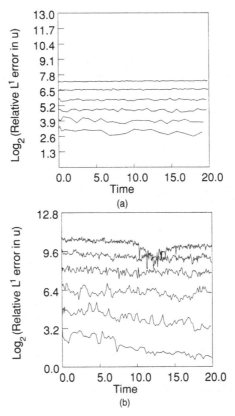

Figure 7.5. L^1 relative errors for the velocity with the adaptive quadrature method for various mesh sizes for (a) $p = 4$ and (b) $p = 6$; $\omega(\mathbf{x}) = (1 - |\mathbf{x}|^2)^3$ (courtesy of J. Strain).

$\omega_0(\mathbf{x}) = (1 - |\mathbf{x}|^2)^3$ when an adaptive quadrature method corresponding to $p = 4$ or $p = 6$ is used. It shows that in this case the adaptive strategy allows us to maintain the desired level of accuracy all along the calculation.

Adaptive quadrature techniques raise two issues. First, the computations of the quadrature weights can be seen as a processing of the volumes of the particles in order to correct the effect of the strain in the particle distribution. A similar interpretation could have been done for the triangulated vortex method, except that this method differs from regular vortex methods by a particular way to desingularize the particles (as a matter of fact, this is the only difference between the triangulated vortex method and the adaptive quadrature method for $p = 2$).

A second comment, which is valid for all the methods described in this section, is that this method performs well as long as the strain does not take apart particles too much (that is, here, as long as H remains of the order of h). This is the case for circular patches, but in more severe situations, such as flows with saddle points (e.g., boundary layers, shear layers), it seems unavoidable to regrid the particles during the simulation.

7.2. Location Processing Techniques

We are interested here in replacing the (occasionally) strained grid distribution with a new Lagrangian regular grid and simultaneously transporting accurately the vorticity from the old grid to the new. As a matter of fact, this is a situation that we have already encountered in two dimensions (the vortex sheet calculations of Section 2.1) and in three dimensions (shear layers and jet calculations of Section 3.2). In both cases it was important to compensate for the strain in the flow by continuous insertion of new elements. The particular geometry of the flow, however, enabled us to do it in a rather simple and efficient way through a monitoring of the distance between adjacent elements. The strength of the new elements in these cases can be obtained by simple, low-order interpolation without any spreading of the vorticity outside the support of the vorticity.

For more general flows or when it is not possible to keep track of the connectivity of the elements (because of the diffusion, for example) one must clearly be more careful in the way circulation is assigned to the new elements. In the following we consider the one-dimensional problem and in particular the interpolation of scalar quantities (such as the vorticity field in two-dimensional flows), bearing in mind that all the techniques described below apply in a straightforward way to any dimension, by using simple tensor product formulas.

We denote by $(\tilde{x}_p)$ and (x_p) respectively the old (distorted) and the new (regular) particle locations, and we denote by $\tilde{\omega}_p$, $\tilde{\Gamma}_p$ and ω_p, Γ_p the local vorticity values and circulations at the old and the new locations. The new particles are assumed to lie on a grid with spacing h. The natural way to compute Γ_p is through a classical interpolation rule:

$$\Gamma_p = \sum_q \tilde{\Gamma}_q W \left(\frac{x_p - \tilde{x}_q}{h} \right). \tag{7.2.1}$$

In the above formula, W is an interpolation kernel whose properties determine the type and the quality of the interpolation. For reasons that will be subsequently clear, we will immediately require that the function W satisfies

$$\sum_p W \left(\frac{x - x_p}{h} \right) \equiv 1. \tag{7.2.2}$$

To measure the discrepancy between the old and the new particle distributions,

$$\sum_p \tilde{\Gamma}_p \delta(x - \tilde{x}_p) - \sum_p \Gamma_p \delta(x - x_p),$$

we multiply the above quantity by a test function ϕ (which in practice may be thought of as a blob function, for the evaluation of the vorticity, or a regularized Biot–Savart kernel for the evaluation of the velocity). We get

$$E = \sum_p \tilde{\Gamma}_p \phi(\tilde{x}_p) - \sum_p \Gamma_p \phi(x_p), \tag{7.2.3}$$

which, by virtue of Eq. (7.2.1), becomes

$$E = \sum_p \tilde{\Gamma}_p \left[\phi(\tilde{x}_p) - \sum_q \phi(x_q) W \left(\frac{x_q - \tilde{x}_p}{h} \right) \right]. \tag{7.2.4}$$

We have thus to evaluate the function

$$f(x) = \phi(x) - \sum_q \phi(x_q) W \left(\frac{x_q - x}{h} \right).$$

Because of Eq. (7.2.2), this can be rewritten as

$$\sum_q [\phi(x) - \phi(x_q)] W \left(\frac{x_q - x}{h} \right).$$

Taylor expansions of ϕ yield

$$f(x) = \sum_\alpha \sum_q [(x_q - x) \cdot \nabla \phi]^\alpha W \left(\frac{x - x_q}{h} \right).$$

From this formula it results that if W satisfies

$$\sum_q (x - x_q)^\alpha W\left(\frac{x - x_q}{h}\right) = 0 \quad \text{for } 1 \le |\alpha| \le m - 1, \tag{7.2.5}$$

then

$$f(x) = O(h^m),$$

and the regridding procedure will be of order m.

It is worth noting that moment conditions (7.2.5) can be seen as the discrete analog of the ones encountered in the definition of cutoff functions. Note also that conditions (7.2.5) are equivalent to the conditions

$$\sum_q x_q^\alpha W\left(\frac{x_q - x}{h}\right) = x^\alpha \quad \text{for } 0 \le |\alpha| \le m - 1 \tag{7.2.6}$$

[this is seen when $(x - x_q)^\alpha$ is developed and moment properties are used at previous orders], which means that the interpolation formula is exact for polynomials of a degree less than or equal to $m - 1$ (we then say that it is exact to the degree $m - 1$ or, equivalently, of order m) and brings us into the classical analysis framework for interpolation formulas.

Finally conditions (7.2.5) can also be interpreted in terms of conservation of moments for the particle distributions. For $m = 2$, for example, one can deduce from Eq. (7.2.5) that

$$\sum_p \Gamma_p(x - x_p) = \sum_{p,q} \tilde{\Gamma}_q W\left(\frac{x_p - \tilde{x}_q}{h}\right)(x - x_p)$$

$$= \sum_{p,q} \tilde{\Gamma}_q [(x - \tilde{x}_q) + (\tilde{x}_q - x_p)] W\left(\frac{x_p - \tilde{x}_q}{h}\right)$$

$$= \sum_q \tilde{\Gamma}_q (x - \tilde{x}_q) \sum_p W\left(\frac{x_p - \tilde{x}_q}{h}\right)$$

$$+ \sum_q \tilde{\Gamma}_q \sum_p (\tilde{x}_q - x_p) W\left(\frac{\tilde{x}_p - x_q}{h}\right)$$

$$= \sum_q \tilde{\Gamma}_q (x - \tilde{x}_q).$$

This identity expresses the conservation of the linear impulse [condition (7.2.2) immediately implies the conservation of total circulation] when one is switching from the old to the remeshed distribution. Conservation of higher moments would follow along the same lines from higher-order moment properties of the interpolation kernel.

We now come to the construction of more efficient interpolation kernels. We note in passing that this work will be useful in defining vortex-in-cell schemes and, more generally, techniques that require exchanging information between a regular grid of finite-difference type and a Lagrangian particle grid (see Chapter 8).

7.2.1. Interpolation Formulas: General Definitions and Fourier Analysis

To facilitate the analysis we restrict our attention to an equispaced regular grid with unit mesh size onto which we map the quantities of interest.

The general interpolation formula in one dimension then is given by:

$$Q(x) = \sum_n q_n \, W(x - n). \qquad (7.2.7)$$

Let us introduce here some vocabulary from interpolation theory. We say that Eq. (7.2.7) is an ordinary interpolation formula if $Q(x)$ interpolates exactly the given ordinates q_n, i.e., if

$$W(0) = 1, \qquad W(n) = 0, \quad (n \neq 0),$$

or, alternatively,

$$Q(n) = q_n.$$

Interpolation formulas that do not have this property spread the value q_n among the neighboring grid points and are thus referred to as smoothing interpolation formulas.

Following Schoenberg [184], we analyze the properties of the interpolating functions through their behavior in the Fourier space. The characteristic function $g(k)$ of the interpolating (even) function $W(x)$ is defined as

$$g(k) = \int_{-\infty}^{+\infty} W(x) \, e^{-ikx} \, dx.$$

We assume that g is a smooth function, which means that W decays fast at infinity, a condition that is always satisfied in practice. A justification of this function lies in the following result, which provides a useful criterion to check the accuracy of interpolation formulas.

Theorem 7.2.1. *Consider the interpolation formula*

$$Q(x) = \sum_{-\infty}^{+\infty} q_n W(x - x_n).$$

Let the interpolating function $W(x)$ decay fast enough to satisfy the condition

$$|W(x)| \le A e^{-B|x|}, \qquad \text{where } A > 0, B > 0.$$

The formula is of degree m if the following two conditions hold simultaneously:

$$g(k) - 1 \text{ has a zero of order } m \text{ at } k = 0, \tag{7.2.8}$$

$$g(k) \text{ has zeros of order } m \text{ at all } k = 2\pi n \ (n \ne 0). \tag{7.2.9}$$

Proof. From the definition of g we may write that

$$\int_{-\infty}^{+\infty} W(y) e^{-i(k+2\pi n)y} \, dy = g(k + 2\pi n).$$

Now, by multiplying both sides by $e^{2\pi inx}$ and summing over all n to take into account all the alias coefficients, we get that

$$\sum_{-\infty}^{+\infty} e^{2\pi inx} \int_{-\infty}^{+\infty} W(y) e^{-i(k+2\pi n)y} \, dy = \sum_{-\infty}^{+\infty} e^{2\pi inx} g(k + 2\pi n).$$

By Poisson's summation formula we obtain

$$\sum_{-\infty}^{+\infty} e^{-ik(x-n)} W(x - n) = \sum_{-\infty}^{+\infty} e^{2\pi inx} g(k + 2\pi n),$$

and finally

$$\sum_{-\infty}^{+\infty} e^{ikn} W(x - n) = e^{ikx} \sum_{-\infty}^{+\infty} e^{2\pi inx} g(k + 2\pi n).$$

Note that for $x = 0$ the above formula provides a classical relation between the Fourier transform of the discrete smoothing formula and the continuous Fourier transform of the smoothing function as $\sum_{-\infty}^{+\infty} e^{ikn} W(n) = \sum_{-\infty}^{+\infty} g(k+2\pi n)$. Now, by fixing x and expanding the left-hand side of the equation, we obtain

$$\sum_{-\infty}^{+\infty} e^{ikn} W(x - n) = \sum_{\nu=0}^{\infty} \frac{i^\nu k^\nu}{\nu!} \sum_{-\infty}^{+\infty} n^\nu W(x - n).$$

On the right-hand side, assumption (7.2.9) implies that the terms $g(k + 2\pi n)$ $e^{2\pi inx}$ for $n \neq 0$, after expansion around $2\pi n$, do not contribute any terms in k of order less than m. So the equation becomes

$$\sum_{\nu=0}^{\infty} \frac{i^\nu k^\nu}{\nu!} \sum_{-\infty}^{+\infty} n^\nu W(x - n) = e^{ikx} g(k) + O(k^m).$$

Assumption (7.2.8) allows us to expand $g(k)$ around $k = 0$ to get

$$g(k) = 1 + O(k^m).$$

Expanding e^{ixk}, we finally may write

$$\sum_{\nu=0}^{\infty} \frac{i^\nu k^\nu}{\nu!} \sum_{-\infty}^{+\infty} n^\nu W(x - n) = \sum_{\nu=0}^{\infty} \frac{i^\nu k^\nu}{\nu!} x^\nu + O(k^m).$$

It remains now to identify the coefficients of the successive powers of k on both sides of the above equation to obtain that the interpolation formula is exact to the order $m - 1$.   □

It should be observed that assumption (7.2.8) translated back in the physical space is just the moment properties

$$\int W(y)\, dy = 1, \quad \int y^\alpha W(y)\, dy = 0, \text{ if } 1 \leq |\alpha| \leq m - 1.$$

This is not surprising, since Eq. (7.2.6) can be rewritten as

$$h \sum_p \left(\frac{n}{h}\right)^\alpha W\left(\frac{nh}{h}\right) = 0.$$

In view of quadrature estimate (A.1.3) proved in Appendix A, the above left-hand side can be estimated as

$$\int x^\alpha W(x)\, dx + O\left(h^m \left|W\left(\frac{\cdot}{h}\right)\right|_{m,1}\right) = \int x^\alpha W(x)\, dx + O(h).$$

Letting h tend to 0 gives the continuous moment properties for W. The additional conditions (7.2.9) are needed as a result of the sampling that has to follow the convolution in the interpolation process.

As a matter of fact, a related point of view on interpolation accuracy is given by Hockney and Eastwood [103], based on a splitting of the interpolation error into a convolution and sampling errors. The sampling, or aliasing error, in the Fourier space can be evaluated as the sum of the contribution of modes that are separated by a multiple of the grid size. More precisely, if q_n are the values taken by a smooth function q and if we define the continuous interpolated quantity as

$$Q_c(x) = \int q(x')\, W(x - x')\, dx',$$

whose values at $x = x_p$ give the mesh-defined function, its discrete and continuous Fourier transforms are linked through

$$\hat{Q}(k) = \sum_{-\infty}^{+\infty} \mathcal{F}\hat{Q}_c(k - nk_g).$$

The contribution from the nth term is called the nth alias contribution and the sum is called the alias sum (these concepts have already been encountered in the discussion of Beale's method in Section 7.1). On sampling values of the continuous function on the grid locations, the mesh has no clear way to distinguish between the principal harmonic and its aliases, thus overestimating or underestimating the correct value of the interpolated function, depending on the phase of the primary harmonic and its aliases. In the physical space the effect of aliasing may be described as loss of information due to the finite size of the grid. This loss of information is dependent on the mesh size and the smoothness of the interpolating function. In the physical space, smoothness is related to the number of continuous derivatives of the interpolating function, while in the transformed space it is translated as the rate of decay of the transform (and hence the order of the low-pass filter). A function W that is continuous in all derivatives up to the nth has a transform $\hat{W}(k)$ that decays as $k^{-(n+1)}$.

Following Hockney and Eastwood [103], we may explain the smoothness constraint in the context of low-pass filters. Consider a smooth function Q_c that has a transform that is bandlimited in the interval $[-(k_c/2), k_c/2]$. Now if we use a grid with size h such that the grid wave number $k_g = 2\pi/h$ is such that $k_g > k_c$, then the smooth function is being oversampled. We are in principle then able to recover the correct function without aliasing effects, although at some excessive computational cost (unnecessarily small h). The function would be critically sampled (at a minimum possible computational cost) when h is such that $k_g = k_c$. Beyond this critical spacing the function would be undersampled and the function we recover would be modified by the contribution of the aliases. So from the above analysis we see that the grid spacing is the main factor that determines the possibility for the accurate interpolation of a function, depending on the smoothness of the interpolated function. One may note here that through the overlap constraint the present analysis is linked to the particle core size. In order to take full advantage of the computational grid we need also interpolation schemes that would act as low-pass filters with a finite bandwidth removing all the harmonics that are larger than k_g. An ideal low-pass filter in the transformed space would then be the function

$$\hat{W}(k) = \Pi(k/k_g) = \begin{cases} 0, & \text{if } k/k_g > 1/2 \\ 1/2, & \text{if } k = k_g \\ 1, & \text{if } k/k_g < 1/2 \end{cases}.$$

It is then easily seen that this function would totally suppress the alias contributions. Conditions (7.2.8) and (7.2.9) are satisfied at any order. Unfortunately, the above bandlimited transform corresponds to a function that is not limited in the physical space as it is the transform of the function

$$W(x) = \frac{1}{h} \frac{\sin(x/2h)}{(x/2h)}.$$

Hence the complete alias elimination would imply interpolation to all mesh points as this function has a very slow decay. Such a dealiasing operation would therefore lead to a computationally impractical scheme. Good interpolation schemes then are those that are bandlimited in the physical space (thus involving a few interpolating points and hence few operations) and at the same time are close approximations of the ideal low-pass filter in the transformed space.

A measure of the effectiveness (ε) of the interpolating scheme in a given computational grid to eliminate the effect of aliasing is given by the relative magnitude of $\hat{W}$ at the principal wave number and its alias, i.e.,

$$\varepsilon = \left| \frac{\hat{W}(k - nk_g)}{\hat{W}(k)} \right|. \tag{7.2.10}$$

The rest of this section is devoted to the derivation of interpolation formulas of practical use. We will adopt successively two points of view. In the first one we will construct interpolation kernels of increasing smoothness, while in the second one we will try to optimize the shape of the kernel with respect to conditions (7.2.8) and (7.2.9).

7.2.2. Smoothing Interpolation Formulas

As the first classical example let us consider the piecewise linear interpolation function given by

$$W(x) = \begin{cases} 0 & \text{if } x \le -1 \text{ or } x \ge -1 \\ 1 + x & \text{if } -1 \le x \le 0 \\ 1 - x & \text{if } 0 \le x \le 1 \end{cases}.$$

The characteristic function of this equation may be expressed as

$$g(k) = \left[\frac{\sin(\pi k)}{\pi k}\right]^2,$$

which has zeros of multiplicity 2 at $k = \pm 1, \pm 2, \pm 3, \ldots$, which means that condition (7.2.9) is satisfied for $m = 2$. Moreover, direct expansion of $g(k)$ for small k gives that

$$g(k) = 1 + \pi^2 k^2/6 + \cdots$$

so that condition (7.2.8) is also satisfied, and the above interpolation formula is then exact for linear functions. The form of this interpolation formula in Fourier space suggests a certain class of splines, the so-called B-splines, that may be used for interpolation with a desired degree of smoothness. Their Fourier transform is given by the formula

$$g(k) = \left[\frac{\sin(k/2)}{k/2}\right]^m.$$

As m increases, the decay of the Fourier transform at large wave numbers gets faster, resulting in smoother functions and less severe aliasing effects. The effectiveness coefficient ε of the mth B-spline, as given by formula (7.2.10), is

$$\varepsilon = \left(1 - \frac{2\pi n}{k}\right)^{-(m+1)}.$$

In the physical space, the smoothing function is obtained through successive convolutions of the top-hat function (which itself corresponds to $m = 1$). For a given m one obtains a piecewise polynomial of degree $m - 1$, which is of class C^{m-2} and with support extending to $2m - 3$ grid points (in two or three dimensions this number should be raised of course to the power 2 or 3). For $m = 2$ we recover the tent function, which is associated with the so-called cloud-in-cell interpolation scheme for vortex-in-cell methods (see Section 8.1):

$$M_2(x) = \begin{cases} 0 & \text{if } |x| > 1 \\ 1 - |x| & \text{if not} \end{cases}.$$

For $m = 3$, the piecewise quadratic function is sometimes referred to as the TSC (for triangular-shaped cloud) interpolation function:

$$M_3(x) = \begin{cases} 0 & \text{if } |x| > 3/2 \\ \frac{1}{2}(-|x| + 3/2)^2 & \text{if } 1/2 \leq |x| \leq 3/2 \\ \frac{1}{2}(x + 3/2)^2 - \frac{3}{2}(x + 1/2)^2 & \text{if } 0 \leq |x| \leq 1/2 \end{cases}.$$

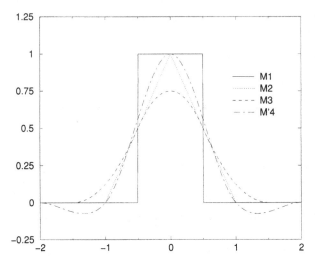

Figure 7.6. B-splines of increasing smoothness and the M_4' function.

These functions are presented in Figure 7.6. For a formula giving B-splines in a systematic way, we refer to the book of Hockney and Eastwood [103].

Note now that, based on Theorem 7.2.1, the B-splines can interpolate only exactly linear functions as the Fourier transform of the mth-order spline has a zero of order m for $k = \pm 1, \pm 2, \ldots$, but has a zero of only multiplicity 2 for $k = 0$. So although with increasing order they improve the smoothness of the interpolating quantity on the grid from the possibly scattered particle locations, their accuracy is limited to second order.

Monaghan [156] has presented a systematic way of increasing the accuracy of the interpolating functions while maintaining the smoothness properties. His idea, which is based on extrapolation, is actually reminiscent of a technique we mentioned in Section 2.3 to construct higher-order kernels through a combination of a cutoff and its radial derivative. If W is an interpolation kernel of second order and class C^{m-1} with $m \geq 3$, consider the new kernel

$$\tilde{W}(x) = \frac{1}{2}(3W + xW').$$

It is readily seen from integration by parts that this kernel has a vanishing second-order moment. Moreover, if W is the B-spline $\chi^{*(m)}$, its Fourier transform has $(\sin \xi/\xi)^m$ in factor and $\mathcal{F}W' = \xi \mathcal{F}W$ has $(\sin \xi/\xi)^{m-1}$ in factor. We can thus conclude that, if $m = 3$ or $m = 4$, the interpolation will be exact for quadratic functions, and the interpolation will be third-order accurate. If $m \geq 5$ it will be fourth-order accurate. For reference we present here the original and the improved interpolating formulas that correspond to $m = 4$:

$$M_4(x) = \begin{cases} 0 & \text{if } |x| > 2 \\ \frac{1}{6}(-|x| + 2)^3 & \text{if } 1 \le |x| \le 2 \\ \frac{1}{6}(-|x| + 2)^3 - \frac{4}{6}(-|x| + 1)^3 & \text{if } |x| \le 1 \end{cases},$$

$$M_4'(x) = \begin{cases} 0 & \text{if } |x| > 2 \\ \frac{1}{2}(2 - |x|)^2(1 - |x|) & \text{if } 1 \le |x| \le 2 \\ 1 - \frac{5x^2}{2} + \frac{3|x|^3}{2} & \text{if } |x| \le 1 \end{cases}. \qquad (7.2.11)$$

This last formula has been used with significantly improved results in SPH (for Smooth Particles Hydrodynamics) simulations by Monaghan [155, 156]. We will see below that it is also a very efficient tool for vortex simulations in two and three dimensions. The function M_4' is plotted in Figure 7.6.

7.2.3. Ordinary Interpolation Formulas

In this type of formula the goal is to treat directly algebraic system (7.2.6), prescribing the exactness of the interpolation for a given degree while somewhat relaxing the smoothness constraint. If we set $W_i = W(x - x_i)$, this system can be rewritten as

$$\sum_{i=1}^{P} W_i = 1,$$

$$\sum_{i=1}^{P} x_i W_i = x.$$

. . .

If P is the number of grid points in the support of W, we have P parameters allowing us, in principle, to satisfy up to P moment conditions. Rewriting W_i in terms of W then yields the desired interpolation function. For $P = 2$ we find (again) the tent function, which allows us to conserve the linear impulse and is second order. For $P = 3$, we obtain the third-order interpolation function

$$\Lambda_2(x) = \begin{cases} 1 - x^2, & \text{if } 0 \le x < 1/2 \\ (1 - x)(2 - x)/2, & \text{if } 1/2 \le x < 3/2 \\ 0, & \text{otherwise} \end{cases}. \qquad (7.2.12)$$

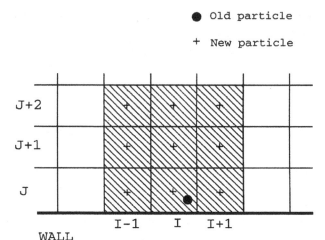

Figure 7.7. Remeshing in a bounded domain.

This function has been successfully used by several authors [125, 187]. For $P = 4$, the result is Everett's fourth-order formula:

$$\Lambda_3(x) = \begin{cases} (1 - x^2)(2 - x)/2, & \text{if } 0 \le x < 1 \\ (1 - x)(2 - x)(3 - x)/6, & \text{if } 1 \le x < 2. \\ 0, & \text{otherwise} \end{cases}$$

One interesting feature of this construction is that it allows some flexibility in dealing with bounded domains (so far all our interpolation formulas were derived in the absence of boundaries).

When boundaries are present the remeshing procedure is complicated as the new mesh points have to be outside the body. It is obvious then that the schemes used for an unbounded domain have to be modified for particles that are located in an (I, J) cell that is adjacent to the boundary. A scheme similar to the Λ_2 scheme described above requires again nine points and conserves the same quantities as for the unbounded case. In Figure 7.7 the nine cells affected by a cell adjacent to the wall are shown. The interpolating kernel is again the product of two one-dimensional forms, but now

$$\Lambda(x, y) = \Lambda_I(x)\Lambda_J(y)$$

with $\Lambda_I = \Lambda_2$, and

$$\Lambda_J = \begin{cases} 1 - 3/2v + 1/2v^2, & \text{for cells } J \\ v(2 - v), & \text{for cells } J + 1 \\ v(v - 1)/2, & \text{for cells } J + 2, \\ 0, & \text{for all other } J \end{cases}$$

where $v = (x - x')/h$, x' is the center of the cell next to the boundary and J here denotes the off-boundary direction.

Note that the kernels derived so far in this section do not have continuous derivatives. The second-order kernel is not even continuous. This implies that when interpolating quantities having large fluctuations for small particle separation they might introduce large interpolation errors from small errors in the actual particle locations. As a consequence one has to be careful when using these interpolation kernels not to allow too important distortions in the particles, which implies frequent regridding. By contrast, the M_4' scheme [formula (7.2.11)] achieves the same order of accuracy, but with a greater smoothness. (It is of class C^1.) This makes it more flexible regarding the frequency of remeshing.

To illustrate this point, we show several calculations of the evolution of an elliptical vortex patch. The dynamics of this particular flow, already considered in Section 2.4, eventually produces very high strain that results in ejection of thin filaments of vorticity. It is impossible to resolve reasonably these filaments without regridding the particles. Figure 7.8 shows vorticity values along the principal axis of the ellipse obtained with the Λ_2 scheme and the M_4' scheme. Both formulas were used with a frequency of one remeshing every four time steps. The Λ_2 scheme does allow us to capture the filaments but produces overshoot and oscillations that are avoided by the M_4' scheme. Figure 7.9 shows the number of particles as a function of time and the enstrophy obtained with the remeshing formulas. They indicate that the smoother results obtained by the M_4' formula do not go with an increased spreading of the vorticity. Both formulas lead to an enstrophy decay of $\sim 4\%$, which is fairly low, given the steep gradients produced by its dynamics and the small number of points in this simulation. Figure 7.10 shows a comparison of how the minimal numerical dissipation introduced by the M_4' formula allows high-resolution vortex simulations, indicating that an initially elliptical vorticity profile may relax in an inviscid flow to a nonaxisymmetric configuration in agreement with the analysis and experiments of Driscoll and Fine [74]. More details on these calculations can be found in Ref. 127.

The case of a circular patch, already considered several times in this chapter, although less severe than the ellipse just considered, allows us to quantify the accuracy of the remeshing procedure. Figure 7.11 compares the L^2 particle velocity error for $\omega_0(\mathbf{x}) = (1 - |\mathbf{x}|^2)^3$ when no-remeshing, the Λ_2 remeshing, and the M_4' remeshing formulas are used. These curves indicate that, when the initial grid is refined, the M_4' scheme does allow us to retain the gain in the velocity accuracy for all times.

A conclusion of these tests and of other simulations for a variety of two- and three-dimensional flows is that the M_4' formula provides a fairly good

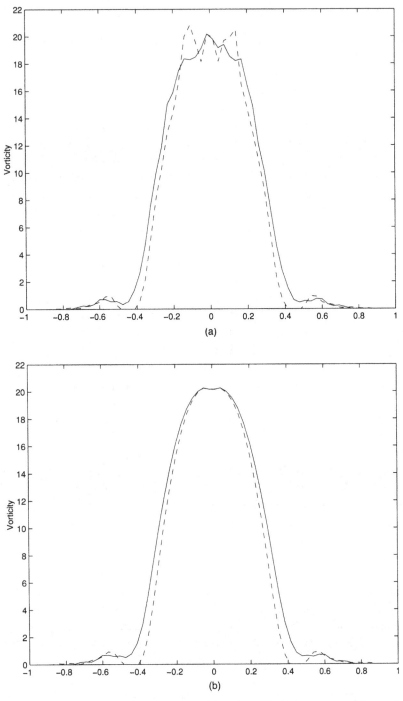

Figure 7.8. Effect of remeshing on the vorticity of an elliptical vortex: vorticity on the axes of (a) an ellipse, remeshing formula Λ_2 (7.2.12); (b) remeshing formula M_4' (7.2.11) (courtesy of M. L. Ould-Salihi).

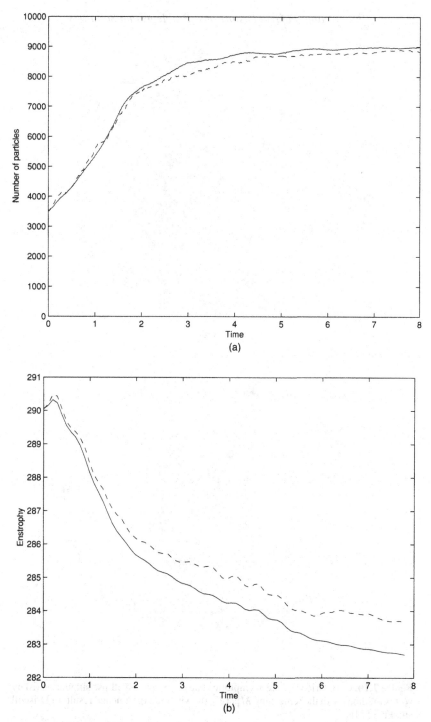

Figure 7.9. Effect of remeshing on the vorticity of an elliptical vortex: (a) number of particles, (b) enstrophy for Λ_2 (dashed curves) and M_4' (solid curves) formulas (courtesy of M. L. Ould-Salihi).

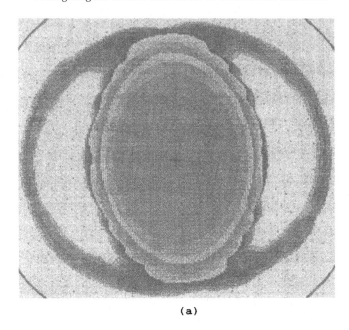

(a)

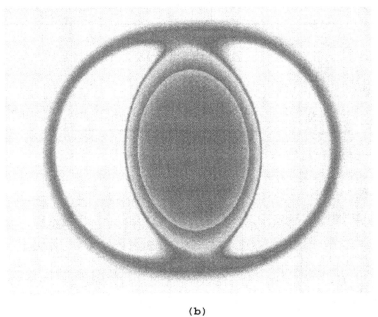

(b)

Figure 7.10. Comparison of the asymptotic state of an inviscid ellipse simulation (b) by vortex methods with the remeshing M_4' [127] (a) with the experimental results of Driscoll and Fine [74].

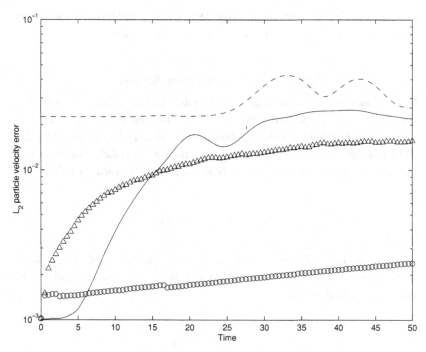

Figure 7.11. Error curves for $\omega_0(\mathbf{x}) = (1 - |\mathbf{x}|^2)^3$: no remesh, $h = 0.1$ (solid curve); no remesh, $h = 0.05$ (dotted curve); remeshing Λ_2 ($\triangle$); and remeshing M'_4 ($\bigcirc$).

compromise between accuracy and smoothness. Near walls, it is possible to combine the M'_4 scheme in the directions parallel to the wall and the Λ_2 scheme in the direction normal to the wall. Note that, in the case of no-slip boundary condition, the sensitivity of the Λ_2 scheme to excessive distortion in the particle distribution has minor effects in this case when it is used in the normal direction: By the continuity condition, the normal derivative of the normal component of the velocity vanishes at the boundary, and thus the strain essentially acts in the directions parallel to the wall.

To conclude this subsection, let us mention several side effects of remeshing techniques. In three-dimensional calculations one observes that, besides maintaining an accurate grid, remeshing produces subgrid dissipation that facilitates underresolved calculations. For viscous calculations, it allows the vorticity to spread its support. The remeshing frequency has then to satisfy $\nu\Delta t_r \leq \varepsilon^2$. Moreover, if blobs of variable size are used (see Subsections 2.6.3 and 5.4.3) regridding is crucial, even in the absence of distortion in the particle distribution, to maintain the overlapping condition for both the Biot–Savart law and the diffusion. Finally, when vortex methods on different grids are combined through domain-decomposition techniques (see Section 8.3) regridding provides a very simple way to transfer vorticity from one domain to another.

7.2.4. The ALE Remeshing Scheme

For completeness, let us describe an alternative technique for the rezoning of the distorted grid associated with the particles. This technique may be devised based on a technique initially applied to the Arbitrary Lagrangian–Eulerian (ALE) method. In this interpolation the regridding procedure may be carried out with an integral formulation for the remapping of the vorticity field, which is

$$\tilde{\Gamma}_i = \int \int_{A_i} \omega(\mathbf{x}) \, d\mathbf{x},$$

where $\tilde{\Gamma}_i$ is the circulation assigned to the particle in cell A_i of the new mesh and $\omega(\mathbf{x})$ is the vorticity of the old (distorted) mesh. The above integral may be expressed in discrete form as

$$\tilde{\Gamma}_i = \sum_j \Gamma_j \frac{A_{ij}}{A_j},$$

where A_{ij} denotes the overlapping area of the old cell i and the new cell j. This remeshing procedure involves then the computation of those overlap areas. An efficient algorithm for a piecewise constant vorticity field is presented by Dukowicz and Kodis [76].

By Stokes formula, the surface integral is translated into a line integral,

$$\tilde{\Gamma}_i = \oint_{C_k} \mathbf{u} \cdot d\mathbf{l},$$

along the contour C_k that defines the overlap region of the old and the new cell. We may efficiently compute this integral by traversing first the cells of the old grid and then the cells of the new grid and accordingly adding the contributions.

In comparison with the methods described in Section 7.2, the ALE technique is more complicated, hard to vectorize, and introduces additional numerical dissipation as it conserves only the circulation of the flow. However, it seems advantageous when one is dealing with remeshing of a vortex field around a complex configuration. There the interpolations described above encounter problems near the boundaries as they have to be modified for each geometry. The ALE method does not care about the specifics of the boundary as it deals directly with the particle cells that may have any geometric configuration.

8

Hybrid Methods

In numerical simulations it is desirable to use numerical methods that are well suited to the physics of the problem at hand. As the dominant physics of a flow can vary in different parts of the domain, it is often advantageous to implement hybrid numerical schemes.

In this chapter we discuss hybrid numerical methods that combine, to various extents, vortex methods with Eulerian grid-based schemes. In these hybrid schemes, Lagrangian vortex methods and Eulerian schemes may be combined in the same part of the domain, in which each method is used in order to discretize different parts of the governing equations. Alternatively, vortex methods and grid-based methods can be combined in the same flow solver, in which each scheme resolves different parts of the domain. In this case we will discuss domain-decomposition formulations. Finally we consider the case of using different formulations of the governing equations in different parts of the domain. In that context we discuss the combination of the velocity–pressure formulation (along with grid-based methods) and the velocity–vorticity formulation (along with vortex methods) for the governing Navier–Stokes equations.

For simplicity, we often use in this chapter the terminology of finite-difference methods but it must be clear that in most cases the ideas can readily be extended to other Eulerian methods, such as finite-element or spectral methods.

One of the attractive features of vortex methods is the replacement of the nonlinear advection terms with a set of ordinary differential equations for the trajectories of the Lagrangian elements, resulting in robust schemes with minimal numerical dissipation. One class of hybrid vortex methods that tries to retain this feature is the Vortex-In-Cell (VIC) scheme, which was introduced in 1973 by Christiansen [55]. In VIC calculations, an Eulerian grid can be implemented in order to compute efficiently the velocity field on the Lagrangian particles. Moreover this Eulerian grid can be used in order to compute diffusion and baroclinic terms in the governing equations. In the first case, the goal is to obtain a fast computation of the particle velocities in regular bounded domains, whereas in the second case, the goal is to take advantage of the efficiency of Eulerian schemes to deal with second-order elliptic or hyperbolic problems.

Another class of hybrid method, which we term Lagrangian–Eulerian domain-decomposition methods, use high-order grid methods and vortex methods in different parts of the domain. For example, a finite-difference scheme can be implemented near solid boundaries, and vortex methods can be implemented in the wake in order to provide the flow solver with accurate far-field conditions. The fact that Eulerian methods can offer more flexibility than vortex methods to deal with viscous boundary conditions, in particular when they are based on a suitable velocity–pressure formulation, can be an attractive feature of Lagrangian–Eulerian domain-decomposition methods.

In the first section we describe various interpolation techniques that are present in all hybrid methods, as they are used to transfer vorticity (or more generally any Lagrangian quantity carried by the particles) between the particles and the fixed grid. In Section 8.2 we discuss VIC methods for inviscid or viscous flows, and finally in Section 8.3, we describe several techniques for particle-grid domain-decomposition algorithms.

8.1. Assignment and Interpolation Schemes

8.1.1. General Setting

In this chapter we denote by $(\mathbf{x}_i)$ the vertices of a fixed Eulerian mesh and by $(\mathbf{x}_p)$ the locations of particles with volumes v_p carrying vorticity values ω_p. The particle approximation of a vorticity field ω is then

$$\omega^h(\mathbf{x}) = \sum_p v_p \omega_p \delta(\mathbf{x} - \mathbf{x}_p).$$

The problem of transferring vorticity values from the particles to the grid can be viewed as a particular case of the general interpolation procedures presented in Chapter 7. However, there are two features in the assignment problem that require some further discussion. First, while in Chapter 7 we were interested in interpolating circulations, particle-grid methods have to deal with pointwise vorticity values. Moreover, remeshing constraints were imposed only by particle considerations while in particle-grid methods one has also to incorporate constraints imposed by the type of the grid-based solver. These in turn depend on the equations and the domain that is discretized by the Eulerian grid.

The definition of assignment schemes starts with the choice of a grid-based family of functions (or filters) ϕ_i satisfying

$$\sum_i \phi_i \equiv 1. \tag{8.1.1}$$

In practice, each function ϕ_i has a small compact support around the grid point $\mathbf{x}_i$. If the grid is constructed through tensor products of one-dimensional meshes, with uniform mesh size ε, we can obtain the functions ϕ_i from a single function ϕ by writing

$$\phi_i(\mathbf{x}) = \phi\left(\frac{\mathbf{x} - \mathbf{x}_i}{\varepsilon}\right).$$

The assignment scheme can then be summarized as

$$\omega_i = \frac{1}{V_i}\sum_p v_p \omega_p \phi_i(\mathbf{x}_p), \tag{8.1.2}$$

where V_i denotes the volume (that is, ε^d in the case of a Cartesian mesh in d dimensions) around the grid point $\mathbf{x}_i$. The first essential requirement concerning the assignment scheme is its conservativity, which results from Eq. (8.1.1):

$$\sum_i V_i \omega_i = \sum_p v_p \omega_p.$$

Conversely, interpolation of a grid quantity onto a distribution of particles located at points $\mathbf{x}_p$ can be achieved through

$$\omega_p = \sum_i \omega_i \phi_i(\mathbf{x}_p). \tag{8.1.3}$$

When the points $\mathbf{x}_i$ are the nodes of a triangulation, a natural choice for the functions ϕ_i is the basis of Lagrange finite elements related to this triangulation. These are functions that are piecewise continuous polynomials and satisfy $\phi_i(\mathbf{x}_j) = \delta_{ij}$.

In the case of a Cartesian mesh, the basis functions ϕ are usually constructed by successive convolutions of the top-hat function (χ) in the square of size ε centered at $\mathbf{x}_i$. These functions constitute the so-called B-splines, already seen in Section 7.2 (see Figure 7.6).

The choice $\phi = \chi \star \chi$ yields a continuous piecewise bilinear function. In this case each particle distributes its weight among the four nearest grid points, with rates proportional to the respective distances from these points. This scheme, which was originally used in Ref. 55, is known as the area-weighting [or Cloud-In-Cell (CIC)] scheme, as the rate of weight assigned from a particle to a grid point can be found by measurement of the amount of overlapping between the volume of the particle and the volume of each grid cell. The next term in this hierarchy of functions is $\phi = \chi \star \chi \star \chi$, which yields a C^1 piecewise continuous quadratic function for each variable. The related assignment and interpolations are generally referred to as the Triangular-Shaped Cloud (TSC) (see for example Ref. 69).

It is important to point out that the assignment schemes related to Cartesian meshes can be easily extended to other mesh configurations, for example polar meshes. Besides, an important case is meshes in rectangular geometries with refinement along one or more directions that can be mapped onto Cartesian meshes. We denote by F the mapping between the physical space and the Cartesian mesh and set

$$\hat{\mathbf{x}}_p = F(\mathbf{x}_p), \quad \hat{\phi}_i = \phi_i \circ F^{-1}.$$

Then formulas (8.1.2) and (8.1.3) become, respectively,

$$\omega_i = \frac{1}{V_i} \sum_p v_p \omega_p \hat{\phi}_i(\hat{\mathbf{x}}_p), \quad \omega_p = \sum_i \omega_i \hat{\phi}_i(\hat{\mathbf{x}}_p).$$

It should be noted that all the schemes described above lead to assignment schemes that conserve, in addition to the circulation, the linear impulse of the vorticity field. From Eq. (8.1.2) we get

$$\sum_i V_i \mathbf{x}_i \omega_i = \sum_p v_p \omega_p \sum_i \mathbf{x}_i \phi_i(\mathbf{x}_p).$$

Next we observe that, for all values of $\mathbf{x}$,

$$\sum_i \mathbf{x}_i \phi_i(\mathbf{x}) = \mathbf{x}, \qquad (8.1.4)$$

which gives

$$\sum_i V_i \mathbf{x}_i \omega_i = \sum_p v_p \omega_p \mathbf{x}_p.$$

Note that, in case of a finite-element formulation, Eq. (8.1.4) is satisfied as soon as the finite-element space contains polynomials of degree at least 1. A numerical comparison of several filters, including the CIC and the TSC filters, is done in Ref. 79.

The conservation of higher moments requires higher-order accurate assignment schemes and the use of finite elements that would contain polynomials at least of degree 2 (in this case the x_i would include, besides the vertices of the triangles, points in the edge of the triangulation), or, in the case of a Cartesian mesh, the use of assignment functions of high order as designed in Section 7.2.

8.1.2. Stability and Accuracy Considerations

Besides accuracy and conservativity, stability is an important requirement for interpolation schemes. Numerical oscillations produced during the interpolation of quantities between grid and particles can accumulate and may result in numerical instabilities. This has indeed been a major difficulty in the original Particle-In-Cell algorithm introduced by Harlow in the context of gas dynamics.

The usual way to avoid oscillations is to ensure that the assignment scheme is enstrophy decreasing. Brackbill and Ruppel [34] (see also Ref. 33) have suggested to combine assignment formula (8.1.2) with a calculation of local cell volumes on the grid by using the same assignment kernel, namely,

$$V_i = \sum_p v_p \phi_i(\mathbf{x}_p). \tag{8.1.5}$$

Replacing the cell volumes by these values is clearly a consistent approximation as it amounts to a quadrature formula for the basis functions ϕ_i. In the case of finite elements or for grids that can be mapped to a Cartesian grid, we can compute volumes by using either the finite-element basis or by mapping first the particles to the Cartesian geometry and then by using the Cartesian assignment basis, leading to volume formulas similar to Eq. (8.1.5). Vorticity values on the grid points are then obtained by Eq. (8.1.2).

Observe that with this formula, modifying the volume evaluation does not affect the circulations $V_i \omega_i$ on the grid. However, volume calculation (8.1.5) has the advantage of guaranteeing a grid enstrophy ($\sum_i V_i \omega_i^2$) that does not exceed the particle enstrophy ($\sum_p v_p \omega_p^2$).

Proposition 8.1.1. *Scheme (8.1.2)–(8.1.5) is conservative and L^2 stable in the sense that*

$$\sum_i V_i \omega_i = \sum_p v_p \omega_p, \tag{8.1.6}$$

$$\sum_i V_i |\omega_i|^2 \leq \sum_p v_p |\omega_p|^2. \tag{8.1.7}$$

Proof. For Eq. (8.1.6) we just observe that, from Eq. (8.1.2),

$$\sum_i V_i \omega_i = \sum_i \sum_p v_p \omega_p \phi_i(\mathbf{x}_p) = \sum_p v_p \omega_p \sum_i \phi_i(\mathbf{x}_p), \qquad (8.1.8)$$

but $\sum_i \phi_i \equiv 1$, so we get the desired identity. For the stability estimate we write

$$\sum_i V_i |\omega_i|^2 = \sum_i V_i^{-1} \sum_{p,q} v_p v_q \omega_p \omega_q \phi(\mathbf{x}_p) \phi(\mathbf{x}_q).$$

Since $\omega_p \omega_q \le 1/2(\omega_p^2 + \omega_q^2)$, this gives

$$\sum_i V_i |\omega_i|^2 \le \sum_i V_i^{-1} \sum_p v_p |\omega_p|^2 \phi(\mathbf{x}_p) \sum_q v_q \phi(\mathbf{x}_q),$$

that is,

$$\sum_i V_i |\omega_i|^2 \le \sum_i V_i^{-1} \sum_p v_p |\omega_p|^2 \phi(\mathbf{x}_p) V_i \le \sum_i \sum_p v_p |\omega_p|^2 \phi_i(\mathbf{x}_p).$$

Using Eq. (8.1.1) again leads now to relation (8.1.7).   □

As a side effect, the volume formula also improves the accuracy of the assignment scheme. Replacing the volume of the cell ε^d by V_i compensates for the computation of grid values when the particle distribution gets distorted. This can be seen by considering the effect of this scheme on a constant vorticity field. With formula (8.1.5), the vorticity field will be recovered exactly, irrespectively of how particles are strained. Another situation in which the improvement is readily seen is near boundaries. A simple-minded implementation of formula (8.1.2) with uniform volume values would lead to $O(1)$ errors, because it misses particles that, in a smooth extension of the vorticity, would lie on the opposite side of the boundary. By not accounting also for these particles, formula (8.1.5) introduces a compensation effect in Eq. (8.1.2). Actually this procedure amounts to an extension of the particle distributions by particles with constant weights outside the computational domain, resulting in a first-order assignment scheme.

In the case of a steep vorticity profile, the accuracy of the assignment scheme can be improved by translating Beale's iterative method, described in Section 7.1, into the particle-grid framework. In this context, this method is designed to ensure that successive assignment–interpolations do not spread the vorticity onto an increasing number of points. It is also efficient to assign vorticity values at grid points near a boundary.

To simplify the forthcoming discussion let us denote by Ξ a particle distribution made of points $\boldsymbol{\xi}_p$ with volumes v_p, and by $A(\Xi)$ and $I(\Xi)$ the assignment and the interpolation operators, respectively. If $\mathcal{V} = (\omega_p)$ is a vector containing the particle values of a function ω, the new particle quantities after an assignment–interpolation sequence can be written as

$$\mathcal{V}' = \mathcal{I}(\Xi)\mathcal{A}(\Xi)(\mathcal{V}). \tag{8.1.9}$$

Of course, since the assignment and the interpolation have been defined in a consistent way, one has $\mathcal{V}' \simeq \mathcal{V}$, that is,

$$I(\Xi)A(\Xi) \simeq Id. \tag{8.1.10}$$

Our goal now is to find new particle quantities $\tilde{\mathcal{V}}$ such that

$$\mathcal{V} = I(\Xi)A(\Xi)(\tilde{\mathcal{V}}). \tag{8.1.11}$$

In view of relation (8.1.10) and rewriting

$$I(\Xi)A(\Xi) = Id + [I(\Xi)A(\Xi) - Id],$$

it is natural to compute $\mathcal{V}'$ as the limit of the following iterations:

$$\begin{aligned} \mathcal{V}^{n+1} &= \mathcal{V} + [Id - I(\Xi)A(\Xi)]\mathcal{V}^n \\ &= \mathcal{V} + \mathcal{V}^n - I(\Xi)A(\Xi)\mathcal{V}^n. \end{aligned} \tag{8.1.12}$$

If the number of particles and the number of grid points are of the order of N, each iteration amounts to $O(N)$ operations. As already noted in Section 7.1, problem (8.1.11) is the discrete analog of a deconvolution process and as such it is ill posed. However, as a finite number of iterations (8.1.12) are performed, this deconvolution affects only a finite range in the Fourier spectrum. In the context of VIC methods, this algorithm is a way to restore information at a subgrid level, depending on the number of iterations that are performed.

It is worthwhile to note in this example that the relevance of Eq. (8.1.12) is based on assumption (8.1.10). Thus, in the presence of boundaries, it is important to use the assignment scheme with volume calculation (8.1.5) and not uniform volume values that would produce an $O(1)$ error at the boundary. Finally, let us mention that in practice the gain obtained through Eqs. (8.1.5) and (8.1.12) is limited to the case when particles are not regridded. When regridding techniques described in Section 7.2 are used, Eq. (8.1.2) with $V_i = \varepsilon^d$ turns out to give a satisfactory accuracy.

8.2. Vortex-In-Cell Methods

In this section we focus on the case in which grid and particle discretizations coexist throughout the computational domain, but are used to solve different parts of the governing equations. We first investigate the case of inviscid flows, in which the grid is used to compute particle velocities, then we address viscous flows, in which a grid solver is also used to resolve the diffusion terms of the equations.

8.2.1. Inviscid Flows

For inviscid flows the grid is used only to compute efficiently the particle velocity field through the computation of the streamfunction. In two dimensions, for every time step the algorithm can be summarized as follows.

- Assign vorticity to the grid by using Eq. (8.1.2).
- Solve $\Delta \Psi = -\omega$ on the grid.
- Differentiate the streamfunction on the grid to get velocities on the grid.
- Interpolate velocity values from the grid to the particles.
- Move the particles.

This scheme can handle efficiently wall boundary conditions, either Dirichlet (for the no-through-flow condition) or periodic, when the geometry of the domain is regular enough so that fast Poisson solvers are available. In order to restore the subgrid information carried by the particles, in the spirit of Particle-Particle, Particle Mesh algorithm, (see Hockney & Eastwood) Anderson [5] corrects this scheme by considering the local interaction of particles. This provides a fast and accurate method for particle velocity calculations.

Note that a CIC assignment scheme does not need any correction at the boundary to correctly assign vorticity from any particle in the computational domain onto the grid. If a smoother assignment scheme is used (e.g., TSC), the vorticity carried by particles in the first half of the cell would spread to a grid point outside the domain. In this case, iterative procedure (8.1.12) can be used to restore accurate values of the vorticity on the first grid points.

The same techniques can be used without modifications in three dimensions with the vortex filament method [69]. For three-dimensional vortex particle schemes, a straightforward extension of the method is to compute the stretching by differentiation of the velocity on the grid and then interpolation on the particles. A conservative variant of this method consists of writing the stretching term in the conservative form div $(\omega : \mathbf{u})$. In this method, grid values of $\mathbf{u}$ and ω are multiplied and then differentiated on the grid and finally interpolated on

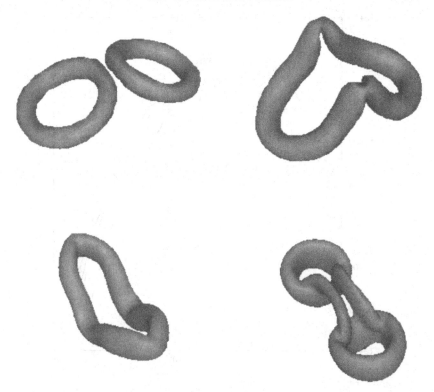

Figure 8.1. Successive stages (from left to right, top to bottom) of the reconnection of two rings of opposite circulation by a VIC method.

the particles. Note that this scheme can be seen as the VIC equivalent of the conservative grid-free scheme described in Chapter 3 [Eq. (3.1.9)]. Figure 8.1 shows a time sequence of vorticity isosurfaces obtained with this method for the classical problem of the reconnection of two vortex rings of opposite circulation in a periodic box. The bridge linking the two rings after the second reconnection ultimately disappears, leaving two rings at an angle of $\sim 90°$ from the original ones. This simulation corresponds to a Reynolds number of 400. Diffusion is solved by a PSE scheme, and remeshing with the M_4' scheme [formula (7.2.11)] is performed at every time step. The particle spacing would correspond to a 64^3 discretization of the computational box. At the end of the calculation (after 300 steps), the vorticity support is covered by $\sim 30, 000$ particles.

It is sometimes believed that particle-grid transfers induce some numerical dissipation in VIC codes. To measure to which extent this is true, we give in Figure 8.2 a comparison, for a classical isotropic homogeneous turbulence experiment, between a VIC and a spectral method at the same 128^3 resolution.

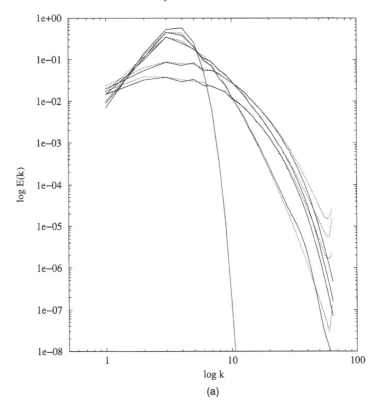

Figure 8.2. Homogeneous isotropic turbulence by a spectral method (solid lines) and a VIC method (dashed lines), (a) spectra; (b) energy curves; (c) enstrophy curves.

The initial turbulent field has a peak at low wave numbers and random phases. The Reynolds number, based on the Taylor microscale, is 75. Figure 8.2(a) shows the spectra at times 0, 1, 2, 6, and 10. Figure 8.2(b) and (c) show the energy and enstrophy evolutions, respectively. All the other statistics, as well as the coherent structures in the vorticity field, are in excellent agreement.

Given the fact that it is in general considered that, for this type of flow, grid-based methods require at least twice as many points in each direction to agree with spectral methods, the VIC method can be considered as rather accurate. Note that the VIC calculation was remeshed at every time-step. These simulations thus serve also as a check of the subgrid-scale effects of remeshing.

The numerical analysis of VIC methods can be done within the framework already used for the numerical analysis of two- and three-dimensional, grid-free vortex methods. If we represent the grid-particle scheme used to compute the particle velocities in the operator form,

$$\mathbf{u}_h = \mathcal{S}_h \boldsymbol{\omega}_h,$$

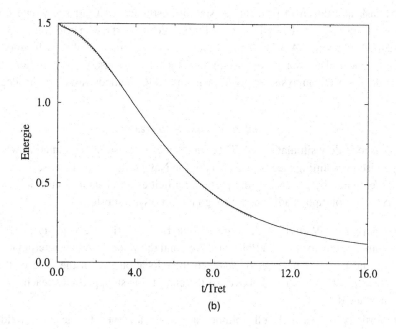

(b)

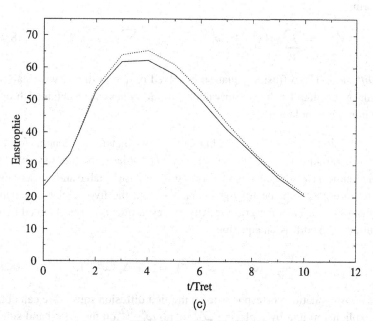

(c)

Figure 8.2. (*Continued*)

the task is reduced to proving L^p stability estimates of Calderon's type for the operator S_h, similar to those valid for the Biot–Savart law in grid-free vortex methods (Theorem A.3.1). Such estimates are in general available by formulation of the grid Poisson solver in a finite-element framework. We refer to Ref. 60 for details on an analysis for two-dimensional VIC methods based on this idea.

8.2.2. Viscous Schemes

In viscous flow simulations, VIC techniques can be used in conjunction with the viscous splitting (see Section 5.1) of the Navier–Stokes equations.

We denote by ω_p^n and $\mathbf{x}_p^n$ the particle vorticities and locations at time $t_n = n\Delta t$. The solution is advanced through the following substeps:

- *Convection*: Vorticity values are assigned to the grid, the velocity field is computed by means of a Poisson solver, and the velocity field is interpolated on the particles in order to advance their locations, as in the case of the inviscid flows described above. At the end of this step, particles reach their new locations $\mathbf{x}_p^{n+1}$.

- Vorticity values and cell volumes are again assigned to the grid, yielding quantities that we denote respectively by $\omega_i^{n+1/2}$ and V_i^{n+1} through the formulas

$$V_i^{n+1} = \sum_p \phi_i\left(\mathbf{x}_p^{n+1}\right), \quad \omega_i^{n+1/2} = \frac{1}{V_i^{n+1}} \sum_p \omega_p^n \phi_i\left(\mathbf{x}_p^{n+1}\right). \qquad (8.2.1)$$

- *Diffusion*: The diffusion equation is solved on the grid, and we obtain the final grid values for the vorticity field. Particle values can be obtained through another interpolation.

The diffusion step of this algorithm can be formulated in a general form by use of a variational methodology. However, in order to be more specific, we will assume a two-dimensional uniform grid and use a stable and conservative, finite-difference scheme for the Laplacian (e.g., the five-point box scheme) denoted by Δ_ε. Then the grid vorticity values at time t_{n+1} are obtained by the solution of the diffusion equation:

$$\frac{V_i^{n+1}}{\varepsilon^2}\left(\omega_i^{n+1} - \omega_i^{n+1/2}\right) = \nu\delta t\, \Delta_\varepsilon\left(\omega_i^{n+1}\right). \qquad (8.2.2)$$

The above equation corresponds to an implicit diffusion solver. We can obtain an explicit formula by replacing ω_i^{n+1} with $\omega_i^{n+1/2}$ on the right-hand side. It remains to update the particle vorticity values. Although it would seem natural to interpolate ω_i^{n+1} onto the particles, it is better computationally to interpolate vorticity increments rather than the values themselves. The final result is then given by

$$\omega_p^{n+1} - \omega_p^n = \sum_i \left(\omega_i^{n+1} - \omega_i^{n+1/2} \right) \phi_i \left(\mathbf{x}_p^{n+1/2} \right). \qquad (8.2.3)$$

We can see the motivation for this formula best by considering the case $v = 0$: in this case, Eqs. (8.2.2) and (8.2.3) do not modify the particle vorticity. On the other hand, the formula $\omega_p^{n+1} = \sum_i \omega_i^{n+1} \phi_i (\mathbf{x}_p^{n+1/2})$ will most likely introduce some numerical dissipation and thus modify the effective viscosity value in the overall algorithm. A variant of this method, used for example in Ref. 42 (see also Ref. 157), is to restart at the end of each diffusion step fresh particles at the grid location, by using the vorticity values ω_i^{n+1}. This technique, however, is still not free of the numerical dissipation introduced by assignment scheme (8.2.1). In addition, it requires using the same spacing for the grid and for the particles, something that violates the overlapping condition necessary to ensure the consistency of Eq. (8.2.1).

We now proceed to show that the algorithm defined by Eqs. (8.2.1)–(8.2.3) is conservative and stable. To check the conservativity we observe that, since the grid solver has been assumed conservative, by Eq. (8.2.2),

$$\sum_i V_i^{n+1} \omega_i^{n+1} = \sum_i V_i^{n+1} \omega_i^{n+1/2},$$

and thus, by Eq. (8.2.3),

$$\sum_p v_p \left(\omega_p^{n+1} - \omega_p^n \right) = \sum_i \left(\omega_i^{n+1} - \omega_i^{n+1/2} \right) \sum_p v_p \phi_i \left(\mathbf{x}_p^{n+1/2} \right)$$

$$= \sum_i V_i^{n+1} \left(\omega_i^{n+1} - \omega_i^{n+1/2} \right) = 0.$$

We now turn to the stability of the algorithm (we follow here the proof given in Ref. 170). We will see that this property heavily relies on the fact that cell volumes are computed on the basis of the particle locations. Let us multiply Eq. (8.2.2) by ω_i^{n+1} and sum over i. We get

$$\sum_i V_i^{n+1} |\omega_i^{n+1}|^2 = \sum_i V_i^{n+1} \omega_i^{n+1/2} \omega_i^{n+1} + \frac{v \Delta t}{\varepsilon^2} \sum_i \Delta_\varepsilon \omega_i^{n+1} \omega_i^{n+1}.$$

We then observe that, by assumption on the finite-difference diffusion solver, $\sum_i \Delta_\varepsilon \omega_i^{n+1} \omega_i^{n+1} < 0$. Writing $2\omega_i^{n+1/2} \omega_i^{n+1} \le |\omega_i^{n+1/2}|^2 + |\omega_i^{n+1}|^2$ then yields

$$\sum_i V_i^{n+1} |\omega_i^{n+1}|^2 \le \sum_i V_i^{n+1} |\omega_i^{n+1/2}|^2. \qquad (8.2.4)$$

Let us now check that the grid-particle interpolation procedure (8.2.3) satisfies

$$\sum_p v_p \left(\left| \omega_p^{n+1} \right|^2 - \left| \omega_p^n \right|^2 \right) \le \sum_i V_i^{n+1} \left(\left| \omega_i^{n+1} \right|^2 - \left| \omega_i^{n+1/2} \right|^2 \right). \qquad (8.2.5)$$

We first rewrite Eq. (8.2.3) as

$$\omega_p^{n+1} = \omega_p^n + \sum_i \phi_i \left(\mathbf{x}_p^{n+1} \right) \delta \omega_i^{n+1},$$

where $\delta \omega_i^{n+1}$ stands for $\omega_i^{n+1} - \omega_i^{n+1/2}$, whence

$$\left| \omega_p^{n+1} \right|^2 = \left| \omega_p^n \right|^2 + 2 \sum_i \phi_i \left(\mathbf{x}_p^{n+1} \right) \omega_p^n \delta \omega_i^{n+1}$$

$$+ \sum_{i,j} \phi_i \left(\mathbf{x}_p^{n+1} \right) \phi_j \left(\mathbf{x}_p^{n+1} \right) \delta \omega_i^{n+1} \delta \omega_j^{n+1}.$$

From Eq. (8.2.1) we deduce that

$$\sum_p v_p \left(\left| \omega_p^{n+1} \right|^2 - \left| \omega_p^n \right|^2 \right) = 2 \sum_i V_i^{n+1} \omega_i^{n+1/2} \delta \omega_i^{n+1}$$

$$+ \sum_{i,j} \phi_i \left(\mathbf{x}_p^{n+1} \right) \phi_j \left(\mathbf{x}_p^{n+1} \right) \delta \omega_i^{n+1} \delta \omega_j^{n+1}.$$

Writing $2 \omega_i^{n+1/2} = \omega_i^{n+1/2} + \omega_i^{n+1} - \delta \omega_i^{n+1}$ yields $2 \omega_i^{n+1/2} \delta \omega_i^{n+1} = \left| \omega_i^{n+1} \right|^2 - \left| \omega_i^{n+1/2} \right|^2 - \left| \delta \omega_i^{n+1} \right|^2$, and thus

$$\sum_p v_p \left(\left| \omega_p^{n+1} \right|^2 - \left| \omega_p^n \right|^2 \right) = \sum_i V_i^{n+1} \left(\left| \omega_i^{n+1} \right|^2 - \left| \omega_i^{n+1/2} \right|^2 \right)$$

$$+ \sum_{i,j} \sum_p v_p \phi_i \left(\mathbf{x}_p^{n+1} \right) \phi_j \left(\mathbf{x}_p^{n+1} \right) \delta \omega_i^{n+1} \delta \omega_j^{n+1}$$

$$- \sum_i V_i^{n+1} \left| \delta \omega_i^{n+1} \right|^2. \qquad (8.2.6)$$

But

$$\sum_{i,j} \sum_p v_p \phi_i \left(\mathbf{x}_p^{n+1} \right) \phi_j \left(\mathbf{x}_p^{n+1} \right) \delta \omega_i^{n+1} \delta \omega_j^{n+1}$$

$$\le \frac{1}{2} \sum_{i,j} \sum_p v_p \phi_i \left(\mathbf{x}_p^{n+1} \right) \phi_j \left(\mathbf{x}_p^{n+1} \right) \left(\left| \delta \omega_i^{n+1} \right|^2 + \left| \delta \omega_j^{n+1} \right|^2 \right)$$

$$= \sum_{i,j} \sum_p v_p \phi_i \left(\mathbf{x}_p^{n+1} \right) \phi_j \left(\mathbf{x}_p^{n+1} \right) \left| \delta \omega_i^{n+1} \right|^2$$

$$= \sum_i \sum_p v_p \phi_i \left(\mathbf{x}_p^{n+1} \right) \left| \delta \omega_i^{n+1} \right|^2 \sum_j \phi_j \left(\mathbf{x}_p^{n+1} \right) = \sum_i V_i^{n+1} \left| \delta \omega_i^{n+1} \right|^2.$$

Combining this last inequality with Eq. (8.2.6) leads to relation (8.2.5). By the stability of grid solver (8.2.4), this in turn implies that

$$\sum_p v_p \left| \omega_p^{n+1} \right|^2 \leq \sum_p v_p \left| \omega_p^n \right|^2,$$

which proves the stability of the overall algorithm.

Note that if an explicit grid solver is used instead of Eq. (8.2.3), to obtain relation (8.2.4) one would have in general to fulfill a stability constraint of the type $v\Delta t \leq C\varepsilon^2$.

VIC methods for inviscid or viscous flows may be viewed as an appealing alternative to pure grid-free vortex methods in simple geometries. In this case, finite-difference methods for the diffusion equation allow more flexibility compared with that of the methods described in Chapter 6 in the treatment of the no-slip boundary conditions. As we already pointed out, the use of a fast Poisson solver also enables fast velocity evaluations. Compared with pure finite-difference methods, VIC methods offer the advantage of a robust and accurate treatment of the convective part of the equations with time steps not constrained by convective CFL conditions. We refer to Ref. 133 for detailed numerical results on VIC methods, including systematic comparisons with high-order finite-difference techniques.

8.3. Eulerian–Lagrangian Domain Decomposition

The motivation for such techniques stems from the observation that the strengths and the weaknesses of grid-based and vortex schemes can be seen as complementary, depending on the physical problem. In certain cases it is advantageous to implement grid-based methods to resolve regions of the flow in which viscous effects are important while using a Lagrangian vortex method for the convection-dominated part of the domain.

For example, for flows around one or several obstacles, one possibility is to rely on existing grid-based codes to resolve the flow around each obstacle while using a vortex method in between the obstacles. Compared with a purely Eulerian method, this approach removes most of the geometrical constraints that would arise from the need to adjust a single mesh to several bodies (if these bodies are in relative motion, this geometrical constraint becomes even more severe, as a new geometry has to be meshed at every time step). Moreover, vortex methods provide an accurate way to implement far-field boundary conditions for the Eulerian scheme that otherwise would have to be modeled.

Compared with a pure vortex method, the hybrid scheme adds flexibility to the treatment of the no-slip boundary conditions, in particular if a velocity–pressure formulation is chosen near the boundaries. As we will see, it also removes the source of inaccuracy in the treatment of kinematic boundary conditions, pointed out in Chapter 4, that result from a vortex blob's overlapping with the solid boundaries. Finally, pushing the particle resolution away from the boundaries simplifies the particle remeshing strategies.

Several options can be considered for the Eulerian solver, depending on the chosen formulation – velocity–pressure or velocity–vorticity – and on the time advancing scheme – explicit or implicit. As for the vortex method, one could consider solving either the full Navier–Stokes equations or their inviscid approximations. This last option has been investigated in Ref. 61 and seems legitimate for high Reynolds numbers, as viscous effects are mostly linked to vorticity generation at the solid boundaries. However, combining viscous and inviscid models through domain decomposition requires some care in the definition of the interface condition. This additional cost does not seem justified by the negligible computational savings that inviscid vortex methods offer compared with that of the viscous schemes.

When designing domain-decomposition techniques, it is customary to distinguish between overlapping and matching domain decomposition. For our purpose, we consider only the first case. As we will see, because of the different nature of the solvers in each subdomain, it will greatly simplify the definition of consistent interface boundary conditions to allow the transfer of information between the subdomains, for a slight computational overhead compared with a matching technique. For a hybrid technique based on matching subdomains, we refer to Ref. 97.

Figure 8.3 sketches a typical domain decomposition with overlapping. For the sake of simplicity we focus on the domain decomposition around a single

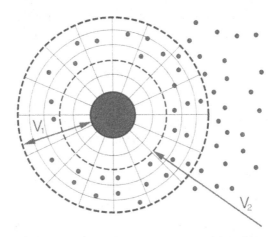

Figure 8.3. Eulerian–Lagrangian domain decomposition with overlapping.

body in a fluid domain V. The Eulerian and Lagrangian domains are denoted by V_1 and V_2, respectively, and S_1 and S_2 are the outer boundary of V_1 and the inner boundary of V_2, respectively. We first consider the case in which velocity–vorticity formulations are used in both domains and then turn to the case in which a velocity–pressure formulation is used in the Eulerian domain.

8.3.1. Velocity–Vorticity Domain Decomposition

In this case, both the Eulerian and the Lagrangian schemes need to handle two equations of a very different nature: an elliptic equation for the calculation of the velocity from the vorticity, and a convection–diffusion parabolic equation to update the vorticity at each time step. Following Ref. 61, we will clearly distinguish between the domain-decomposition techniques that are required for the solution to these two equations.

Computation of the Velocity from the Vorticity

In this section, we assume that the vorticity field is known both on the grid in V_1 and on the particles in V_2. The computation of the velocity then amounts to solving the elliptic system

$$\Delta \Psi = -\omega \qquad (8.3.1)$$

for the streamfunction Ψ. A popular domain decomposition procedure for this problem is the Schwarz alternating method. It consists in solving alternatively Eq. (8.3.1) in V_1 and V_2, each domain using as boundary condition the stream-function value obtained in the other domain at the preceding iteration. In the present case we are faced with the particular additional feature that the elliptic solvers are of a different nature in each domain: a grid solver in V_1, an integral solver in V_2, associated with the use of the Biot-Savart law. As seen in Chapter 4, an integral representation of the solution of Eq. (8.3.1) in V_2 is given by

$$\Psi(\mathbf{x}) = \int_{V_2} G(\mathbf{x} - \mathbf{y})\omega(\mathbf{y}) \, d\mathbf{y} + \int_{S_2} \frac{\partial G}{\partial \mathbf{n}_x}(\mathbf{x} - \mathbf{y})\Psi(\mathbf{y}) \, d\mathbf{y}$$
$$+ \int_{S_2} G(\mathbf{x} - \mathbf{y})\frac{\partial \Psi}{\partial \mathbf{n}}(\mathbf{y}) \, d\mathbf{y}. \qquad (8.3.2)$$

In principle this would require to determine the single layer potential $\partial \Psi / \partial \mathbf{n}$ at the boundary by solving an integral equation. In the context of a domain decomposition the algorithm can be further simplified by using the values of this potential, along with the values of Ψ, obtained from the finite-difference solution in V_1. This procedure was used in [61] and proved to converge in [3].

The integrals in the right-hand side of Eq. (8.3.2) are then discretized on the particles lying in V_2 and on fixed source points located on S_2.

One iteration of the Schwarz alternative method can be summarized as follows:

- Given Ψ on S_1, solve Eq. (8.3.1) in V_1.
- Deduce values for Ψ and $\partial \Psi / \partial n$ on source terms on S_2.
- Use these values to evaluate the right-hand side of Eq. (8.3.2) and compute Ψ on S_1.

After convergence of these iterations, the final velocity is obtained through the grid values in V_1 and by formula (8.3.2) in $V_2 - V_1$. Particle velocities in $V_2 \cap V_1$ are obtained by interpolation of the grid values.

Some comments are in order concerning the expression of the vorticity field to be used in Eq. (8.3.2). As we have seen in Chapter 4, the vortex method solution in V_2 can be either represented by particles or blobs. The blob representation can lead to inconsistency because of possible overlapping with the boundary S_2. In the present case, this inconsistency is irrelevant, provided the two interfaces are separated by at least a distance of the order of ε, because the difference between the velocity fields created by a point particle and a blob of size ε located in the neighborhood of S_2 vanishes away from S_2. This feature illustrates the flexibility that the domain-decomposition approach offers to the vortex method in the treatment of boundary conditions. This flexibility will be further apparent below in the treatment of the vorticity boundary condition and in regridding techniques.

Solution to the Vorticity Convection–Diffusion Equation

We now assume that the velocity is known everywhere at a given time step t_n, and our goal is to update the vorticity in both domains. The method relies on two main ingredients: the PSE scheme described in Section 5.4 and the regridding techniques discussed in Section 7.2.

Let us denote by ω_1^n the grid solution in V_1 and by $\omega_2^n = \sum_p v_p \omega_p^n \delta(\mathbf{x} - \mathbf{x}_p^n)$ the particle solution in V_2 at time t_n. We first assume that an explicit scheme is used in V_1. To update ω_1 on the interior nodes in V_1 requires the knowledge of the vorticity at time t_n on the grid points lying on S_1. Likewise, to update the particle solution in ω_2 on the basis of the particle equations

$$\frac{d\mathbf{x}_p}{dt} = \mathbf{u}(x_p, t), \quad \frac{d\omega_p}{dt} = \nu \Delta \omega_p,$$

one has to know the circulation and locations at time t_n for those particles that may affect the vorticity in V_2. If ρ is the size of the support of the kernel η used in PSE formula (5.4.6), one may easily estimate that these particles lie in a layer around S_2 of width $U \Delta t + \rho \varepsilon$, where U is an estimate of the maximum absolute value of the velocity. We subsequently call V_{12} this layer.

To obtain the missing information on these grid points on S_1 and particles in V_{12}, we have to resort to the particle-grid operators described in Section 8.1 in the following way: We obtain values of ω_1^n on S_1 by assigning to these grid points the circulation of nearby particles in V_2 (this of course requires that the distance between S_1 and S_2 be larger than the width of the assignment function). Conversely, vorticity values on particles in V_{12} are computed by interpolating grid values ω_1^n. It is worthwhile to emphasize that, by allowing these particles to play the role of the interface boundary condition for the vortex method, one simplifies the treatment of vorticity boundary conditions for the vortex method. The overlapping assumption allows us to take advantage of a natural extension of the vortex solution outside its computational domain, something that of course would not be possible if vortex methods were used all the way to the boundary. It allows us, for example, to bypass the integral formulation described in Subsection 6.3.3.

The method can be summarized as follows. For each time-step

- compute the velocity on the grid as well as on particles by using the method described in the previous paragraph,
- project particles onto the grid to get vorticity grid values on S_1 and interpolate from the grid to get particle circulations in V_{12},
- update particles in V_2 with a PSE scheme, and grid values in V_1.

If we consider the case in which the Eulerian method is based on an implicit treatment of the diffusion, the only difference with the explicit case is that the vorticity boundary condition on S_1 will be obtained by projection from the vortex solution obtained at the end, rather than at the beginning, of the time-step. In this case, the algorithm proceeds as follows:

- Compute velocity in the whole domain with the method of Subsection 8.3.1.
- Interpolate vorticity values at the particles in V_{12}.
- Move particles and solve for the diffusion in V_2.
- Project new particle vorticity to get vorticity boundary condition on S_1.
- Solve the implicit Eulerian scheme with this boundary condition.

Note that of course both the explicit and the implicit method require a vorticity boundary condition on the obstacle reflecting the no-slip condition. For a discussion of this topic in the context of finite-difference schemes we refer to Refs. 78, 110.

Regridding for Accurate Vorticity Transfer

A clear consistency requirement for the procedure just outlined is that particle locations yield accurate quadrature formulas in $V_2 \cup V_{12}$, so that grid information can be accurately translated onto particles and vice versa. In the original method, as described in Ref. 61, particles were continuously injected upstream of the obstacle, and the consistency of the interface conditions relied on the assumption that the flow would by itself carry particles in the vortical regions of the flow.

A more robust way to ensure a consistent particle discretization in the interface zone is by periodically remeshing the particle distribution. By using one of the interpolation formulas indicated in Section 7.2, one obtains a new consistent particle solution in V_2. That is due to the fact that V_2 does not contain blobs that would intersect S. After N time steps, particles have moved but still allow an accurate description of the vorticity in V_2 and transfer of vorticity through the procedure above outlined, provided that

$$|U|N\Delta t < d(S, S_1) - \rho\varepsilon. \qquad (8.3.3)$$

Inequality (8.3.3) allows us to estimate the frequency of remeshing needed. It is actually very likely that preserving a regular particle grid for the Navier–Stokes solver in V_2 would impose a more drastic restriction on the remeshing frequency.

The above argument leads to the same conclusion as for the wall boundary conditions: Using an Eulerian solver near the body allows us to bypass the technical difficulties that remeshing would meet in this region.

Figure 8.4 demonstrates that the domain-decomposition method ensures a smooth transfer of vorticity between the Eulerian and the Lagrangian subdomains. In the Eulerian domain a fourth-order compact finite-difference method was used [78] on a polar grid of 51×161 points. Grid-particle interpolation and particle regridding are done with the M_4' formula. The vortex domain starts at $r = 1.32$, and the finite-difference domain ends at $r = 2$. In the present simulation, the vorticity remains mostly attached to the obstacle, and the vortex solution merely provides an accurate far-field solution for the finite-difference solver.

To conclude this section let us emphasize that the techniques just described apply with little modifications to particle–particle domain decomposition. In this case one wishes to use particle discretizations of both the boundary layer and the wake, but with a different resolution. Typically particles would be remeshed,

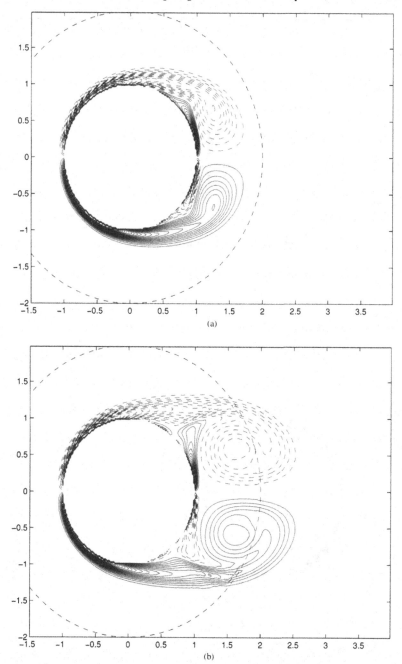

(a)

(b)

Figure 8.4. Time evolution (isovorticity lines) for the flow behind an impulsively started cylinder by the domain-decomposition method of Subsection 8.3.1. The Reynolds number is 550. The dashed circles indicate the outer limit of the finite-difference domain.

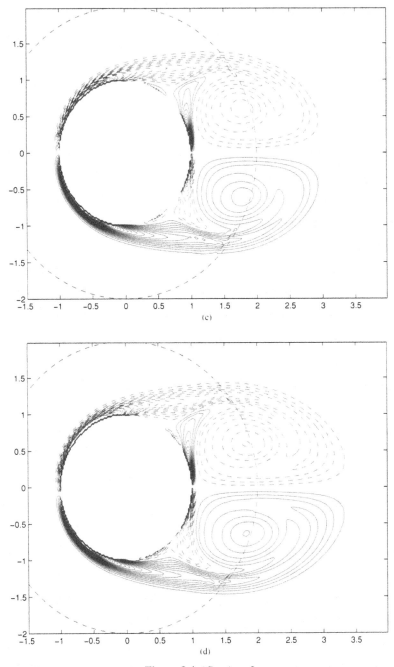

Figure 8.4. (*Continued*)

by means of local mappings, on a polar grid with refinement in the normal direction around the obstacles, on a uniform Cartesian grid in between the obstacles and on another polar grid in their wakes. Each subdomain is resolved by means of a vortex method with variable blobs, as discussed in Section 2.2 and Subsection 5.4.3 (see Figure 5.8). We achieve the transfer of boundary conditions from one subdomain to another simply by remeshing the particles in the overlapping zone on grids by using local mappings corresponding to the desired resolutions. A more detailed discussion on this topic can be found in Ref. 133.

8.3.2. Velocity–Pressure and Velocity–Vorticity Domain Decomposition

Although it involves different formulations of the fluid equations, this case is not more difficult to deal with than the previous one. In particular, the pressure term in the Eulerian domain does not require specific interface boundary conditions.

Here again we have to distinguish between explicit and implicit solvers in the Eulerian zone. The choice between these two options is in general dictated by the type of discretization chosen there (for example a finite-element method will be based on an implicit treatment of the diffusion, whereas finite-volume methods favor fully explicit solvers).

Let us first discuss the case of an implicit Eulerian solver. In this case, to advance the solution from time t_n to time t_{n+1} in V_1, one has to solve a system of the form

$$(Id + \nu \Delta t \Delta)\mathbf{u}^{n+1} + \nabla p^{n+1} = F^n, \quad \nabla \cdot \mathbf{u}^{n+1} = 0, \tag{8.3.4}$$

where F^n depends on $\mathbf{u}^n$ and the nonlinear term evaluated at time t^n. One thus needs a boundary condition for $\mathbf{u}^{n+1}$ at the interface S_1. On the other hand, the (explicit) vortex scheme allows us to update the vorticity field in S_2 to time t_{n+1} by using velocity values and vorticity values on particles around S_2 at time t_n (we assume for the time being that forward Euler time stepping is used to push particles and calculate circulations). It is thus natural to recover the value of $\mathbf{u}^{n+1}$ at S_1 through a Schwarz algorithm similar to the one outlined in Subsection 8.3.1. The difference is that the Eulerian solver has to work on the full Navier–Stokes equations (8.3.4) instead of dealing with only the elliptic system (8.3.1).

For each time step, the coupling scheme can be summarized as follows:

- Differentiate velocity grid values to obtain vorticity in V_1.
- Interpolate vorticity grid values onto particles in V_{12}.

- Update particle location and circulations in V_2 with a PSE scheme.
- Iterate the velocity values on S_1 with the Schwarz algorithm, iterating between the Biot–Savart law and Eqs. (8.3.4).

In case a multilevel time-advancing method is chosen in one of or both subdomains, the above procedure has to be adapted to provide also velocity boundary conditions at the interface at intermediate time levels.

Of course the precise way Eqs. (8.3.4) are implemented depends on the kind of Eulerian method used in V_1. For a finite-element method, pressure and velocities have to be discretized in compatible finite-element spaces, and no boundary conditions need to be prescribed on the pressure. Finite-difference methods in general use projection algorithms. In summary, Eqs. (8.3.4) are first solved without the solenoidal constraint; then the velocity is projected onto divergence-free fields by the addition of the pressure term; the pressure is determined through the solution of a Poisson equation with a Neumann boundary condition to ensure the correct through flow.

If now an explicit scheme is used in the Eulerian domain, a finite-difference scheme can advance the velocity values at the first substep of a projection algorithm on all interior points in V_1 by use of only velocity values at time t_n. However, the value of $\mathbf{u}^{n+1}$ on S_1 is required for providing the boundary condition for the calculation of the pressure in the projection step. This boundary condition is computed through a Schwarz iterative algorithm. More precisely, if we set

$$\tilde{\mathbf{u}}^n = \mathbf{u}^n + \Delta t [\nu \Delta \mathbf{u}^n + (\mathbf{u}^n \cdot \nabla) \mathbf{u}^n],$$

the Schwarz algorithm consists of calculating a velocity $\mathbf{a}$ on S_1 through the following iterations:

- Solve $\Delta p = \nabla \cdot \tilde{\mathbf{u}}^n$ with $\partial p / \partial \mathbf{n} = \mathbf{a} \cdot \mathbf{n}$ on S_1.
- Set $\mathbf{u} = \tilde{\mathbf{u}}^n + \nabla p$ and input this value on S_2 to compute the velocity in V_2 through the Biot–Savart law.
- Compute new value of the velocity $\mathbf{a}$ on S_1.

Compared with that of the implicit scheme, the advantage of the present method is that each iteration of the Schwarz algorithm involves the inversion of only one Laplacian, instead of three in two dimensions or four in three dimensions. We refer to Refs. 132 and 133 for numerical illustrations of these methods.

Appendix A

Mathematical Tools for the Numerical Analysis of Vortex Methods

The goal of this appendix is to provide the mathematical background needed in the numerical analysis of vortex methods and, more generally, particle methods.

The two first sections are devoted to particle approximations of the solutions to advection equations. The third section summarizes some mathematical features of the Navier–Stokes equations. The numerical analysis carried out in Chapters 2 and 3 results from a combination of the results hereafter derived.

As we have seen in several occasions, the basic feature of vortex methods is that the data, that is the initial vorticity and the source terms at the boundary, are discretized on Lagrangian elements where the circulation is concentrated. These elements are termed particles, and mathematically they consist in delta functions. In Section A.1 we answer the following question: in which sense can a set of particles be used to approximate a given smooth function? We then proceed in Section A.2 to demonstrate that particles moving along a given flow are explicit exact weak solutions, in a sense that we precisely define, of the corresponding advection equation. This is the mathematical reason why particle methods are suitable for the numerical approximation of transport equations. We then show stability properties for the weak form of the transport equation. Together with the results of Section A.1, these stability estimates are the central tool for the numerical analysis of particle methods for linear equations.

For the numerical analysis of vortex methods, it then remains to handle the non-linear features of the flow equations. This is the purpose of Section A.3, where in particular we give the Calderon theorem which allows to control in an optimal way the nonlinear coupling resulting from the calculation of the velocity in terms of the vorticity.

In all this appendix, our computational domain, which we will denote by Ω, will be either the whole space $\mathbf{R}^d$, or a square box with periodic boundary conditions, or a bounded domain. In this later case we will assume

261

no-through-flow-type boundary conditions, that is the velocity field (**a**) satisfies **a** · **n** = 0 at the boundary, where **n** denotes the outward normal to the boundary. In the periodic case all coefficients will of course be supposed periodic.

The functional framework will be based on the Sobolev spaces $W^{m,p}(\Omega)$ of functions which are, together with their derivatives of order up to m, in $L^p(\Omega)$. If one wishes to deal with periodic geometries, then the functions will in addition be assumed to be periodic. If Ω is bounded and no-through-flow-type boundary conditions are prescribed, we will assume the functions to vanish at the boundary, together with their derivatives (that is we will work with the spaces $W_0^{m,p}(\Omega)$).

In all three cases ($\mathbf{R}^d$, periodic box, or bounded domains and functions vanishing at the boundary), to simplify the notations, when there is no ambiguity we will denote by $W^{m,p}$ the resulting Sobolev spaces. We will denote by $\| \cdot \|_{m,p}$ the natural norms and by $| \cdot |_{m,p}$ the associated semi-norms of these spaces. We will finally denote by $W^{-m,p}$ the distribution space made by all continuous linear form on W^{m,p^*}, where p^* denotes the conjugate exponent to p ($1/p + 1/p^* = 1$), and by $\| \cdot \|_{-m,p}$ its norm. We refer to [2] for the mathematical properties of these spaces.

Finally if X is a functional Banach space and T is a positive time, we will use the spaces $L^p(0, T; X)$ of functions of **x** and t, which are for all time in X and are in L^p as vector valued functions of time.

A.1. Data Particle Approximation

If one looks for a functional framework in which one could quantify the degree of approximation of a function by a Dirac mass (or particle), it is clear that the space **M** of bounded measures is not appropriate: consider 2 particles x_1 and x_2 arbitrarily close to each other, but distinct; then the difference has norm 2 in **M** (take the duality of $\delta(\mathbf{x} - \mathbf{x}_1) - \delta(\mathbf{x} - \mathbf{x}_2)$ against a continuous test function ϕ having the value 1 at x_1 and -1 at x_2), and thus does not tend to 0 when $x_1 - x_2$ tends to 0. Therefore a bigger functional space has to be found, or, by duality, a space which acts on test functions in a space which is smaller than the space of continuous function. Good candidates for the test function spaces are the Sobolev spaces $W^{m,p}(\Omega)$, which by Sobolev inclusions are made of continuous functions if $m \geq d$.

A.1.1. Deterministic Initializations

Let us assume that u is a smooth function and that particles are laid on a regular lattice: for $h > 0$ given, $\mathbf{x}_p = \mathbf{p}h, \mathbf{p} \in \mathbf{Z}^d$, if $\Omega = \mathbf{R}^d$. If Ω is a d-cube we divide it in cells of size h and take for the particles the center of each cell.

Then the particle discretization of u defined by

$$u^h(\mathbf{x}) = \sum_p h^d u(\mathbf{x}_p)\,\delta(\mathbf{x} - \mathbf{x}_p), \tag{A.1.1}$$

is an approximation of u in the sense of

Theorem A.1.1. *Let $m \geq d$. Assume $u \in W^{m,\infty}(\Omega) \cap W^{m,1}(\Omega)$. Then the following error estimate holds, for all $p \in]1, +\infty[$*

$$\|u - u^h\|_{-m,p} \leq Ch^m. \tag{A.1.2}$$

Proof. Let ϕ a test function in W^{m,p^*}. If we denote by B_p the cell centered at $\mathbf{x}_p$ with side of length h, we have

$$\langle u - u^h, \phi \rangle = \sum_p \left\{ \int_{B_p} u(\mathbf{x})\phi(\mathbf{x})\,d\mathbf{x} - |B_p| u(\mathbf{x}_p)\phi(\mathbf{x}_p) \right\}.$$

Our estimate thus amounts to a quadrature estimate related to the midpoint quadrature rule for the function $u\phi$. Observe that, depending on the situation under consideration, this function is either a periodic function or a function vanishing at the boundary (in particular, for the no-flow case, under our assumptions on u, $u\phi$ is indeed in $W_0^{m,1}(\Omega)$).

This quadrature rule is actually of infinite order:

Lemma A.1.2. *If $g \in W^{m,1}$ for some $m \geq d$ then*

$$\sum_p \left\{ \int_{B_p} g(\mathbf{x})\,d\mathbf{x} - |B_p| g(\mathbf{x}_p) \right\} \leq Ch^m |g|_{m,1}. \tag{A.1.3}$$

This assertion can be proved by using either Bramble–Hilbert lemma or Poisson's summation formula [8]. However the second proof is not optimal in the sense that it requires more regularity than really necessary and we only sketch the proof based on Bramble–Hilbert lemma. One first checks that there exist coefficients c_α such that for all g which are polynomials of order less than m

$$\int_{\hat{B}} g(\mathbf{x})\,d\mathbf{x} - g(0) = \sum_{1 \leq |\alpha| \leq m-1} c_\alpha \int_{\hat{B}} \partial^\alpha g(\mathbf{x})\,d\mathbf{x},$$

where $\hat{B}$ denotes the unit d-dimensional square. By Bramble–Hilbert lemma and the Sobolev embedding $W^{m,1} \subset C^0$, for $m \geq d$, this implies that

$$\left| \int_{\hat{B}} g(\mathbf{x}) \, d\mathbf{x} - g(0) - \sum_{1 \leq |\alpha| \leq m-1} c_\alpha \int_{\hat{B}} \partial^\alpha g(\mathbf{x}) \, d\mathbf{x} \right| \leq C |g|_{m,1}.$$

It remains now to map $\hat{B}$ into B_p and to add up all the resulting contributions (this is the technique also used to control projection errors in finite-element methods). A classical scaling argument, produces a factor of the order of $h^{|\alpha|}$ in front of the contribution of the derivatives of order α. To conclude, it remains to notice that the contribution of the sum of the integrals of derivatives on Ω vanishes (recall that if Ω is bounded, we deal either with periodic functions or functions vanishing at the boundary, so that this property is true), leaving us only with the highest order derivatives in the estimate, with a factor $O(h^m)$.

With the estimate (A.1.3) it is now easy to prove (A.1.2), since one can write

$$|u\phi|_{m,1} \leq \|u\|_{m,p} \|\phi\|_{m,p^*}$$

and $u \in W^{m,p}(\Omega)$ for all values of p.   □

Remark.

1. The above result clearly applies to geometries which can be smoothly mapped onto rectangles (for instance circular domains through the use of polar coordinates). In this case, the uniform quadrature weights in the mapped space, once multiplied by the Jacobean of the mapping give the actual values of the volumes around the particles in the physical space.

2. If one wishes to allow test functions that do not vanish at the boundary, the accuracy of the midpoint quadrature formula breaks down to second order; if one seeks higher order particle approximations it is necessary to use Gauss-type quadrature points. We do not pursue on this matter which is classical in numerical integration.

A.1.2. Random Initializations

We now analyze two quadrature formulas of Monte Carlo type, that is, formulas which are based on the use of random number sequences.

The first method we wish to deal with is the genuine Monte Carlo method, where particles are initialized using independent random variables in the support of the function to approximate. More precisely if Ω is the unit square $[0, 1]^d$, let us consider the space Ξ of all sequences $\xi = (\xi_1, \xi_2, \ldots)$ with values in Ω. A sequence of particle initializations consists in the choice of a particular sequence within this space, and one specific initialization is given by the N first elements in this sequence, for a finite value of N. In practice it requires the call of a random generator number. Note that this formula is also the basic ingredient underlying random walk methods for the simulation of viscous flows.

One can prove that this formula is consistent, for almost all sequences ξ, in the following sense

Lemma A.1.3. *Let* $g \in L^2(\Omega)$*. Then*

$$\int_\Xi \left| \int_\Omega g(\mathbf{x})\, d\mathbf{x} - N^{-1} \sum_{i=1}^N g(\xi_i) \right|^2 d\xi \le N^{-1} \int_\Omega |g(\mathbf{x})|^2\, d\mathbf{x} \qquad (A.1.4)$$

Proof. By setting $\bar{g} = g - \int_\Omega g\, d\mathbf{x}$, we are brought back to the case when $\int_\Omega g\, d\mathbf{x} = 0$. In this case, if we develop the left-hand side of (A.1.4), we find

$$\frac{1}{N^2} \int_\Xi d\xi \sum_{i,j} g(\xi_i) g(\xi_j)$$

If $i \ne j$, Fubini's theorem implies that the corresponding term in the above sum will be the square of the integral of g, that is zero. We are thus left with

$$\frac{1}{N^2} \int_\Xi d\xi \sum_i |g(\xi_i)|^2 = \frac{1}{N} \int_\Omega |g(\mathbf{x})|^2\, d\mathbf{x}$$

which yields the expected estimate. $\square$

The interest of this quadrature estimate is clearly that it requires only weak regularity of g. However its convergence is rather slow and we now present a method which can be seen as a compromise between mid-point formulas, which are potentially high-order accurate but require smoothness, and the fully random technique just analyzed.

This quadrature formula consists in splitting the computational box in cells of uniform size h; then, instead of taking the center of the cell, one chooses randomly a point inside these boxes. If we denote by $\mathbf{x}_i$, $i \in [1, N]$, the left-bottom corner of the cell number i, the particle will have location $\mathbf{x}_i + \xi_i h$, where ξ_i is chosen randomly in $[0, 1]^d$. One can prove

Lemma A.1.4. *Assume* $g \in W^{1,\infty}(\Omega)$*. Then*

$$\int_\Xi \left| \int_\Omega g(\mathbf{x})\, d\mathbf{x} - N^{-1} \sum_{i=1}^N g(\mathbf{x}_i + \xi_i h) \right|^2 d\xi \le h N^{-1} |g|^2_{1,\infty} \qquad (A.1.5)$$

Proof. It goes along the same lines as for the formula (A.1.4). If we assume that g has mean value 0, on expanding the left-hand side of (A.1.5), one gets

$$\frac{1}{N^2} \int_\Xi d\xi \sum_{i,j} g(\mathbf{x}_i + \xi_i h) g(\mathbf{x}_j + \xi_j h)$$

$$= \frac{1}{h^d N^2} \left[h^{-d} \sum_{i \neq j} \int_{B_i} g(\mathbf{x}) \, d\mathbf{x} \int_{B_j} g(\mathbf{x}) \, d\mathbf{x} + \sum_i \int_{B_i} |g(\mathbf{x})|^2 \, d\mathbf{x} \right]$$

where B_i denotes the cell number i. Next we observe that

$$\left| \sum_i \int_{B_i} g(\mathbf{x}) \, d\mathbf{x} \right|^2 = 0$$

or, equivalently,

$$\sum_{i \neq j} \int_{B_i} g(\mathbf{x}) \, d\mathbf{x} \int_{B_j} g(\mathbf{x}) \, d\mathbf{x} + \sum_i \left| \int_{B_i} g(\mathbf{x}) \, d\mathbf{x} \right|^2 = 0.$$

Thus

$$h^{-d} \sum_{i \neq j} \int_{B_i} g(\mathbf{x}) \, d\mathbf{x} \int_{B_j} g(\mathbf{x}) \, d\mathbf{x} + \sum_i \int_{B_i} |g(\mathbf{x})|^2 \, d\mathbf{x}$$

$$= \sum_i \left[\int_{B_i} |g(\mathbf{x})|^2 \, d\mathbf{x} - h^{-d} \left| \int_{B_i} g(\mathbf{x}) \, d\mathbf{x} \right|^2 \right]$$

We next note that, by a straightforward expansion of g:

$$\int_{B_i} |g(\mathbf{x})|^2 \, d\mathbf{x} - h^{-d} \left| \int g(\mathbf{x}) \, d\mathbf{x} \right|^2 = h^d O\left(h |g|_{1,\infty}^2 \right)$$

and (A.1.5) follows. $\qquad\qquad\square$

For $d = 2$ we get a second-order accuracy if the function g is differentiable. This must be compared with the mid-point formula which yields the same accuracy but for a function which is twice differentiable. However in the case of the Monte Carlo method, the order of accuracy is to be taken in a statistical sense. Also observe that the amelioration of this formula over the genuine Monte Carlo obviously deteriorates as the dimension increase.

The quadrature formulas that we have seen use uniform distributions of points. The particle weights that result are given by the local value of the function to approximate. Alternatively, one may also consider quadrature formulas where the weights are constant, but the distribution of points adjust to the function to approximate. This type of quadrature formulas can be useful in some particular situations (they are actually a well-known tool in applications of particle methods to plasma physics) but are clearly of more delicate use.

A.2. Classical and Measure Solutions to Linear Advection Equations

In this section we summarize the mathematical background needed for the definition and the numerical analysis of particle methods. The outline of this section is as follows; we first recall some basic facts related to classical solutions of linear advection equations, including a precise regularity result. Then we use the notion of measure solution to give a mathematical meaning to particle approximations as it was first described in [169]. Finally we exploit further this notion of weak solutions to derive error estimates in distribution spaces for particle approximations. Our analysis will be based on techniques described in [60] then further simplified in [44, 45]. We will be working in d-dimension (in practice $d = 2$ or 3), $\mathbf{a} = (a_1, \ldots, a_d)$ will denote a vector field and a_0 a scalar function. We will denote by L the operator defined by

$$Lu = \frac{\partial u}{\partial t} + \operatorname{div}(\mathbf{a}u) + a_0 u$$

and by L^* its formal adjoint:

$$L^*u = -\frac{\partial u}{\partial t} - (\mathbf{a} \cdot \nabla)u + a_0 u.$$

A.2.1. Classical Solutions: Explicit Expression and Regularity Properties

Let us first recall how smooth solutions of the advection equation can be explicitly written, using the notion of characteristics.

If f and u_0 are given functions in Ω we will call u a classical solution of the advection equation

$$L^*u = f; \quad u(\cdot, 0) = u_0 \tag{A.2.1}$$

a distribution solution of this equation which is continuous with respect to $\mathbf{x} \in \Omega$ and t. We will denote by $\mathbf{X}(t; \mathbf{x}, s)$ the characteristics associated to the vector field **a**, passing through **x** at time s, that is the solution to the system:

$$\frac{d\mathbf{X}}{dt} = \mathbf{a}(\mathbf{X}, t); \quad \mathbf{X}(s; \mathbf{x}, s) = \mathbf{x}. \tag{A.2.2}$$

It is well known that this system has a unique solution as soon as **a** is Lipschitz continuous with respect to **x**.

The existence and uniqueness of classical solutions are given by

Theorem A.2.1. *Assume $u_0 \in C^0$, $f \in L^1(0, T; C^0)$. Assume further that $\mathbf{a} \in L^\infty(0, T; W^{1,\infty}(\Omega))$ with $\mathbf{a} \cdot \mathbf{n} = 0$ at the boundary if we deal with bounded domains with no-through-flow boundary conditions, and $a_0 \in L^1(0, T; C^0(\Omega))$. Then the problem (A.2.1) has a unique classical solution given by*

$$u(\mathbf{x}, t) = u_0(\mathbf{X}(0; \mathbf{x}, t)) \exp\left(\int_0^t a_0(\mathbf{X}(s; \mathbf{x}, t), s) \, ds \right)$$

$$- \int_0^t f(\mathbf{X}(s; \mathbf{x}, t), s) \exp\left(\int_s^t a_0(\mathbf{X}(\sigma; \mathbf{x}, t), \sigma) \, d\sigma \right) ds. \tag{A.2.3}$$

If $a_0 = f = 0$ the above result means that the solution merely propagates along the characteristics. Actually writing u in the corresponding Lagrangian coordinates reduces our PDE to a first-order linear ODE for the function of time $t \to u(\mathbf{X}(t; \mathbf{x}, s), t)$. Upon inverting the flow map, the explicit integration of this ODE gives the formula (A.2.3).

Observe that, in the case of no through-flow boundary conditions, if u_0 and f vanish at the boundary, the same property is true for $u(\cdot, t)$ at later times. This clearly results from (A.2.3), once it is observed that, due to the no through-flow condition, characteristics which intersect the boundary, must remain for all times on the boundary. Let us now deal with the so-called conservative form

$$Lu = f; \quad u(\cdot, 0) = u_0 \tag{A.2.4}$$

of the advection equation. Although it is obviously equivalent to the form (A.2.1), up to some changes of signs and to the addition of the term div **a** to a_0 in the operator L^*, it is worthwhile to rewrite the solution of (A.2.4) in a slightly different way, emphasizing its conservative features. Let us denote by $J(t; \mathbf{x}, s)$ the Jacobian of the transform $\mathbf{x} \to \mathbf{X}(t; \mathbf{x}, s)$

$$J(t; \mathbf{x}, s) = \left| \det \frac{\partial \mathbf{X}}{\partial \mathbf{x}} \right|$$

Then we have

Proposition A.2.2. *Under the same assumptions as for Theorem A.2.1, if $a_0 = 0$ the classical solution of (A.2.4) can be written*

$$u(\mathbf{x}, t) = u_0(\mathbf{X}(0; \mathbf{x}, t))J(0; \mathbf{x}, t) + \int_0^t f(\mathbf{X}(s; \mathbf{x}, t), s)J(s; \mathbf{x}, t)\, ds.$$

If $u_0 \in L^1(\Omega)$ and $f \in L^1(0, T; L^1(\Omega))$ then $u \in L^\infty(0, T; L^1(\Omega))$ and

$$\int_\Omega u(\mathbf{x}, t)\, d\mathbf{x} = \int_\Omega u_0(\mathbf{x})\, d\mathbf{x} + \int_0^t \int_\Omega f(\mathbf{x}, s)\, d\mathbf{x}\, ds.$$

This result follows easily from theorem A.2.1 and the well known fact that the Jacobean is solution to the ODE

$$\dot{J} = J \operatorname{div} \mathbf{a}.$$

The second equality above can be seen as the integral counterpart of the point-wise conservation expressed in Theorem A.2.1. In the important particular case of a divergence free velocity field $\mathbf{a}$, the Jacobian remains equal to 1 for all time, and both conservations, pointwise and in the mean, hold.

We now state a regularity result for equation (A.2.1). This result will be crucial in the stability estimates for weak solutions needed in the numerical analysis of the vortex methods.

Theorem A.2.3. *Let $m \geq 1$, $p > 1$, $T > 0$ and assume $a_0, \mathbf{a} \in L^\infty(0, T; W^{m,p})$. Assume further that there exist 2 constants A and M such that the following estimates are valid for $t \in [0, T]$*

$$\sum_{k=1}^m A^{1-k} |\mathbf{a}(\cdot, t)|_{k,\infty} \leq M \qquad (A.2.5)$$

$$\|a_0(\cdot, t)\|_{0,\infty} + \sum_{k=1}^m A^{1-k} |a_0(\cdot, t)|_{k,\infty} \leq M \qquad (A.2.6)$$

Then, if $u_0 \in W^{m,p}$ and $f \in L^1(0, T; W^{m,p})$, the classical solution u to (A.2.1) belongs to $L^\infty(0, T; W^{m,p})$. Moreover there exist constants C_k only depending on M and T such that, for $k \leq m$

$$|u(\cdot, t)|_{k,p} \leq C \left\{ \sum_{i=1}^k A^{k-i} \left(|u_0|_{i,p} + \int_0^T |f(\cdot, t)|_{i,p}\, dt \right) \right.$$

$$\left. + A^{k-1} \left(\|u_0\|_{0,p} + \int_0^T \|f(\cdot, t)\|_{0,p}\, dt \right) \right\} \qquad (A.2.7)$$

Proof. For a given m, we proceed by induction on $k \leq m$. For $k = 0$, we write
the solution on the form (A.2.3) and we estimate the L^p norm of the right-hand
side using the change of variables along the characteristics $\mathbf{y} = \mathbf{X}(0; \mathbf{x}, t)$ for the
contribution of u_0 and $\mathbf{y} = \mathbf{X}(s; \mathbf{x}, t)$ for the contribution of f. We observe that,
due to (A.2.5), $|a|_{1,\infty} \leq M$, and thus the jacobians of these change of variables
are bounded by $\exp MT$ and the L^p estimate (A.2.7) for $k = 0$ follows.

We next differentiate k times the original equation with respect to all possible
combination of variables. This gives a system of equations which can be written
in the following general form

$$-\frac{\partial D^k u}{\partial t} - (a \cdot \nabla) D^k u = D^k f + B(Da, D^k u)$$

$$+ \sum_{i=2}^{k} P_i(D^i a, D^{k+1-i} u) + \sum_{i=1}^{k} Q_i(D^i a_0, D^{k-i} u)$$

where $D^i v$ denotes the vector made up of all derivatives of order i of v and
B, P_i, Q_i denote various bilinear forms. If we denote by F the above right-hand
side, formula (A.2.3) yields

$$D^k u(\mathbf{x}, t) = D^k u_0(\mathbf{X}(0; \mathbf{x}, t)) - \int_0^t F(\mathbf{X}(s; \mathbf{x}, t), s) \, ds.$$

Arguing as for $k = 0$ we get

$$\|D^k u(\cdot, t)\|_{0,p} \leq C(M, T) \left[\|D^k u_0\|_{0,p} + \int_0^t \|F(\cdot, s)\|_{0,p} \, ds \right].$$

It remains now to develop F to obtain the following bound

$$|u(\cdot, t)|_{k,p} \leq C(M, T) \left\{ |u_0|_{k,p} + \int_0^T \left[|f(\cdot, s)|_{k,p} + |a_0|_{k,\infty} |u(\cdot, s)|_{0,p} \right. \right.$$

$$\left. \left. + \sum_{i=1}^{k-1} |u(\cdot, s)|_{i,p} (|a|_{k+1-i,\infty} + |a_0|_{k-i,\infty}) \right] ds \right\}$$

We now use (A.2.5)–(A.2.6) to estimate the derivatives of $\mathbf{a}$ and a_0, and we
assume that (A.2.7) is true for all derivatives of u up to order $k - 1$. We obtain

$$|u(\cdot, t)|_{k,p} \leq C(M, T) \left\{ |u_0|_{k,p} + \int_0^T |f(\cdot, s)|_{k,p} \, ds + A^{k-1} |u_0|_{0,p} \right.$$

$$\left. + A^{k-1} \int_0^T \|f(\cdot, s)\|_{0,p} \, ds + \sum_{i=1}^{k-1} A^{i-k} \left[\sum_{j=1}^{i} A^{j-i} (|u_0|_{j,p} \right. \right.$$

$$+ \int_0^T |f(\cdot, t)|_{j,p} \, dt + A^{i-1} (\|u_0\|_{0,p} + \int_0^T \|f(\cdot, t)\|_{0,p} \, dt) \Big] \Big\}$$

$$\leq C(M, T) \Big\{ |u_0|_{k,p} + \int_0^T |f(\cdot, s)|_{k,p} \, ds + A^{k-1} |u_0|_{0,p}$$

$$+ A^{k-1} \int_0^T \|f(\cdot, s)\|_{0,p} \, ds + \sum_{j=1}^{k-1} A^{j-k} (|u_0|_{j,p}$$

$$+ \int_0^T |f(\cdot, t)|_{j,p} \, dt) \Big\}$$

which ends our proof. $\qquad\qquad\qquad\qquad\qquad\qquad\qquad\qquad\qquad\qquad\qquad$ □

A.2.2. Measure Solutions and Justification of Particle Approximations

To give a mathematical meaning to particle approximations, we now need to define measure solutions to the advection equation in its conservative form (A.2.4). We will have to deal with continuous functions either vanishing at infinity, or at the boundary of the computational domain Ω, or periodic in Ω, depending on the case we wish to focus on. To avoid unnecessary notational complications, we will denote by C^0 the corresponding space. $\mathbf{M}$ will be the space of Radon measures on Ω, and $\langle \cdot, \cdot \rangle$ the duality pairing between $\mathbf{M}$ and C^0.

Definition A.2.4. *For $u_0 \in \mathbf{M}$ and $f \in L^1(0, T; \mathbf{M})$, $u \in L^\infty(0, T; \mathbf{M})$ will be called a measure solution to (A.2.4) if it satisfies*

$$\int_0^T \langle u(\cdot, t), L^* \phi(\cdot, t) \rangle \, dt = \langle u_0, \phi(\cdot, 0) \rangle + \int_0^T \langle f(\cdot, t), \phi(\cdot, t) \rangle \, dt \quad \text{(A.2.8)}$$

for all test function ϕ such that $\phi(\cdot, T) = 0$ and

$$\phi \in C^0([0, T]; C^0); \quad L^* \phi \in L^1(0, T; C^0). \qquad\qquad \text{(A.2.9)}$$

In view of (A.2.8), it is clear that the space of test functions should at least fulfill the regularity requirements given in (A.2.9). The existence and uniqueness theorem which follows confirms that these conditions actually define the right test function space.

Theorem A.2.5. *Assume $\mathbf{a}$ and a_0 satisfy the same regularity assumptions as in Theorem A.2.1. Then for all $u_0 \in \mathbf{M}$ and $f \in L^1(0, T; \mathbf{M})$, problem*

(A.2.8)–(A.2.9) has a unique measure solution, which is given by:

$$\langle u(\cdot, t), \phi \rangle = \left\langle u_0, \phi(\mathbf{X}(t; \cdot, 0)) \exp\left(-\int_0^t a_0(\mathbf{X}(s; \cdot, 0), s)\, ds\right) \right\rangle$$
$$+ \int_0^t ds < f(\cdot, s), \phi(\mathbf{X}(t; \cdot, s)) \exp\left(-\int_s^t a_0(\mathbf{X}(\sigma; \cdot, s), \sigma)\, d\sigma\right)$$

$$\text{(A.2.10)}$$

for all continuous test function ϕ.

Proof. Let us first check the uniqueness. Since the problem is linear we have to show that 0 is the only solution to the homogeneous problem

$$\int_0^T \langle u(\cdot, t), L^\star \phi(\cdot, t) \rangle = 0, \quad \text{for all admissible } \phi.$$

Let $\psi \in L^1(0, T; C^0)$. In view of Theorem A.2.1, the problem

$$L^\star \phi = \psi; \quad \psi(\cdot, T) = 0$$

has a unique classical solution on $[0, T]$. This solution obviously satisfies the properties required for the test function in (A.2.9). We therefore have

$$\int_0^T \langle u(\cdot, t), \psi(\cdot, t) \rangle\, dt = 0, \quad \text{for all } \psi \in L^1(0, T; C^0),$$

which implies $u \equiv 0$.

Let us now prove that the formula (A.2.10) actually gives a solution to our problem. If we replace ϕ by $L^\star \phi$ in the right-hand side of (A.2.10) and if we denote by A_1 the term in front of u_0 and by A_2 the contribution of f, before integration in time, we observe that

$$A_1(\cdot, t) = L^\star \phi(\mathbf{X}, t) \exp\left(-\int_0^t a_0(\mathbf{X}, s)\, ds\right)$$
$$= -\frac{d}{dt}\left\{\phi(\mathbf{X}, t) \exp\left(-\int_0^t a_0(\mathbf{X}, s)\, ds\right)\right\}$$

where to simplify the notations we have replaced $(\mathbf{X}(s; \mathbf{x}, 0), s)$ by $(\mathbf{X}, s)$. Therefore

$$\int_0^T \langle u_0, A_1(\cdot, t) \rangle\, dt = -\left\langle u_0, \left\{\phi(\mathbf{X}, T) \exp\left(-\int_0^T a_0(\mathbf{X}, s)\, ds\right) - \phi(\mathbf{X}, 0)\right\}\right\rangle.$$

But $\phi(\cdot, T) = 0$ and, with our notational convention, $(\mathbf{X}, 0) = (\mathbf{x}, 0)$. Thus

$$\int_0^T \langle u_0, A_1(\cdot, t)\rangle dt = \langle u_0, \phi(\cdot, 0)\rangle.$$

For the contribution of f we similarly observe that

$$L^\star\phi(\mathbf{X}(t; \mathbf{x}, s), t) \exp\left(-\int_s^t a_0(\mathbf{X}(\sigma; \mathbf{x}), \sigma)\, ds\right)$$

$$= -\frac{d}{dt}\left\{\phi(\mathbf{X}(t; \mathbf{x}, s), t) \exp\left(-\int_s^t a_0(\mathbf{X}(\sigma; \mathbf{x}, s), \sigma)\, ds\right)\right\},$$

so that we have to compute

$$I = -\int_0^T dt \int_0^t ds$$

$$\times \left\langle f(\cdot, s), \frac{d}{dt}\left\{\phi(\mathbf{X}(t; \cdot, s), t) \exp\left(-\int_s^t a_0(\mathbf{X}(\sigma; \cdot, s), \sigma)\, ds\right)\right\}\right\rangle.$$

Integrating first over t then over s yields

$$I = -\int_0^T ds \int_s^T dt \frac{d}{dt}\left\langle f(\cdot, s), \{\phi(\mathbf{X}(t; \cdot, s), t)\right.$$

$$\times \exp\left(-\int_s^t a_0(\mathbf{X}(\sigma; \cdot, s), \sigma)\, ds\right)\}\bigg\rangle$$

$$= \int_0^T ds \langle f(\cdot, s), \phi(\cdot, s)\rangle,$$

which finally proves that u satisfies (A.2.8). □

Notice that the definition as well as the formula (A.2.10) is the most natural extension of classical solutions: starting from the formula giving the classical solution, if one takes the duality of the resulting function with a test function, one obtains the formula (A.2.10) by using a change of variables along the characteristics in the integrals. Let us also observe that the above theory can be readily extended to forcing terms f which are time measure instead of integrable time functions. Within this framework we can now define particle solutions.

Corollary A.2.6 (Fundamental example of particle solutions).

1) Assume $u_0 = \alpha_0\delta(\mathbf{x} - \mathbf{x}_0)$ and $f = 0$. Then the measure solution of (A.2.4) is given by

$$u(\mathbf{x}, t) = \alpha(t)\delta(\mathbf{x} - \mathbf{x}(t)) \tag{A.2.11}$$

where $\mathbf{x}(t) = \mathbf{X}(t; \mathbf{x}, 0)$ *and* α *satisfies*

$$\frac{d\alpha}{dt} + a_0(\mathbf{x}(t), t)\alpha = 0; \quad \alpha(0) = \alpha_0.$$

2) *If* $u_0 = 0$ *and* $f = \beta_0 \delta(\mathbf{x} - \mathbf{x}_0) \otimes \delta(t - t_0)$ *then*

$$u(\mathbf{x}, t) = \beta(t)\delta(\mathbf{x} - \mathbf{X}(t; \mathbf{x}_0, t_0))$$

where $\beta(t) = 0$ *for* $t < t_0$ *and, for* $t \geq t_0$,

$$\frac{d\beta}{dt} + a_0(\mathbf{X}(t; \mathbf{x}_0, t_0), t)\beta(t) = 0; \quad \beta(t_0) = \beta_0.$$

Proof. These formulas easily follow from the general formula (A.2.10). Let us only check the first one. If u is given by (A.2.11), then we have

$$\langle u(\cdot, t), \phi \rangle = \alpha(t)\phi(\mathbf{X}(t; \mathbf{x}, 0)),$$

which is precisely (A.2.10), once the ODE giving α has been explicitly integrated.   □

If $a_0 \equiv 0$, the conservation of the weights of the particles in the solution formula (A.2.11) can be seen as the result of a compensation effect between the evolution of the local volumes (in case the flow is not incompressible) and the modification of the local values of the underlying solution to the advection equation.

Before going further let us point out that the above framework applies to the case of systems as well. In this case the operator L takes the form

$$L\mathbf{U} = \frac{\partial \mathbf{U}}{\partial t} + \mathrm{div}(\mathbf{a} : \mathbf{U}) + [A_0][\mathbf{U}],$$

$\mathbf{U}$ is a vector solution, $\mathbf{a}$ is a vector field and $\mathbf{a} : \mathbf{U}$ is the vector of components $(\sum_j a_j U_i)_i$ and A_0 is a square matrix. All the results and definitions above remain unchanged, except for the definition of L^* were A_0 has to be replaced by its transpose matrix. In particular, formula (A.2.11) remains valid (α is then a vector-valued coefficient). This important point is useful for the definition and analysis of vortex methods in three dimensions.

We now are in a position to address the second question raised in the preamble of this appendix, in the case of a deterministic initialization of the particles (we do not consider here the case of random initializations, for which things are far more subtle). From Theorem A.1.1, we know that, for this type of initialization, the initial discrepancy between the exact initial data and its particle approximation is naturally measured in a distribution space of the kind $W^{-m,p}(\Omega)$. We have also seen that both the exact and the particle solutions are explicit solutions to the advection equation. We therefore have to look for stability properties of the advection equation in these spaces.

Regarding the smoothness of the flow field **a** and of the coefficient a_0, we will keep the assumptions (A.2.5)–(A.2.6) already made to establish the regularity result for the classical solutions. For a sake of simplicity in the notations, the following stability result is stated for scalar problems. Its extension to systems is straightforward.

Theorem A.2.7. *Let* $m \geq 1$. *Under the assumptions (A.2.5)–(A.2.6), if* $\phi_0 \in W^{-m,p}$ *and* $f \in L^1(0, T; W^{-m,p})$, *the Cauchy problem*

$$L\phi = f; \quad \phi(\cdot, 0) = \phi_0$$

has a unique weak solution $\phi \in L^\infty(0, T; W^{-m,p})$.

Moreover this solution can be written $\phi = \sum_{k=0}^{m} \phi^k$, *where* $\phi^k \in W^{-k,p}$ *for* $0 \leq k \leq m$ *and there exists a constant* $C(M, T)$ *only depending on* M *and* T *such that for* $t \in [0, T]$

$$A^{1-m}\|\phi^0(\cdot, t)\|_{0,p} + \sum_{k=1}^{m} A^{k-m}\|\phi^k(\cdot, t)\|_{-k,p}$$

$$\leq C(M, T) \left(\|\phi_0\|_{-m,p} + \int_0^t \|f(\cdot, s)\|_{-m,p} \, ds \right) \quad \text{(A.2.12)}$$

This theorem will be the main ingredient for a unified convergence theory for the two- and three-dimensional Euler equations. Before giving its proof let us comment briefly on its meaning. First of all, if one thinks of the constant A involved in the regularity assumptions for **a**, a_0 as an $O(1)$ term, then (A.2.12) is just the natural stability result one could expect in the distribution spaces. Next, if one allows A to take large values, A is then a scale factor which measures the regularity of the flow field (in a viscous flow one can for instance think of A as the square root of the Reynolds number), the splitting of the solution into components of increasing smoothness and the estimates of these components indicate that there is a compensation effect which makes that the largest part of the solution concerns smooth (compared to the data) components of the solution. In vortex methods these solutions will have to be regularized, and the smoother the error components are, the less one will have to pay in the error estimate for this regularization.

Let us also mention that it is possible to generalize this result by allowing the exponent p to take different values for the different components of the solution (see Ref. 60 for this more general setting). This flexibility turns out to be useful for the analysis of vortex methods in unbounded domains.

To establish our result, we need a distribution representation result, which is nothing but a precise rephrasing of a classical result in distribution theory

Lemma A.2.8 [60]. *Let T be a distribution that satisfies*

$$|\langle T, \phi \rangle| \leq \sum_{k=0}^{m} c_k |\phi|_{k, p^*} \qquad (A.2.13)$$

for some positive constants c_k and all functions ϕ that are C^∞ with compact support in Ω (or periodic if we are in this situation). Then we can write

$$T = \sum_{\alpha \in \mathbf{N}^2, |\alpha| \leq m} T_\alpha, \quad T_\alpha \in W^{-|\alpha|, p},$$

$$\|T_\alpha\|_{-|\alpha|, p} \leq c_{|\alpha|}.$$

Proof. Denoting by p^* the conjugate exponent to p, we classically begin by viewing T as a linear continuous form on the subspace $H \in (L^{p^*})^N$ for an appropriate value of N, containing the N-uplets made up by all derivatives of order up to m of functions of W^{m, p^*}.

We then provide $(L^{p^*})^N$ with the following norm:

$$(f_\alpha)_{|\alpha| \leq m} \rightarrow \sum_{|\alpha| \leq m} c_\alpha \|f_\alpha\|_{0, p^*}.$$

By virtue of relation (A.2.13), T has a norm less than or equal to 1 on H. Thus we can extend T to the whole space $(L^{p^*})^N$, with a norm still less than or equal to 1. But we know the forms of all such continuous linear mappings: there exist L^p functions g_α such that the extended mapping $\bar{T}$ can be given by the following expression:

$$\langle \bar{T}, \phi \rangle = \sum_{|\alpha| \leq m} \int g_\alpha \phi_\alpha \, d\mathbf{x}.$$

It is now not difficult to check that this implies

$$T = \sum_{|\alpha| \leq m} \frac{\partial^{|\alpha|} g_\alpha}{\partial x_1^{\alpha_1} \cdots \partial x_d^{\alpha_d}}.$$

By the choice we have made for the norm in H, we also have

$$\sum_{|\alpha| \le m} \int g_\alpha \phi_\alpha \, dx \le \sum_{|\alpha| \le m} c_\alpha \|\phi_\alpha\|_{0,p^*}$$

for all $(\phi_\alpha) \in (L^{p^*})^N$. This proves that

$$\|g_\alpha\|_{0,p} \le c_{|\alpha|}, \quad |\alpha| \le m$$

and it remains to set $\mathcal{T}_\alpha = [(\partial^{|\alpha|} g_\alpha)/(\partial x_1^{\alpha_1} \cdots \partial x_d^{\alpha_d})]$ to obtain the desired decomposition. $\qquad\square$

We can now give the Proof of Theorem A.2.7.

Proof of Theorem A.2.7. We will prove only estimate (A.2.12) (see Ref. 60 for the existence and uniqueness of weak solutions in $W^{-m,p}$). From Definition A.2.4, the natural notion of weak solutions of Eq. (A.2.4) in $W^{-m,p}$ is

$$\int_0^T \langle \phi(\cdot, t), L^* \psi(\cdot, t) \rangle \, dt = \langle \phi_0, \psi(\cdot, 0) \rangle + \int_0^T \langle f(\cdot, t), \psi(\cdot, t) \rangle \, dt.$$

$$\text{(A.2.14)}$$

for all admissible test function (in the sense of (A.2.9) with C^0 replaced by W^{m,p^*}), where $\langle \cdot, \cdot \rangle$ denotes now the duality pairing between $W^{-m,p}$ and W^{m,p^*}.

Let ξ be a function in $L^1(0, T; W^{m,p^*})$ and ψ be the (classical) solution to

$$\psi(\cdot, T) = 0; \quad L^* \psi = \xi.$$

In view of Theorem A.2.3 (up to changing t into $-t$) we know that $\psi \in L^\infty(0, T; W^{m,p^*})$ and, more precisely, for $k \le m$ and $t \le T$,

$$|\psi(\cdot, t)|_{k,p^*} \le C \left[A^{k-1} \int_0^T \|\xi(\cdot, t)\|_{0,p^*} \, dt + \sum_{1 \le i \le k} A^{k-i} \int_0^T |\xi(\cdot, t)|_{k,p^*} \, dt \right].$$

$$\text{(A.2.15)}$$

We next choose the function ξ of the form $\lambda(t)\xi$, where $\lambda \in L^1(0, T)$ and $\xi \in W^{m,p^*}$. Combining relations (A.2.14) and (A.2.15) yields

$$\int_0^T \lambda(t) \langle \phi(\cdot, t), \xi \rangle \, dt \le C \left[\|\phi_0\|_{-m,p} + \int_0^T \|f(\cdot, t)\|_{-m,p} \right]$$

$$\times \left[\int_0^T |\lambda(t)| \, dt \right] \left(A^{m-1} \|\xi\|_{0,p^*} + \sum_{1 \le i \le m} A^{m-i} |\xi|_{i,p^*} \right).$$

This relation, valid for all $\lambda \in L^1(0, T)$, implies that for all $t \in [0, T]$ and all $\xi \in W^{m,p^*}$

$$|\langle \phi(\cdot, t), \xi \rangle| \leq C \left[\|\phi_0\|_{-m,p} + \int_0^T \|f(\cdot, t)\|_{-m,p} \right]$$

$$\times \left(A^{m-1} \|\xi\|_{0,p^*} + \sum_{1 \leq i \leq m} A^{m-i} |\xi|_{i,p^*} \right).$$

Our claim follows then immediately from the representation Lemma A.2.8. □

A.3. Mathematical Facts about the Flow Equations

This section is devoted to some results concerning the incompressible Navier–Stokes equations. We focus on the statement of the results, without proofs, that are necessary in the convergence studies in Chapters 2 and 3. For a thorough mathematical discussion with emphasis on the vorticity form of the equations we refer to a review article [142], books [193, 143], and the references therein.

As we have seen on several occasions, the flow equations couple two problems of a different nature: a kinematic part (the calculation of the velocity in function of the vorticity) and a dynamic part (the transport-diffusion equation for the vorticity). It is thus convenient to give a separate account of the mathematical properties related to these problems.

A.3.1. Kinematics

The kinematic part of the flow equations consists of solving the system

$$\text{div } \mathbf{u} = 0,$$

$$\nabla \times \mathbf{u} = \omega,$$

supplemented by boundary conditions. For the sake of simplicity, in the notation we consider here only periodic boundary conditions or equations in the whole space $\mathbf{R}^2$ or $\mathbf{R}^3$ that vanish at infinity, but the estimates below extend to domains with solid boundaries and the no-through-flow boundary condition. If the vorticity is divergence free (a condition that of course applies only in three dimensions) and is integrable, the solution this problem can be written as

$$\mathbf{u} = \mathbf{K} \star \omega. \tag{A.3.1}$$

In the case of periodic boundary condition, the above convolution should be taken on an extension by periodicity of the vorticity to the whole space. Since, roughly speaking, one obtains the velocity by integrating the vorticity, one may expect that it is more regular. The following theorem summarizes the smoothing property of the Biot–Savart kernel.

Theorem A.3.1 (Calderon). *Let* $1 < p < \infty$. *For periodic boundary conditions, one has*

$$\|\mathbf{K} \star \omega\|_{1,p} \leq C\|\omega\|_{0,p}. \qquad (A.3.2)$$

For peridodic boundary conditions or solutions in $\mathbf{R}^d$, $d = 2, 3$, *one has*

$$|\mathbf{K} \star \omega|_{1,p} \leq C\|\omega\|_{0,p}, \qquad (A.3.3)$$

and, if $p \leq d$ *and* $1/q = 1/p - 1/d$ ($q = \infty$ *if* $p = d$),

$$\|\mathbf{K} \star \omega\|_{0,q} \leq C\|\omega\|_{0,p}. \qquad (A.3.4)$$

Note that, although it is natural to expect derivatives of the velocity to have the same regularity as the vorticity, estimates (A.3.2) and (A.3.3) are not true for $p = \infty$. L^∞ estimates, however, can be recovered by use of the following interpolation estimates, valid for $p > d$:

$$\|u\|_{0,\infty} \leq C\|u\|_{0,p}^{s}|u|_{1,p}^{1-s}, \quad s = d/p. \qquad (A.3.5)$$

Relation (A.3.4) results from estimate (A.3.3) and Sobolev inequalities. The fact that, in relation (A.3.4), q is larger than p reflects the fact that the velocity decays at infinity at a slower rate than the velocity (typically, for a vorticity with zero circulation and compact support, the velocity decays no faster than $1/r^d$).

Finally it is important to stress that the above estimates can be rephrased within the framework of distribution introduced in Section A.1. In the case of boundary conditions, estimates (A.3.2) equivalently is given by

$$\|\mathbf{K} \star \omega\|_{0,p} \leq C\|\omega\|_{-1,p}. \qquad (A.3.6)$$

We refer to Ref. 60 for similar results in the case $\Omega = \mathbf{R}^2$, $\mathbf{R}^3$. The regularization properties of $\mathbf{K}$ can be further strengthened if one uses mollified kernels. For simplicity we assume here that we are in the periodic case. If ζ is a smooth cutoff function with a unit integral let us set $\zeta_\varepsilon = \varepsilon^{-d}\zeta(\mathbf{x}/\varepsilon)$ and $\mathbf{K}_\varepsilon = \mathbf{K} \star \zeta_\varepsilon$.

Let $T \in W^{-m,p}(\Omega)$ with $m \geq 1$. By the representation Lemma A.2.8, we can write

$$T = \sum_{|\alpha| \leq m-1} \partial^\alpha f_\alpha,$$

with $f_\alpha \in W^{-1,p}(\Omega)$ and $\|f_\alpha\|_{-1,p} \leq C\|T\|_{-m,p}$. Therefore

$$\|\mathbf{K}_\varepsilon \star T\|_{0,p} \leq \sum_{|\alpha| \leq m-1} \|\partial^\alpha \zeta_\varepsilon\|_{0,1} \|\mathbf{K} \star f_\alpha\|_{0,p}.$$

It is readily seen by a change of variables that $\|\partial^\alpha \zeta_\varepsilon\|_{0,1} \leq C\varepsilon^{-|\alpha|}$, and, in view of relation (A.3.6), we obtain

$$\|\mathbf{K}_\varepsilon \star T\|_{0,p} \leq C\varepsilon^{1-m}\|T\|_{-m,p}. \tag{A.3.7}$$

Let us finally sketch the proof of the Poincaré identity [Eqs. (4.2.12) and (4.3.19)]. We will focus on the three-dimensional case. Let us start with scalar formula (4.2.12). We consider a volume of fluid, as shown in Figure A.1. The volume of fluid may contain bodies, but it can be rendered simply connected if thin tubes are drawn that join the surface of the body with the outer surface. Since $1/|\mathbf{x} - \mathbf{x}'|$ does not satisfy Laplace's equation at $\mathbf{x}$ we exclude this point from the volume by surrounding it with a sphere with a surface Σ and of radius ε.

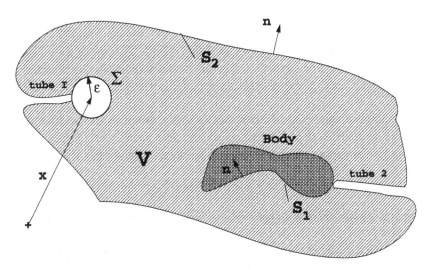

Figure A.1. Geometrical definitions of a domain that includes a body.

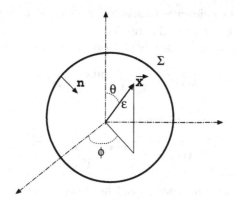

Figure A.2. Infinitesimal volume around a point $\mathbf{x}$ (see Eq. A.3.9).

By using the Green's theorem we obtain

$$\int_{S+\Sigma(\mathbf{x},\varepsilon)} \left(\frac{\mathbf{n}\cdot\nabla\Phi}{|\mathbf{x}-\mathbf{x}'|} - \Phi\mathbf{n}\cdot\nabla\frac{1}{|\mathbf{x}-\mathbf{x}'|} \right) d\mathbf{x}' = \int_{V-B(\mathbf{x},\varepsilon)} \frac{1}{|\mathbf{x}-\mathbf{x}'|}\Delta\Phi(\mathbf{x}')\,d\mathbf{x}'.$$

$$(A.3.8)$$

Integration over the infinitesimal tubes joining Σ and S_2 and S_1 and S_2 vanishes by continuity of Φ. On the surface Σ we have that (see Fig. A.2)

$$\mathbf{x}-\mathbf{x}' = -\varepsilon\,\mathbf{e}_r, \quad \mathbf{n} = -\mathbf{e}_r, \quad d\mathbf{x}' = \varepsilon^2\sin(\theta)\,d\theta\,d\phi, \quad (A.3.9)$$

$$\nabla\frac{1}{|\mathbf{x}-\mathbf{x}'|} = \frac{\mathbf{x}-\mathbf{x}'}{|\mathbf{x}-\mathbf{x}'|^3} = -\varepsilon\frac{\mathbf{e}_r}{\varepsilon^3} \qquad (A.3.10)$$

Hence

$$\int_{\Sigma(\mathbf{x},\varepsilon)} \left[\frac{\mathbf{n}\cdot\nabla\Phi(\mathbf{x}')}{|\mathbf{x}-\mathbf{x}'|} - \Phi(\mathbf{x}')\mathbf{n}\cdot\nabla\frac{1}{|\mathbf{x}-\mathbf{x}'|} \right] d\mathbf{x}' \qquad (A.3.11)$$

$$= -\Phi(\mathbf{x})\int_0^{2\pi} d\phi \int_0^{\pi} \varepsilon^2\frac{\mathbf{e}_r\mathbf{e}_r}{\varepsilon^2}\sin(\theta)\,d\theta + O(\varepsilon) \qquad (A.3.12)$$

$$= 4\pi\,\Phi(\mathbf{x}) + O(\varepsilon). \qquad (A.3.13)$$

On the other hand

$$\int_{B(\mathbf{x},\varepsilon)} \frac{1}{|\mathbf{x}-\mathbf{x}'|}\Delta\Phi(\mathbf{x}')\,d\mathbf{x}' = \Delta\Phi(\mathbf{x}) + O(\varepsilon^2). \qquad (A.3.14)$$

Hence, taking the limit as $\varepsilon \to 0$, we obtain Eq. (4.2.12).

In the case of a two-dimensional domain we use $\psi = \log(|\mathbf{x}|)$ in the Green's theorem and we can obtain similar results.

Let us now turn to vector identity (4.3.19). We define

$$\mathbf{A}(\mathbf{x}) = \frac{1}{4\pi} \int_{V_i} \frac{\mathbf{u}(\mathbf{x}')}{|\mathbf{x} - \mathbf{x}'|} \, d\mathbf{x}'. \tag{A.3.15}$$

Using Poisson's formula, we obtain that

$$\Delta \mathbf{A}(\mathbf{x}) = \begin{cases} -\mathbf{u}(\mathbf{x}) & \text{if } \mathbf{x} \text{ in } V_i \\ 0 & \text{if } \mathbf{x} \text{ in } V_e \end{cases}. \tag{A.3.16}$$

A relationship between the vorticity and the velocity field can be obtained with the vector identity:

$$\nabla \times \nabla \times \mathbf{A} = \nabla(\nabla \cdot \mathbf{A}) - \Delta \mathbf{A}. \tag{A.3.17}$$

Taking the curl and the divergence of the vector quantity $\mathbf{A}$ we have that

$$\nabla \times \mathbf{A} = \frac{1}{4\pi} \int_{V_i} \frac{(\nabla \times \mathbf{u})(\mathbf{x}')}{|\mathbf{x} - \mathbf{x}'|} \, d\mathbf{x}' + \frac{1}{4\pi} \int_{S_i} \frac{(\mathbf{n} \times \mathbf{u})(\mathbf{x}')}{|\mathbf{x} - \mathbf{x}'|} \, d\mathbf{x}', \tag{A.3.18}$$

$$\nabla \cdot \mathbf{A} = \frac{1}{4\pi} \int_{V_i} \frac{(\nabla \cdot \mathbf{u})(\mathbf{x}')}{|\mathbf{x} - \mathbf{x}'|} \, d\mathbf{x}' + \frac{1}{4\pi} \int_{S_i} \frac{(\mathbf{n} \cdot \mathbf{u})(\mathbf{x}')}{|\mathbf{x} - \mathbf{x}'|} \, d\mathbf{x}'. \tag{A.3.19}$$

Finally Poincaré formula (4.3.19) is obtained when Eqs. (A.3.16), (A.3.18), and (A.3.19) are substituted in Eq. (A.3.17).

A.3.2. Dynamics

We now turn to the full, nonlinear flow equations. We quickly review some important features of these equations, and distinguish between inviscid and viscous flows.

Inviscid Case

In two dimensions, the incompressible Euler equations consist of a transport equation with the vorticity coupled with system (A.3.1) for the velocity calculation. As for the linear transport equation, the vorticity is advected with the flow: With the notations introduced in Section A.1 for the characteristics,

$$\omega(\mathbf{x}, t) = \omega_0[\mathbf{X}(0; \mathbf{x}, t)], \tag{A.3.20}$$

where ω_0 is the initial vorticity field. Note that this formula is valid for flows in the $\mathbf{R}^2$ as well as in domains with solid walls and either periodic or no-through-flow boundary conditions. Of course it requires that the characteristics be well defined, which implies some kind of regularity in the flow. One can prove that a bounded vorticity is enough to ensure this minimal smoothness. If the initial vorticity is bounded, formula (A.3.20) forces the vorticity to remain bounded for all time, and a boot-strap argument yields a bounded solution to the flow Euler equations.

If the initial vorticity is not bounded, one can still prove the existence of weak solutions to the Euler equations as long as $\omega_0 \in L^p(\mathbf{R}^2)$ for some $p > 1$, and in this case it results from the incompressibility of the flow that $\|\omega(\cdot, t)\|_{0,p} = \|\omega_0\|_{0,p}$. For $p = 2$, in particular, this is the conservation of the enstrophy. An important case of even weaker regularity is the case of a vortex sheet. As we have said in Chapter 2, if the strength of the sheet is an analytic function, it remains so for a short time. Without this analyticity assumption, weak solutions can be obtained if the strength has a definite sign.

In three dimensions the vorticity transport equation is augmented by the stretching term $(\omega \cdot \nabla)\mathbf{u}$, which causes the enstrophy to change in time. The vorticity and the velocity gradients involve derivatives of the same order, so the stretching term acts a quadratic forcing on the vorticity equation. The possibility of a blow up of the solution in a finite time, even starting from a very smooth initial vorticity, cannot be excluded.

Viscous Equations

In the presence of viscosity, the molecular dissipation has a mollifying effect on the fields. In two dimensions and for periodic boundary conditions or flows in the whole space, this results in a decay of the enstrophy (for wall-bounded flows with no-slip boundary conditions, this is not true due to the vorticity generation at the walls). It also guarantees the well posedness of the equations for less smooth initial vorticity (for example, an initial vorticity consisting of a vortex sheet or even Dirac masses).

In three dimensions, it gives the existence of finite-energy solutions for all times. However, the existence of solutions that remain smooth for all time is an open problem.

Appendix B

Fast Multipole Methods for Three-Dimensional N-Body Problems

A fundamental issue in the use of vortex methods is the ability to use efficiently large numbers of computational elements for simulations of viscous and inviscid flows.

The traditional cost of the method scales as $O(N^2)$ as the N computational elements and particles induce velocities at each other, making the method unacceptable for simulations involving more than a few tens of thousands of particles. We reduce the computation cost of the method by making the observation that the effect of a cluster of particles at a certain distance may be approximated by a finite series expansion. When the space is subdivided in uniform boxes it is straightforward to construct an $O(N^{3/2})$ algorithm [189]. In the past decade faster methods have been developed that have operation counts of $O(N \log N)$ [17] or $O(N)$ [91], depending on the details of the algorithm. In these algorithms the particle population is decomposed spatially into clusters of particles (see, for example, Figure B.1) and we build a hierarchy of clusters (a tree data structure) – smaller neighboring clusters combine to form a cluster of the next size up in the hierarchy and so on. The hierarchy allows one to determine efficiently where the multipole approximation of a certain cluster is valid.

The N-body problem appears in many fields of engineering and science ranging from astrophysics to micromagnetics and computer animation. In the past few years these N-body solvers have been implemented and applied in simulations involving vortex methods. Koumoutsakos and Leonard [125] implemented the Greengard–Rohklin (GR) scheme in two dimensions for vector computer architectures, allowing for simulations of bluff-body flows with millions of particles. Winckelmans et al. [201] presented three-dimensional, viscous simulations of interacting vortex rings by using an implementation of a Barnes–Hut (BH) scheme for parallel computer architectures. Bhatt et al. [30] presented a vortex filament method to perform inviscid vortex ring interactions, with an

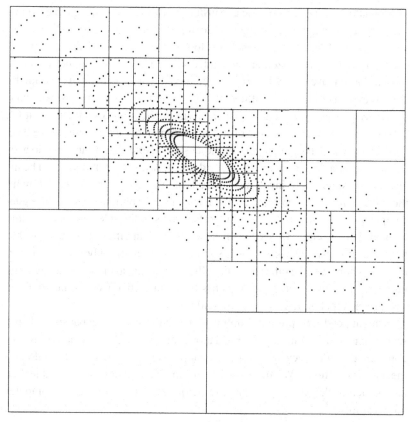

Figure B.1. A spiral distribution of particles decomposed into a hierarchy of clusters.

alternative implementation of a BH scheme for a Connection machine parallel computer architecture.

Historically the BH method was first implemented for large-scale astrophysical simulations. Several N-body algorithms are extensions of tree codes originally developed for gravitational interactions [201]. This has been motivated not only by the ease of implementation to a variety of applications, but also by the fact that in three dimensions the GR scheme suffers from the large computational cost of $O(p^4)$ associated with the translation of a pth-order multipole expansion. This cost has prohibited the use of large numbers of terms in the multipole expansions and has led to the general adoption of the BH method for N-body solvers. In a related effort to simplify the multipole expansion schemes, Anderson [7] presented a computational scheme based on the Poisson integral formula (hereafter referred to as PI). Greengard [92] presented a strategy to reformulate the translation of the expansions of the original GR scheme as convolution operators, thus enabling the use of fast Fourier transforms (FFTs) and the reduction of the computational cost to $O(p^2)$ operations. This strategy has been concisely summarized and extended in the work of Epton and Dembart [80]. If one can overcome the $O(p^4)$ difficulty, it may be beneficial to adopt the GR strategy, as by using a sufficiently large number of expansions, we avoid the costly pairwise interactions. The pairwise interactions usually account for most of the computational cost of the algorithm, and we try to minimize their number by using large numbers of expansions. Note that a large number of pairwise interactions may lead to algorithms of, say, $O(N^{1.1})$ that would be inefficient for simulations involving hundreds of millions of particles [30]. Finally the particulars of the problem at hand and the distribution of the Lagrangian computational elements can play important roles in the selection of a suitable method. For example, Brady [35] has constructed an efficient $N^{3/2}$ algorithm for three-dimensional vortex sheet simulations, keeping only a few terms in the expansions. On the other hand the efficiency of the approach of using large numbers of terms in the expansions has been quite effective in simulations of two-dimensional bluff-body and inviscid flows.

In this appendix we present a summary of the GR multipole expansion scheme with efficient $O(p^2)$ multipole translations for general N-body problems. We compare the efficiency of computing the expansions based on the GR and the PI formulations. We discuss also the implementation of a suitable tree data structure that has been shown to be quite effective for vector computer architectures.

B.1. Multipole Expansions

We present a summary of the multipole expansions technique as presented by Greengard and Rohklin [91] and Anderson [7]. We conduct some computations to assess the accuracy and the efficiency of the two techniques and describe the fast multipole translation theory following Epton and Dembart [80]. Finally we describe our tree data structure and its implementation so as to take advantage of vector computer architectures.

B.1.1. The Greengard–Rohklin Formulation

In order to introduce the subject of multipole expansions, we consider a unit source at a point $\mathbf{Q}(\mathbf{x}')$ (Figure B.2). This unit source induces a potential at point $\mathbf{P}(\mathbf{x})$ given by

$$\Psi(\mathbf{P}; \mathbf{Q}) = \Psi(\mathbf{x}; \mathbf{x}') = \frac{1}{|\mathbf{x} - \mathbf{x}'|}, \qquad (\text{B.1.1})$$

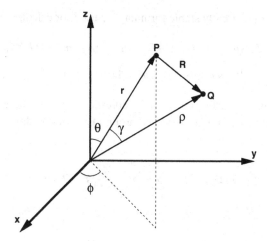

Figure B.2. Coordinate definition for multipole expansion.

where the spherical coordinates of $\mathbf{x}$ and $\mathbf{x}'$ are given by (r, θ, ϕ) and (ρ, α, β) respectively. The distance between the two points is denoted by $\mathbf{R}$ and the angle between vectors $\mathbf{x}$ and $\mathbf{x}'$ is denoted as γ. If we define $\mu = \rho/r$ and $\mathbf{u} = \cos \gamma$ then the potential at point $\mathbf{Q}$ may be expressed as

$$\Psi(\mathbf{P}; \mathbf{Q}) = \frac{1}{\mathbf{R}} = \frac{1}{r\sqrt{1 - 2u\mu + \mu^2}}. \qquad \text{(B.1.2)}$$

For $\mu = \rho/r < 1$, we use the generating formula for Legendre polynomials P_n so that the potential is expressed as

$$\Psi(\mathbf{P}; \mathbf{Q}) = \sum_{n=0}^{\infty} \frac{\rho^n}{r^{n+1}} P_n(u). \qquad \text{(B.1.3)}$$

This equation describes the far-field potential at a point $\mathbf{P}$, that is due to a charge of unit strength centered at $\mathbf{Q}$. To obtain a computationally tractable formulation we proceed to express the Legendre functions in terms of spherical harmonics,

$$P_n(\cos \gamma) = \sum_{m=-n}^{n} Y_n^{-m}(\alpha, \beta) Y_n^m(\theta, \varphi) \qquad \text{(B.1.4)}$$

and the spherical harmonics in terms of the associated Legendre polynomials,

$$Y_n^m(\theta, \varphi) = \sqrt{\frac{(n - |m|)!}{(n + |m|)!}} P_n^{|m|} \cos(\theta) e^{im\varphi}. \qquad \text{(B.1.5)}$$

The following numerically stable formulas are used for calculations:

$$(n - m) P_n^m(u) = (2n - 1) u P_{n-1}^m(u) - (n + m - 1) P_{n-2}^m(u) \quad \text{(B.1.6)}$$

$$P_m^m(u) = (-1)^m (2m - 1)! (1 - u^2)^{\frac{m}{2}}. \quad \text{(B.1.7)}$$

Summarizing, we see that the far-field representation of the potential induced by a collection of N_v sources centered around $\mathbf{Q}$ with coordinates $(\rho_i, \alpha_i, \beta_i)$ is expressed as

$$\Psi(\mathbf{P}; \{q_i\}) = \sum_{n=0}^{\infty} \sum_{m=-n}^{n} \frac{M_n^m}{r^{n+1}} Y_n^m(\theta, \varphi), \quad \text{(B.1.8)}$$

where:

$$M_n^m = \sum_{i=1}^{N_v} q_i \rho_i^n Y_n^{-m}(\alpha_i, \beta_i). \quad \text{(B.1.9)}$$

Note that if we wish to form a local expansion of the field around the origin then we express $1/\mathbf{R}$ as

$$\frac{1}{\mathbf{R}} = \frac{1}{\rho \sqrt{1 - 2\frac{r}{\rho} u + \left(\frac{r}{\rho}\right)^2}} \quad \text{(B.1.10)}$$

$$= \sum_{n=0}^{\infty} \frac{r^n}{\rho^{n+1}} P_n(u). \quad \text{(B.1.11)}$$

B.1.2. Translation of Multipole Expansions

Once the multipole expansions due to a collection of sources have been computed, one is usually interested in computing the far-field coefficients of the same collection expanded about some other point, say $\mathbf{S}$, so that the potential would be represented as

$$\Psi(\mathbf{S}; \mathbf{P}) = \sum_{n=0}^{\infty} \sum_{m=-n}^{n} \frac{L_n^m}{\sigma^{n+1}} Y_n^m(\Theta, \Phi) \quad \text{(B.1.12)}$$

where (σ, Θ, Φ) are the spherical coordinates of the distance between points $\mathbf{P}$ and $\mathbf{S}$.

This defines the translation problem for multipole expansions for fast multipole methods. Following Refs. 92 and 80 we present a concise summary of the theory underlying the translation of multipole expansions. We make use of the following definitions of harmonic outer functions O_n^m and inner functions I_n^m:

$$O_n^m(\mathbf{x}) = O_n^m(r, \theta, \varphi) = \frac{(-1)^n i^{|m|}}{A_n^m} \frac{Y_n^m(\theta, \varphi)}{r^{n+1}}, \qquad (\text{B.1.13})$$

$$I_n^m(\mathbf{x}) = I_n^m(r, \theta, \varphi) = i^{-|m|} A_n^m r^n Y_n^m(\theta, \varphi). \qquad (\text{B.1.14})$$

More specifically for $|\mathbf{x}| > |\mathbf{x}'|$ we obtain

$$O_n^m(\mathbf{x} - \mathbf{x}') = \sum_{n'=0}^{\infty} \sum_{m'=-n'}^{n'} (-1)^{n'} I_{n'}^{-m'}(\mathbf{x}') O_{n+n'}^{m+m'}(\mathbf{x}), \qquad (\text{B.1.15})$$

and for the inner expansions we get that

$$I_n^m(\mathbf{x} - \mathbf{x}') = \sum_{n'=0}^{\infty} \sum_{m'=-n'}^{n'} (-1)^{n'} I_{n'}^{m'}(\mathbf{x}') I_{n-n'}^{m-m'}(\mathbf{x}), \qquad (\text{B.1.16})$$

so now we may express the equation for the potential induced at point $\mathbf{x}$ from a unit source at point $\mathbf{x}'$ as

$$\frac{1}{|\mathbf{x} - \mathbf{x}'|} = O_0^0(\mathbf{x} - \mathbf{x}'). \qquad (\text{B.1.17})$$

In order to exhibit further the formulation of these translation operators we consider the configuration shown in Figure B.3. We wish to determine the potential induced by a collection of sources within a sphere centered at $\mathbf{x}_0$ and having radius R_0 [denoted as $s(\mathbf{x}_0, R_0)$] to a collection of points/sources within a sphere $s(\mathbf{x}_3, R_3)$. This is achieved in the following steps:

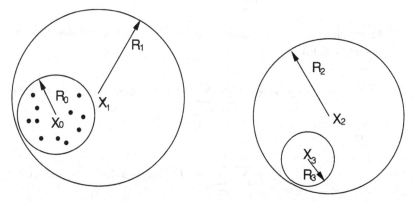

Figure B.3. Sketch for the translation of the multipole expansions.

(i) We compute a set of multipole expansion coefficients C_n^m [by using Eq. (B.1.8)] for the far-field representation of a set of sources distributed within $s(\mathbf{x}_0, R_0)$. Then the far-field representation of the field induced by this cluster of particles at a location $\mathbf{x}$ is given by

$$\Psi(\mathbf{x}) = \sum_{n=0}^{\infty} \sum_{m=-n}^{n} C_n^m O_n^{-m}(\mathbf{x} - \mathbf{x}_0). \qquad (B.1.18)$$

(ii) We translate the outer expansion about $\mathbf{x}_0$ to an outer expansion about $\mathbf{x}_1$ (child to parent):

$$\psi(\mathbf{x}) = \sum_{l=0}^{\infty} \sum_{j=-l}^{l} D_l^j O_l^{-j}(\mathbf{x} - \mathbf{x}_1), \qquad (B.1.19)$$

where

$$D_l^j = \sum_{n=0}^{l} \sum_{m=-n}^{n} C_n^m I_{l-n}^{j-m}(\mathbf{x}_1 - \mathbf{x}_0). \qquad (B.1.20)$$

(iii) We translate the outer expansion about $\mathbf{x}_1$ to an inner expansion about $\mathbf{x}_2$ (box–box interaction):

$$\psi(\mathbf{x}) = \sum_{l=0}^{\infty} \sum_{j=-l}^{l} E_l^j I_l^j(\mathbf{x} - \mathbf{x}_2), \qquad (B.1.21)$$

where

$$E_l^j = \sum_{n=0}^{l} \sum_{m=-n}^{n} D_n^m O_{l+n}^{-j-m}(\mathbf{x}_2 - \mathbf{x}_1). \qquad (B.1.22)$$

(iv) We translate the inner expansion about $\mathbf{x}_2$ to an inner expansion about $\mathbf{x}_3$ (parent to child):

$$\psi(\mathbf{x}) = \sum_{l=0}^{\infty} \sum_{j=-l}^{l} F_l^j I_l^j(\mathbf{x} - \mathbf{x}_3), \qquad (B.1.23)$$

where

$$F_l^j = \sum_{n=0}^{l} \sum_{m=-n}^{n} E_n^m I_{n-l}^{m-j}(\mathbf{x}_3 - \mathbf{x}_2). \qquad (B.1.24)$$

(v) Once the coefficients of the multipole expansions have been computed in the sphere $s(\mathbf{x}_3, R_3)$ we perform a local expansion by using Eq. (B.1.11) to compute the potential at the individual points.

The above representations for the translation operations of spherical harmonics reveal that they require the evaluation of double summations that are essentially convolution operations over the coefficients of the expansions and can essentially be computed with two-dimensional FFTs.

B.2. The Poisson Integral Method

In order to approximate the potential due to a collection of particles, Anderson [7] proposed an alternative simplified technique. This technique is based on the observation that a harmonic function (Ψ) external to a sphere of radius R may be described with its boundary values $g(R\mathbf{s})$ on the surface of the sphere. So given a point $\mathbf{x}$ we define $\mathbf{x}_p$, a point on the unit sphere that points in the direction of $\mathbf{x}$, so that:

$$\Psi(\mathbf{x}) = \frac{1}{4\pi} \int_{S^2} \left[\sum_{n=0}^{\infty} (2n+1) \left(\frac{R}{r}\right)^{n+1} P_n(\mathbf{s} \cdot \mathbf{x}_p) \right] g(R\mathbf{s}) \, d\mathbf{s}, \qquad \text{(B.2.1)}$$

where S^2 denotes the surface of the unit sphere and P_n is the nth-order Legendre polynomial. We use a quadrature formula to integrate the function on the surface of the sphere with K integration points $\mathbf{s}_i$ and weights w_i to obtain an approximation of the form

$$\Psi(\mathbf{x}) \approx \frac{1}{4\pi} \sum_{i=1}^{K} \left[\sum_{n=0}^{M} (2n+1) \left(\frac{R}{r}\right)^{n+1} P_n(\mathbf{s}_i \cdot \mathbf{x}_p) \right] g(R\mathbf{s}_i) w_i. \qquad \text{(B.2.2)}$$

In order to compute the far-field multipole expansion of a cluster of particles based on this formulation, the function $g(R\mathbf{s})$ is determined on certain quadrature points on the surface of a sphere by use of the direct interactions of the potential induced by the sources onto the evaluation points. Subsequently these coefficients are used in approximation (B.2.2) to compute the potential induced at distances sufficiently large compared with the radius of the cluster. In order to translate the expansions, we may use the above formula again by considering the coefficients of $g(R\mathbf{s}_i)$ on the inner sphere to be sources themselves. We complete the method by observing that a local expansion approximation may

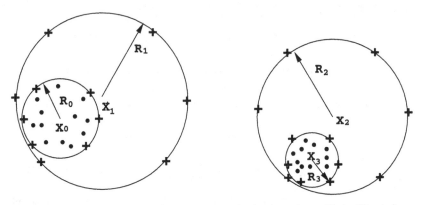

Figure B.4. Sketch for the translation of the multipole expansions with the PI technique.

be constructed by using the following formula.

$$\Psi(\mathbf{x}) \approx \frac{1}{4\pi} \sum_{i=1}^{K} \left[\sum_{n=0}^{M} (2n+1) \left(\frac{r}{R} \right)^n P_n(\mathbf{s}_i \cdot \mathbf{x}_p) \right] g(R\mathbf{s}_i) w_i. \qquad \text{(B.2.3)}$$

Note that the expansions are in terms of r/R in this formula.

Anderson [7] showed that approximations with $M = 2p + 1$ terms may be compared with multipole schemes that have p terms retained in the expansions. The strength of this method relies on its simplicity and its easy extension from two to three dimensions. This is exhibited if the implementation of this technique is considered in the context of an $O(N)$ algorithm for the computation of the potential field induced by a set of particles within a sphere $s(\mathbf{x}_0, R_0)$ to a cluster of points in $s(\mathbf{x}_0, R_0)$ (Figure B.4). This interaction is performed in the following steps:

(i) The potential induced by the particles (denoted in Figure B.4 by dots) is computed on quadrature points properly selected on the surface of the enclosing sphere (denoted in Figure B.4 by +), thus constructing the function $g(R_0\mathbf{s}_i)$

(ii) The potential induced by the quadrature points on $s(\mathbf{x}_0, R_0)$ is computed on the quadrature points of $s(\mathbf{x}_1, R_1)$ with approximation (B.2.2).

(iii) The coefficients $g(R_1\mathbf{s}_i)$ are considered to be sources themselves so that the coefficients $g(R_2\mathbf{s}_i)$ are computed with approximation (B.2.2).

(iv) The coefficients $g(R_3\mathbf{s}_i)$ are computed subsequently by performing a local expansion from the quadrature points on sphere $s(\mathbf{x}_2, R_2)$ to the quadrature points on the sphere $s(\mathbf{x}_3, R_3)$ with approximation (B.2.3).

(v) By use of approximation (B.2.3) the potential on the particles inside the sphere $s(\mathbf{x}_3, R_3)$ is computed with the coefficients $g(R_3 \mathbf{s}_i)$.

The simplicity of the formulas implemented in this technique make it an attractive alternative to the multipole expansion method of GR. We consider below a comparison of the two methods in terms of their accuracy and computational cost.

B.3. Computational Cost

The computational cost, associated with the multipole expansions of the GR scheme, scales as $O(p^2 N)$ for the multipole–particle operations and as $O(\log N \, p^2)$ for the multipole translation operations that use the convolution formulation discussed above. For the PI formulation, assuming K integration points and M terms in the Poisson kernel, the cost scales as $O(K \times M \times N)$ for the particle–multipole interactions and as $O(K \times K \times M)$ for the multipole translation. Both algorithms have error terms that scale as $(R/r)^{L+1}$, where L corresponds to p terms in the multipole method and to $M + 1$ for the PI scheme. The number of the required integration points (K) corresponding to 5th-, 7th-, 9th-, and 14th-order quadrature formulas require 12, 24, 32, and 72 points, respectively, or approximately $K \approx 2 \, m^2/5$ points for an mth-order integration formula. So in order to achieve the same accuracy with the PI method as with the multipole scheme we require that $m \approx M \approx P$. This then implies a computational cost of $O(p^3 N)$ for the particle–multipole interactions while it implies an $O(\log N \, p^5)$ cost associated with the multipole translations.

Although such estimates depend on the particular implementation of the method, it is evident that for the same order of accuracy the multipole method with fast translation operations would lead to much faster computations, especially for large expansion orders. Moreover, in the particular implementation of the methods it was easier to unroll the computational loops involving the particle–multipole interactions for the GR scheme than the respective operations in the PI scheme. In (Figure B.5) we present also the relative error as computed by the two methods by using $P = 8$ and different values of m in the PI method.

In Figure B.6 we present some representative timing results for the construction of the multipole expansions as well as for the particle–multipole interactions. A number of N particles were distributed randomly inside a cube of side 2 and the potential was evaluated on L points uniformly distributed along a line extending from the center of the cube and along one of its sides. The first

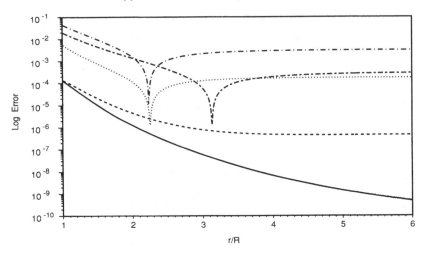

Figure B.5. Relative errors of the GR and the PI multipole expansion schemes: Solid curves, GR, $P = 8$; dashed curve, PI, $m = 14$; dotted curve, $m = 9$; dashed–dashed curve, $m = 7$; dashed–dotted curve, $m = 5$.

N	L	$m = 5$	$m = 7$	$m = 9$	$m = 14$	Multipoles $P = 8$
10^5	10^5	0.0708	0.1397	0.1863	0.4191	0.1023
		0.5268	0.5844	0.6395	1.0870	0.0959
10^6	10^6	0.7111	1.4219	1.8968	4.2680	1.0251
		5.2724	5.8516	6.4145	10.8925	0.9585

Figure B.6. Computational cost of the fast multipole method and the PI method.

line corresponds to the CPU time required for the construction of the multipole expansions while the second line corresponds to the CPU time required for performing the particle–multipole expansions. It is observed then that when the GR multipole expansion scheme is used, 1-order-of-magnitude faster calculations are achieved for the same order of accuracy.

B.4. Tree Data Structures

A key issue in the implementation of fast multipole methods is the associated data structure and the computer platform. We present a methodology that has been successfully implemented for the GR algorithm on vector computer architectures. Several of its features could be carried over to parallel platforms involving large numbers of vector processors. The GR scheme relies on a predetermined tree data structure and a large number of terms to be kept in the expansions while the BH scheme determines the interaction list while traversing the tree data structure, and a trade-off is made between the number of terms kept in the expansions (usually two to four) and the distance at which the expansions are favored over direct interactions.

In order to exploit the observation that the effect of a cluster of particles at a certain distance may be approximated by a finite sum of series expansions by using the equations described above, we need to organize the particles into a hierarchy of clusters. This hierarchy of clusters allows us to determine efficiently when the approximation is valid. In order to establish the particle clusters we may resort to a tree-building algorithm.

The contribution of a cluster of particles to the potential of a given vortex can then be computed to desired accuracy if the particle is sufficiently far from the cluster in proportion to the size of the cluster and a sufficiently large number of terms in the multipole expansion are taken. This is the essence of the particle–box method, requiring $O(N \log N)$ operations. One then tries to minimize the work required by maximizing the size of the cluster used while keeping the number of terms in the expansion within a reasonable limit and maintaining a certain degree of accuracy.

The box–box scheme goes one step further as it accounts for box–box interactions as well. These interactions are in the form of shifting the expansions of a certain cluster with the desired accuracy to the center of another cluster. Then those expansions are used to determine the velocities of the particles in the second cluster. This has as the effect of minimizing the tree traversals for the individual particles requiring only $O(N)$ operations.

In order to construct the tree data structure, the three-dimensional space is considered to be a cube enclosing all computational elements. We apply the operation of continuously subdividing a cube into eight identical cubes until each cube has only a certain maximum number of particles in it or a maximum allowable level of subdivisions has been reached.

The hierarchy of boxes defines a tree data structure that is common for both algorithms. The tree construction proceeds level by level starting at the finest level of the particles and proceeding upward to coarser box levels. Because of the simplicity of the geometry of the computational domain, the addressing of the elements of the data structure is facilitated significantly. As the construction proceeds, pointers are assigned to the boxes so that there is direct addressing of the first and the last particle index in them as well as direct access to their children and parents. This facilitates the computation of the expansion coefficients of the children from the expansions of the parents for the box–box algorithm and the expansions of the parents from those of the children for the particle–box.

The data structure is used to determine when the expansions are to be used and when pairwise interactions have to be calculated. It helps in communicating to the computer the geometric distribution of the particles in the computational domain. The particles reside at the finest level of the structure. Clusters of particles form the interior nodes of the tree, and hierarchical relations are established. The data structure adds to the otherwise minimal memory requirements of the vortex method.

The tree has to be reconstructed at every step as the particles change positions in the domain. There are several ways that nearby particles could be clustered together and some of the decisions to be made are

The center of expansions. Usually the geometric center of the cells is used as it facilitates the addressing of the data structure.

The cluster size. In our experience we find beneficial to follow a hybrid strategy as we keep at least L_{min} particles per box until we reach a predetermined finest level of boxes. The number L_{min} may be chosen by the user, depending on the particle population and configuration so as to achieve an optimal computational cost.

Addressing the clusters. As particles are usually associated with a certain box, it is efficient to sort the particle locations in the memory so that particles that belong to the same box occupy adjacent locations in the memory devoted to the particle arrays. Such memory allocation enhances the vectorization significantly as very often we loop over particles of the same box (e.g., to construct the expansions at the finest level, or to compute interactions) and the loops have an optimum stride of one.

We now describe two algorithms, a particle–box (PB) and a box–box (BB), in which we may distinguish three stages:

- Building the data structure (tree).
- Establishing the interaction lists (by nonrecursively descending the tree).
- Computing pairwise interactions for all particles in the domain.

The building of the data structure is common for PB & BB algorithms, but they differ in the tree descent and the pairwise interactions. Care has been exercised at all stages to maximize vectorization. In our respective two-dimensional implementation, the building of the data structure consumes $\sim$5%–7% of the time, the descent consumes another 5%–10%, so that the largest amount is spent in computing the pairwise interactions.

B.4.1. The Particle–Box Algorithm

Step 1: Building the Data Structure (Tree). *Step 1a*: For each of the cubes at each level that is not further subdivided, we compute the p terms of the multipole expansions. These expansions are used to describe the influence of the particles at locations that are well separated from their cluster.

Step 1b: We construct the expansions of all parent boxes by shifting the expansion coefficients of their children. The tree is traversed upward in this stage. Rather than constructing the expansions of all the members of a family (that is, traverse each branch until the root is reached) we construct the expansions of all parent boxes at each level simultaneously. This enables long loops over the parent boxes at each level. Care is taken so that the procedure is fully vectorized by taking advantage of the regularity of the data structure and the addressing of the boxes in the memory. Moreover, the regularity of the data structure allows us to precompute many coefficients that are necessary for the expansions. Straightforward implementation of these translations leads to a computational cost of $O(p^4)$. This has been the major reason that several implementations of the algorithm have used only up to $p = 3$ terms in the multipole expansions. However, such an approach results in large numbers of particle–particle interactions and hence a large computational cost. We can implement the technique proposed in Ref. 92 (see above) to reduce the computational cost of this translation to $O(p^2)$ operations by observing that this translation amounts to a convolution, and we use FFTs.

Step 2: Establishing of Interaction Lists. In the present algorithm a breadth-first search is performed at each level to establish the interaction lists of each particle (cell). This search is facilitated by the regularity of the data structure and the identification arrays of the cells in the tree. At each level interaction lists are established for the particles (cells) by looping across the cells of a certain level.

Note that this depth-first search for interaction lists is further facilitated by the fact that every particle belongs to a childless box. It is easy then to observe that all particles in the same box share the same interaction list, which comprises members of the tree that belong to coarser levels. In this way the tree is traversed upward for all particles in a childless box and downward separately for each particle. It is evident that this procedure is more efficient for uniformly clustered configurations of particles because there would be more particles that belong to childless boxes at the finest level.

Step 3: Computation of the Interactions. Once the interaction lists have been established, we compute the velocities of the particles by looping over the elements of the lists. For particles that have the same boxes in their interaction list, this is performed simultaneously so that memory referencing is minimized. Moreover, by systematically traversing the tree, we make the particle–particle interactions symmetric so that the cost of this computation is halved. The cost of this step is $O(Np^2)$.

B.4.2. The Box–Box Algorithm

This scheme is similar to the particle–box scheme except that here every node of the tree assumes the role of a particle. In other words, interactions are not limited to particle–particle and particle–box but interactions between boxes are considered as well. Those interactions are in the form of shifting the expansion coefficients of one box into another and the interaction lists are established with respect to the locations of every node of the tree.

The scheme distinguishes five categories of interacting elements of the tree with respect to a cell denoted by **c**.

- **List 1:** All childless boxes neighboring **c**.
- **List 2:** Children of colleagues (boxes of the same size) of **c**'s parents that are well separated from **c**. All such boxes belong to the same level with **c**.
- **List 3:** Descendants (not only children) of **c**'s colleagues whose parents are adjacent to **c** but are not adjacent to **c** themselves. All such boxes belong to finer levels.
- **List 4:** All boxes such that box **c** belongs to *their* list 3. All such boxes are childless and belong to coarser levels.
- **List 5:** All boxes well separated from **c**'s parents. Boxes in this category do not interact directly with the cell **c**.

If the cell **c** is childless it may have interacting pairs that belong to all four lists. However, if it is a parent it is associated with boxes that belong to lists 2 and 4, as described above. These observations are directly applied in our algorithm and we may distinguish again the following 4 steps.

Step 1: Building the Data Structure. This procedure is the same as that for the particle–box scheme. This fact enables us to compare directly the two algorithms and assess their efficiency.

Step 2: Construction of Interaction Lists. To establish the interaction lists we proceed again level by level, starting at the coarsest level. For each level we distinguish childless and parent boxes. In establishing lists 1 and 3 we need only loop over childless boxes, whereas to establish lists 2 and 4 we loop over all cells that are active in a certain level.

Step 2a: Here we establish lists 1 and 3. We start at the level of the parents of box **c** and we proceed level by level, examining again breadth first, until we reach the finest level of the structure (the particles). The elements of list 1 are basically the particles and account for the particle–particle interactions. Care is exercised so that this computation is symmetric, and we need to traverse the tree downward only, thus minimizing the search cost. The elements of list 3 are the boxes and are accountable for the particle–box interactions in this scheme.

Step 2b: Here we establish interaction lists 2 and 4 for all boxes in the hierarchy. We start at the coarsest possible level and proceed downward until reaching the level of box **c** to establish the interaction lists. To do so for a certain box, we start by examining boxes that are not well separated from their parents (otherwise they would have been dealt with at the coarser level). Subsequently the children of those boxes are examined to establish interaction lists.

Step 3: Computations of the Interactions. In this scheme we consider three kinds of interactions: the box–box, particle–box, and particle–particle interactions. The latter two categories were discussed in the previous section. For the box–box interactions, once the respective interaction lists have been established (with members of lists 2 and 4), we need to transfer those expansions down to those of the children and add them to the existing ones. We vectorize this procedure by looping over the number of boxes at each level. The use of pointers to access the children of each box enhances this vectorization. Note that we can calculate an arbitrarily high number of expansions efficiently by unrolling the loop over the number of expansions into the previously mentioned loop.

In summary, a fast N-body solver can be implemented for a number of different cases in vortex methods (such as fast velocity evaluation among vortices, solution of the systems arising in panel methods, fast Poisson solver for the streamfunction–vorticity formulation in complex geometries, etc.). The speed and the efficiency of the final computer code may depend strongly on the physical problem as well as on the programmer and the particular computer architecture that is being implemented.

Bibliography

[1] Abramowitz, M. and Stegan, I.A. 1964, *Handbook of Mathematical Functions*, Dover, New York.

[2] Adams, R. 1975, *Sobolev Spaces*, Academic, New York.

[3] Albuquerque, C. 1999, Ph.D. thesis, University of Lisbon, Portugal.

[4] Anderson, C. 1985, *A vortex method for flows with slight density variation*, J. Comput. Phys. **61**, 417–444.

[5] Anderson, C. 1986, *A method of local corrections for computing the velocity field due to a distribution of vortex blobs*, J. Comput. Phys. **62**, 111–123.

[6] Anderson, C. 1989, *Vorticity boundary conditions and boundary vorticity generation for two-dimensional viscous incompressible flows*, J. Comput. Phys. **80**, 72–97.

[7] Anderson, C. 1992, *An implementation of the fast multipole method without multipoles*, SIAM J. Sci. Stat. Comput. **13**, 923–947.

[8] Anderson, C. and Greengard, C. 1985, *On vortex methods*, SIAM J. Num. Anal. **22**, 413–440.

[9] Anderson, C. and Greengard, C., eds. 1988, *Vortex Methods*, Vol. 1360 of Lecture Notes in Mathematics Series, Springer-Verlag, Berlin.

[10] Anderson, C. and Greengard, C. 1988, *Vortex ring interaction by the vortex method*, in Ref. 9, pp. 23–36.

[11] Anderson, C. and Greengard, C., eds. 1991, *Vortex Dynamics and Vortex Methods*, Vol. 28 of Lectures in Applied Mathematics Series, American Mathematical Society, New York.

[12] Anderson, C. and Reider, M. 1994, *Investigation of the Prandtl Navier–Stokes equation procedures for two-dimensional incompressible flows*, Phys. Fluids **6**, 2380–2389.

[13] Appel, A. 1985, *An efficient program for many-body simulation*, SIAM J. Sci. Stat. Comput. **6**, 85–103.

[14] Aref, H. 1983, *Integrable, chaotic and turbulent vortex motion in two-dimensional flows*, Ann. Rev. Fluid Mech. **15**, 345–389.

[15] Baker, G.R. and Shelley, M.J. 1986, *Boundary integral techniques for multi-connected domains*, J. Comput. Phys. **64**, 112–132.

[16] Bar-Lev, M. and Yang, H.T. 1975, *Initial flow field over an impulsively started circular cylinder*, J. Fluid Mech. **72**, 625–647.

[17] Barnes, J.E. and Hut, P. 1986, *A hierarchical $O(N \log N)$ force-calculation algorithm*, Nature (London) **324**, 446–449.

[18] Batchelor, G.K. 1967, *An Introduction to Fluid Dynamics*, Cambridge University Press, Cambridge.

[19] Beale, J.T. 1986, *A convergent 3d vortex method with grid-free stretching*, Math. Comput. **46**, 401–424.

[20] Beale, J.T. 1988. *On the accuracy of vortex methods at large time*, in *Computational Fluid Dynamics and Reacting Gas Flows*, B. Engquist et al., eds., Springer-Verlag, New York, pp. 19–32.

[21] Beale, J.T. and Greengard, C. 1993, *Convergence of Euler–Stokes splitting of the Navier–Stokes equations*, Commun. Pure Appl. Math. **46**, 1083–1115.

[22] Beale, J.T., Cottet, G.-H., and Huberson, S., eds. 1993, *Vortex Flows and Related Numerical Methods*, NATO ASI Series, Kluwer Academic, Dordrecht, The Netherlands.

[23] Beale, J.T., Eydeland, A., and Turkington, B. 1991, *Numerical Tests of 3-D Vortex Methods Using a Vortex Ring with Swirl*, Ref. 11, 1–10.

[24] Beale, J.T. and Majda, A. 1981, *Rates of convergence for viscous splitting of the Navier–Stokes equations*, Math. Comput. **37**, 243–259.

[25] Beale, J.T. and Majda, A. 1982, *Vortex methods II: high order accuracy in 2 and 3 dimensions*, Math. Comput. **32**, 29–52.

[26] Beale, J.T. and Majda, A. 1985, *High order accurate vortex methods with explicit velocity kernels*, J. Comput. Phys. **58**, 188–208.

[27] Belotserkovskii, S.M., Skripach, B.K., and Tabachnikov, V.G. 1971, *Wing in Unsteady Gas Flow* Nauka, Moscow.

[28] Belotserkovskii, S.M. and Lifanov, I.K.H. 1993, *Methods of Discrete Vortices*, CRC, Boca Raton, FL.

[29] Benfatto, G. and Pulvirenti, M. 1986, *Convergence of Chorin–Marsden product formula in the half-plane*, Commun. Math. Phys. **106**, 427–458.

[30] Bhatt, S., Liu, P., Fernandez, V., and Zabusky, N. 1995, *Tree codes for vortex dynamics: Application of a programming framework*, Workshop on Solving Irregular Problems on Parallel Machines, Santa Barbara, CA, April 25–28.

[31] Blasius, H. 1908, *Grenzchichten in Flüssigkeiten mit Kleiner Reibug*. Z. Angew. Math. Phys. **56,** 1–20.

[32] Bouard, R. and Coutanceau, M. 1980, *The early stage of development of the wake behind an impulsively started cylinder for* $40 \leq Re \leq 10^4$, J. Fluid Mech. **101**, 583–607.

[33] Brackbill, J. and Khote, D. 1988, *FLIP: a method for low-dissipation particle in cell method for fluid flow*, Comput. Phys. Commun. **48**, 25–38.

[34] Brackbill, J. and Ruppel, H. 1986, *Flip: a method for adaptively zoned, particle in cell calculations of fluid flows in two dimensions*, J. Comput. Phys. **65**, 314–343.

[35] Brady, M. 1998, Ph.D. dissertation, Caltech.

[36] Brenier, Y. 1985, *Une méthode particulaire pour les équations non linéaires de convection–diffusion en dimension 1*, Rapp. Rech. 0422, INRIA.

[37] Brenier, Y. and Cottet, G.-H. 1995, *Convergence of vortex methods with random rezoning for the 2-D Euler and Navier–Stokes equations*, SIAM J. Num. Anal. **32**, 1080–1097.

[38] Buntine, J. and Pullin, D. 1989, *Merger and cancellation of strained Vortices*, J. Fluid Mech. **205**, 263–295.

[39] Caflish, R., ed. 1989, *Mathematical Aspects of Vortex Dynamics*, Society for Industrial and Applied Mathematics, Philadelphia, PA.

[40] Casciola, C.M., Piva, R., and Bassanini, P. 1996, *Vorticity generation on a flat surface in 3D flows*, J. Comput. Phys. **129**, 345–356.

[41] Chacon-Rebollo, T. and Hou, T.Y. 1990, *A Lagrangian finite element method for the 2-D Euler equations*, Commun. Pure Appl. Math. **43**, 735–767.

[42] Chang, C.C. and Chern, R.L. 1991, *A numerical study of flow around an impulsively started cylinder by a deterministic vortex method*, J. Fluid Mech. **233**, 243–263.

[43] Cheer, A. 1989, *Unsteady separated wake behind an impulsively started cylinder flow*, J. Fluid Mech. **201**, 485–505.

[44] Choquin, J.-P. 1987, *Simulation numérique d'écoulements tourbillonnaires de fluides incompressibles par des méthodes particulaires*, Thèse de l'Université Paris VI.

[45] Choquin, J.-P. and Cottet, G.-H. 1988, *Sur l'analyse d'une classe de méthodes de vortex tri-dimensionnelles*, C. R. Acad. Sci. Paris **306**, 739–42.

[46] Choquin, J.-P., Cottet, G.-H. and Mas-Gallic, S. 1988, *On the validity of vortex methods for nonsmooth flows*, in Ref. 9, 56–67.

[47] Choquin, J.-P. and Lucquin-Desreux, B. 1988, *Accuracy of a deterministic particle method for the Navier–Stokes equations*, Int. J. Num. Fluids **8**, 1439–1458.

[48] Choquin, J.P. and Huberson, S. 1989, *Particle simulation of viscous flow*, Comput. Fluids **17**, 397–410.

[49] Chorin, A.J. 1973, *Numerical study of slightly viscous flow*, J. Fluid Mech. **57**, 785–796.

[50] Chorin, A.J. 1978, *Vortex sheet approximation of boundary layers*, J. Comput. Phys. **27**, 428–442.

[51] Chorin, A.J. 1980, *Vortex models and boundary layer instability*, SIAM J. Sci. Stat. Comput. **34**, 1–21.

[52] Chorin, A.J. 1990, *Hairpin removal in vortex interactions*, J. Comput. Phys. **91**, 1–21.

[53] Chorin, A.J. 1993, *Hairpin removal in vortex interactions II*, J. Comput. Phys. **107**, 1–9.

[54] Chorin, A.J., Hugues, T.J., McCracken, M.F., and Marsden, J. 1978, *Product formulas and numerical algorithms*, Commun. Pure Appl. Math. **31**, 205–256.

[55] Christiansen, J.P. 1973, Numerical solution of hydrodynamics by the method of point vortices, J. Comput. Phys. **13**, 363–379.

[56] Cohen, A. and Perthamme, B. 1998, *Optimal approximation of transport equations by particle and pseudo particle methods*, submitted.

[57] Collins, W.M. and Dennis, S.C.R. 1973, *Flow past an impulsively started circular cylinder*, J. Fluid Mech. **60**, 105–127

[58] Cottet, G.-H. 1982, Ph.D. thesis, Université Paris 6, Paris.

[59] Cottet, G.-H. 1988, *Vorticity boundary conditions and the deterministic vortex method for the Navier–Stokes equation*, in Ref. 39.

[60] Cottet, G.-H. 1988, *A new approach for the analysis of vortex methods in 2 and 3 dimensions*, Ann. Inst. Henri Poincaré **5**, 227–285.

[61] Cottet, G.-H. 1991, *Particle-grid domain decomposition methods for the Navier–Stokes equations in exterior domains*, in Ref. 11, 100–118.

[62] Cottet, G.-H. 1991, *Large time behavior of deterministic particle approximations to the Navier–Stokes equations*, Math. Comput. **56**, 45–59.

[63] Cottet, G.-H. 1994, *A vorticity creation algorithm for the Navier–Stokes equations in arbitrary domain*, in Navier–Stokes equations and related non-linear problems, A. Sequeira (ed).

[64] Cottet, G.-H. 1996, *Artificial viscosity models for vortex and particle methods*, J. Comput. Phys. **127**, 299–308.

[65] Cottet, G.-H., Goodman, J., and Hou, T.Y. 1991, *Convergence of the grid-free point vortex method for the three-dimensional Euler equations*, SIAM J. Num. Anal. **28**, 291–307.

[66] Cottet, G.-H. and Mas-Gallic, S. 1983, *Une méthode de décomposition pour une équation de type convection-diffusion combinant résolution explicite et méthode particulaire*, C. R. Acad. Sci. Paris **297**, 133–136.

[67] Cottet, G.-H. and Mas-Gallic, S. 1987, *A particle method to solve transport-diffusion equations; Part II: the Navier–Stokes system*, Internal Report 158, Ecole Polytechnique, Paris.

[68] Cottet, G.-H. and Mas-Gallic, S. 1990, *A particle method to solve the Navier–Stokes system*, Num. Math. **57**, 1–23.

[69] Couet, B., Buneman, O., and Leonard, A. 1981, *Simulation of three-dimensional flows with vortex-in-cell methods*, J. Comput. Phys. **39**, 305.

[70] Crighton, D.G. 1985, *The Kutta condition in unsteady flow*, Ann. Rev. Fluid Mech. **17**, 411–445.

[71] Da Rios, L.S. 1906, *Sul moto 'un liquido indefinito con un filetto vorticoso di forma qualunque*, Rend. Circ. Mat. Palermo, **22**, 117–135.

[72] Daube, O. 1992, *Resolution of the 2D Navier–Stokes equations in velocity–vorticity form by means of an imfluence matrix technique*, J. Comput. Phys. **103**, 402–414.

[73] Degond, P. and Mas-Gallic, S. 1989, *The weighted particle method for convection–diffusion equations*, Math. Comput. **53**, 485–526.

[74] Driscoll, C.F., and Fine, K. 1990, *Experiments on vortex dynamics in pure electron plasmas*, Phys. Fluids B **6**, 1359–1366.

[75] Dritschel, D.G. 1989, *Contour dynamics and contour surgery: numerical algorithms for extended, high-resolution modelling of vortex dynamics in two dimensional, inviscid, incompressible flows*, Comput. Phys. Rep. 1077–146.

[76] Dukowicz, J.K. and Kodis, J.W. 1987, *Accurate conservative remapping (rezoning) for arbitrary Lagrangian–Eulerian computations*, SIAM J. Sci. Stat. Comput. **8**, 305–321.

[77] E, W. and Hou, T.Y. 1987, *Convergence of the vortex method for oscillatory vorticity field*, CAM Report 10, Department of Mathematics, UCLA.

[78] E, W. and Liu J.G. 1996, *Finite-difference schemes for incompressible flows in vorticity formulation*, in *Vortex Flows and Related Numerical Methods II*, ESAIM Proc., 181–195, *http://www.emath.fr/proc/Vol1/*.

[79] Ebiana, A.B. and Bartholomew, R.W. 1996, *Design consideration for numerical filters used in vortex-in-cell*, Comput. Fluids **25**, 61–75.

[80] Epton, M.A. and Dembart, B. 1995, *Multipole translation theory for the three-dimensional Laplace and Helmholtz equations*, SIAM J. Sci. Stat. Comp. **16**, 865–897.

[81] Falkner, V.M. 1953, *The solution of lifting-plane problems by vortex lattice theory*, Aeronaut. Res. Counc. Rep. Mem. 25911-30.

[82] Feynman, R.P. 1957, *Application of Quantum Mechanics to Liquid Helium*, Prog. Low temp. Phys. **I**, 16–53.

[83] Fishelov, D. 1990, *A new vortex scheme for viscous flows*, J. Comput. Phys. **86**, 211–224.

[84] Fischer, P.F. 1997, *An overlapping Schwarz method for spectral element solution of the incompressible Navier–Stokes equations*, J. Comput. Phys. **133**, 84–101.

[85] Friedmann, A. 1964, *Partial Differential Equations of Parabolic Type*, Prentice-Hall, Englewood Cliffs, NJ.

[86] Fritts, M.J., Crowley, W.P., and Trease, H.E., eds. 1985, *The Free Lagrange Method*, Vol. 238 of Lecture Notes in Physics Series, Springer-Verlag, Berlin, New York.

[87] Gagnon, Y., Cottet, G.-H., Dritschel, D., Ghoniem, A., and Meiburg, E. eds. 1996, *Vortex flows and related numerical methods II*, ESAIM Proc., *http://www.emath.fr/proc/Vol1/*.

[88] Gharakhani, A and Ghoniem, A. 1998, *Simulation of the piston driven flow inside a cylinder with an eccentric port*, J. Fluids Eng. **120**, 319–326.

[89] Greengard, C. 1985, *The core spreading vortex method approximates the wrong equation*, J. Comput. Phys. **61**, 345–348.

[90] Greengard, C. 1986, *Convergence of the vortex filament method*, Math. Comput. **47**, 387–398.

[91] Greengard, L. 1988, *On the efficient implementation of fast multipole algorithms*, Dept. of Computer Science Report 602, Yale University, New Haven, CT.

[92] Greengard, L. and Rokhlin, V. 1987, *A fast algorithm for particle simulation*, J. Comput. Phys. **73**, 325–348.

[93] Greengard, L. and Strain, J. 1990, *A fast algorithm for the evaluation of heat potentials*, Commun. Pure Appl. Math. **XLIII**, 949–963.

[94] Goodman, J. 1987, *The convergence of random vortex methods*, Commun. Pure Appl. Math. **40**, 189–220.

[95] Goodman, J., Hou, T.Y., and Lowengrub, J. 1990, *Convergence of the point vortex method for the 2-D Euler equations*, Commun. Pure Appl. Math., **43**, 415–430.

[96] Gray, A. and Mathews, G.B. 1952, *A Treatise on Bessel Functions and Their Applications to Physics*, MacMillan, London.

[97] Guermond, J.L., Huberson, S., and Shen, W.S. 1993, *Simulation of 2D external viscous flows by means of a domain decomposition method*, J. Comput. Phys. **108**, 343–352.

[98] Hald, O. 1979, *Convergence of vortex methods II*, SIAM J. Num. Anal. **16**, 726–755.

[99] Hasimoto, H. and Kambe, T., eds. 1987, *Fundamental Aspects of Vortex Motion*, IUTAM Proc., North-Holland, Amsterdam.

[100] Henshaw, W.D., Kreiss, H.-O., and Reyna, L.G. 1989, *On the smallest scale for the compressible Navier–Stokes equations*, Theor. Comput. Fluid Dyn. **1**, 65–95.

[101] Hess, J.L. 1975, *Review of integral-equation techniques for solving potential flow problems with emphasis on the surface-source method*, Comput. Methods Applied Mech. Eng. **5**, 145–196.

[102] Hess, J.L. 1990, *Panel methods in computational fluid dynamics*, Ann. Rev. Fluid Mech. **22**, 255–274.

[103] Hockney, R.W. and Eastwood, J.W. 1981, *Computer Simulation Using Particles*, McGraw-Hill, New York.

[104] Hoeijmakers, H.W.M. 1983, *Computational Vortex Flow Aerodynamics*, AGARD Conf. Proc. **342-18**, 1–35.

[105] Hornung, H. 1989, *Vorticity generation and transport*, in *Proceedings of the 10th Australasian Fluid Mechanics Conference*, University of Melbourne, Australia.

[106] Hou, T.Y. 1990, *Convergence of a variable blob vortex method for the Euler and Navier–Stokes equations*, SIAM J. Num. Anal. **27**, 1387–1404.

[107] Hou, T.Y. 1991, *Vortex methods with general variable blob shape function*, private communication.

[108] Hou, T.Y. and Lowengrub, J. 1990, *Convergence of the point vortex method for the 3-D Euler equations*, Commun. Pure Appl. Math. **43**, 965–981.

[109] Hou, T.Y., Lowengrub, J., and Krasny, R. 1991, *Convergence of a point vortex method for vortex sheets*, SIAM J. Num. Anal. **28**, 308–320.

[110] Hou, T.Y. and Wetton, B. 1992, *Convergence of a finite difference scheme for the Navier–Stokes equations using vorticity boundary conditions*, SIAM J. Num. Anal. **29**, 615–639.

[111] Huberson, S., Jollès, A., and Shen, W. 1991, *Numerical simulation of incompressible viscous flows by means of particle methods*, in Ref. 11, 369–384.

[112] Huberson, S. 1986, *Modélisation asymptotique et numérique de noyaux tourbillonaires enroulés*, Thèse d'état, Université Pierre et Marie Curie.

[113] Katz, J. and Plotkin, A. 1991, *Low-Speed Aerodynamics*, McGraw-Hill, New York.

[114] Katzenelson, J. 1989, *Computational structure of the N-body problem*, SIAM J. Sci. Stat. Comput. **11**, 787–815.

[115] Kinney, R.B. and Paolino, M.A. 1974, *Flow transient near the leading edge of a flat plate moving through a viscous fluid*, ASME J. Appl. Mech. **41**, 919–924.

[116] Kinney, R.B. and Cielak, Z.M. 1977, *Analysis of unsteady viscous flow past an airfoil: Part I – theoretical development*, AIAA J. **15**, 1712–1717.

[117] Knio, O. and Ghoniem, A. 1990, *Numerical study of three-dimensional vortex methods*, J. Comput. Phys. **86**, 75–106.

[118] Knio, O. and Ghoniem, A. 1991, *Three dimensional vortex simulation of rollup and entrainment in a shear layer*, J. Comput. Phys. **97**, 172–223.

[119] Knio, O. and Ghoniem, A. 1992, *The 3-dimensional structure of periodic vorticity layers under nonsymmetrical conditions*, J. Fluid Mech. **242**, 353–392.

[120] Knio, O. and Ghoniem, A. 1992, *Vortex simulation of a 3-dimensional reacting shear-layer with infinite-rate kinetics*, AIAA J. **30**, 105–116.

[121] Koumoutsakos, P. 1992, *Simulation of unsteady separated flows using vortex methods*, Ph.D. dissertation, Caltech.

[122] Koumoutsakos, P. and Leonard, A. 1988, *Improved boundary integral method for inviscid boundary condition applications*, AIAA J. **31**, 401–404.

[123] Koumoutsakos, P. and Leonard, A. 1993, *Direct numerical simulations using vortex methods*, in Ref. 22, 179–190.

[124] Koumoutsakos, P., Leonard, A., and Pepin, F. 1994, *Viscous boundary conditions for vortex methods*, J. Comput. Phys. **113**, 52–56.

[125] Koumoutsakos, P. and Leonard, A. 1995, *High resolution simulations of the flow around an impulsively started cylinder using vortex methods*, J. Fluid Mech. **296**, 1–38.

[126] Koumoutsakos, P. and Shiels, D. 1996, *Simulations of the viscous flow normal to an impulsively started and uniformly accelerated flat plate*, J. Fluid Mech. **328**, 177–226.

[127] Koumoutsakos, P. 1997, *Inviscid axisymmetrization of an elliptical vortex*, J. Comput. Phys. **138**, 821–857.

[128] Koumoutsakos, P. 1997, *Active control of vortex–wall interactions*, Phys. Fluids **9**, 3808–3816.

[129] Krasny, R. 1986, *Desingularization of periodic vortex sheet roll-up*, J. Comput. Phys. **65**, 292–313.

[130] Krasny, R. 1986, *A study of singularity formation in a vortex sheet by the point vortex approximation*, J. Fluid Mech. **167**, 65–93.

[131] Krasny, R. 1991, *Vortex sheet computations: roll-up, wake separation*, in Ref. 11, 385–402.

[132] Lemine, M. 1997, *Méthodes de couplage particules-grille pour les écoulements incompressibles*, Colloque national d'analyse numérique, Saint-Etienne.

[133] Lemine, M. 1998, Ph.D. thesis, Université Joseph-Fourier, Grenoble, France.

[134] Leonard, A. 1975, *Numerical simulation of interacting vortex filaments*, in *Proceedings of the IV Intl. Conference on Numerical Methods of Fluid Dynamics*, Springer-Verlag, New York.

[135] Leonard, A. 1980, *Vortex methods for flow simulation*, J. Comput. Phys. **37**, 289–335.

[136] Leonard, A. 1985, *Computing three-dimensional incompressible flows with vortex elements*, Ann. Rev. Fluid Mech. **17**, 523–559.

[137] Lifschitz, A., Suters, W.H., and Beale, J.T. 1996, *The onset of instability in exact vortex rings with swirl*, J. Comput. Phys. **129**, 8–29.

[138] Lighthill, M.J. 1963, *Introduction. Boundary Layer Theory*, J. Rosenhead, ed. Oxford U. Press, New York, 54–61.

[139] Long, D.G. 1988, *Convergence of the random vortex method in two dimensions*, J. Am. Math. Soc. **1**, 779–804.

[140] Lugt, H.J. 1983, *Vortex Flow in Nature and Technology*, Wiley, New York.

[141] Lundgren, T.S. and Mansour, N.N. 1988, *Oscillations of drops in zero gravity with weak viscous effects*, J. Fluid Mech. **194**, 479–510.

[142] Majda, A. 1986, *Vorticity and the mathematical theory of incompressible fluid flow*, Commun. Pure Appl. Math. **39**, S179–S187.

[143] Marchioro, C. and Pulvirenti, M. 1994, *Mathematical Theory of Incompressible Nonviscous Fluids*, Springer-Verlag, Berlin.

[144] Martensen, E. 1959, *Berechnung der Druckverteilung an Gitterprofilen in ebener Potentialströmung mit einer Fredhölmschen Integralgleichung*, Arch. Ration. Mech. Anal. **3**, 235–270.

[145] Martin, J.E. and Meiburg, E. 1991, *Numerical investigation of three-dimensional evolving jets subject to axisymmetric and azimuthal perturbation*, J. Fluid Mech. **230**, 271–318.

[146] Martin, J.E. and Meiburg, E. 1992, *Numerical investigation of three-dimensionally evolving jets under helical perturbations*, J. Fluid Mech. **243**, 457–487.

[147] Mas-Gallic, S. 1987, *Contribution à l'analyse numérique des méthodes particulaires*, Thèse d'Etat, Université Paris VI.

[148] Mas-Gallic, S. 1991, *Deterministic particle method: diffusion and boundary conditions*, in Ref. 11, 433–465.

[149] Meiburg, E. and Lasheras, J.C. 1988, *Experimental and numerical investigation of the three-dimensional transition in plane wakes*, J. Fluid Mech. **190**, 1–37.

[150] Meiburg, E. 1989, *Incorporation and test of diffusion and strain effects in the two-dimensional vortex blob technique*, J. Comput. Phys. **82**, 85–93.

[151] Merriman, B. 1991, *Particle approximation*, in Ref. 11, 481–546.

[152] Mikhlin, S.G. 1964, *Integral equations and their applications to certain problems of mechanics*, in *Mathematical Physics and Engineering*, 2nd ed., Macmillan, New York.

[153] Milizzano, F. and Saffman, P.G. 1977, *The calculation of large Reynolds Number two-dimensional flow using discrete vortices with random walk*, J. Comput. Phys. **23**, 380–392.

[154] Monaghan, J.J. 1982, *Why particle methods work*, SIAM J. Sci. Stat. Comput. **3**, 422.

[155] Monaghan, J.J. 1985, *Particle methods for hydrodynamics*, Comput. Phys. Rep. **3**, 71–124.

[156] Monaghan, J.J. 1985, *Extrapolating B-splines for interpolation*, J. Comput. Phys. **60**, 253–262.

[157] Najm, H. 1993, *A hybrid vortex method with deterministic diffusion*, in Ref. 22, 207–222.

[158] Nitsche, M. and Krasny, R. 1994, *A numerical study of vortex ring formation at the edge of a circular tube*, J. Fluid. Mech. **276**, 139–161.

[159] Orlandi, P. 1990, *Vortex dipole rebound from a wall*, Phys. Fluids A **1429**, 75–110.

[160] Panton, R.L. 1994, *Incompressible Flow*, Wiley, New York.

[161] Pépin, F. and Leonard, A. 1990, *Concurrent implementation of a fast vortex method*, in *Proceedings of the Fifth Distributed Memory Computing Conference I*, pp. 453–462.

[162] Perlman, M. 1985, *On the accuracy of vortex methods*, J. Comput. Phys. **59**, 200–223.

[163] Prager, W. 1928, *Die Druckverteilung an Körpern in ebener Potentialströmung*, Phys. Z. **29**, 865–869.

[164] Prandtl, W. 1904, *Über Flüssigkeitbewegung bei sehr kleiner Reibung*, in Proceedings of the Fourth Mathematics Congress, Heidelberg, pp. 484–491.

[165] Prandtl, W. 1925, *The Magnus effect and windpowered ships*, Wissenschaften **13**, 93–108.

[166] Puckett, E.G. 1989, *A study of the vortex sheet method and its rate of convergence*, SIAM J. Sci. Stat. Comput. **10**, 298–327.

[167] Pullin, D.I. 1992, *Contour dynamics methods*, Ann. Rev. Fluid Mech. **24**, 89–115.

[168] Quartapelle, L. 1981, *Vorticity conditioning in the computation of two-dimensional viscous flows*, J. Comput. Phys. **40**, 453–477.

[169] Raviart, P.A. 1983, *An analysis of particle methods*, in *Numerical Methods in Fluid Dynamics*, Vol. 1127 of Lecture Notes in Mathematics Series, Springer-Verlag, Berlin, 243–324.

[170] Raviart, P.A. 1987, *Méthodes particulaires, Ecole d'été CEA-INRIA-EDF*, unpublished lecture notes.

[171] Rehbach, C. 1978, *Numerical calculation of three-dimensional unsteady flows with vortex sheets*, AIAA Paper 111-1978.

[172] Rosenhead, L. 1931, *The Formation of vortices from a surface of discontinuity*, Proc. R. Soc. London Ser. A **134**, 170.

[173] Rosenhead, L. 1932, *The point vortex approximation of a vortex sheet*, Proc. R. Soc. London Ser. A **134**, 170–192.

[174] Russo, G. 1991, *A Lagrangian vortex method for the incompressible Navier–Stokes equations*, in Ref. 11, pp. 585–596.

[175] Russo, G. 1993, *A deterministic vortex method for the Navier–Stokes equations*, J. Comput. Phys. **108**, 84–94.

[176] Russo, G. and Strain, J.A. 1994, *Fast triangulated vortex methods for the 2D Euler equations*, J. Comput. Phys. **111**, 291–323.

[177] Saffman, P.G. 1981, *Dynamics of vorticity*, J. Fluid Mech. **106**, 49.

[178] Saffman, P.G. and Meiron, D.I. 1986, *Difficulties with three-dimensional weak solutions for inviscid incompressible flow*, Phys. Fluids **29**, 2373–2375.

[179] Saghbini, J.C. and Ghoniem, A. 1997, *Numerical simulation of the dynamics and mixing in a swirling flow*, AIAA Paper 97-0507.

[180] Salmon, J.K. 1991, *Parallel hierarchical N-body methods*, Ph.D. dissertation, Caltech.

[181] Salmon, J.K. and Warren, M. 1994, *Skeletons from the treecode closet*, J. Comput. Phys. **111**, 136–155.

[182] Salmon, J.K., Warren, M.S., and Winckelmans, G. 1994, *Fast parallel tree codes for gravitational and fluid N-body problems*, Int. J. Super. **8**, 129–142.

[183] Sarpkaya, T. 1989, *Computational methods with vortices*, J. Fluids Eng. **11**, 15–52.

[184] Schoenberg, I.J. 1973, *Cardinal Spline Interpolation*, Society for Industrial and Applied Mathematics, Philadelphia, PA.

[185] Sethian, J. and Ghoniem, A. 1984, *Validation study of vortex methods*, J. Comput. Phys. **54**, 425–456.

[186] Shankar, S. and van Dommelen, L. 1996, *A new diffusion procedure for vortex methods*, J. Comput. Phys. **127**, 88–109.

[187] Shiels, D. 1998, *Simulation of controlled bluff body flow with a viscous vortex method*, Ph.D. dissertation, Caltech.

[188] Smith, P.A. and Stansby, P.K. 1988, *Impulsively started flow around a circular cylinder by the vortex method*, J. Fluid Mech. **194**, 45–77.

[189] Spalart, P.R. and Leonard, A. 1981, *Computation of separated flows by a vortex-tracing algorithm*, AIAA Paper 81-1246.

[190] Strain, J.A. 1996, *2D vortex methods and singular quadrature rules*, J. Comput. Phys. **124**, 131–145.

[191] Strain, J.A. 1997, *Fast adaptive 2D vortex methods*, J. Comput. Phys. **132**, 108–122.

[192] Ta Phuoc, L. and Bouard, R. 1985, *Numerical solution of the early stage of the unsteady viscous flow around a circular cylinder: a comparison with experimental visualization and measurements*, J. Fluid Mech. **160**, 93–117.

[193] Temam, R. 1977, *Navier–Stokes Equations*, North-Holland, Amsterdam.

[194] Teng, Z.J. 1982, *Elliptic vortex method for incompressible flow at high Reynolds number*, J. Comput. Phys. **46**, 54–68.

[195] Truesdell, C. 1954, *The Kinematics of Vorticity*, Indiana U. Press, Bloomington, IN.

[196] van Dommelen, L. and Rundesteiner, E.A. 1989, *Fast, adaptive summation of point forces in the two-dimensional Poisson equation*, J. Comput. Phys. **83**, 126–148.

[197] Winckelmans, G. and Leonard, A. 1988, *Weak solutions of the three-dimensional vorticity equation with vortex singularities*, Phys. Fluids **31**, 1838–1839.

[198] Winckelmans, G. 1989, *Topics in vortex methods for the computation of three- and two-dimensional incompressible unsteady flows*, Ph.D. dissertation, Caltech.

[199] Winckelmans, G. and Leonard, A. 1989, *Improved vortex methods for three-dimensional flows*, in Ref. 39, 25–35.

[200] Winckelmans, G. and Leonard, A. 1993, *Contributions to vortex particle methods for the computation of three dimensional incompressible unsteady flows*, J. Comput. Phys. **109**, 247–273.

[201] Winckelmans, G.S., Salmon, J.K., Warren, M.S., Leonard, A., and Jodoin, B. 1995, *Application of fast parallel and sequential tree codes to computing three-dimensional flows with the vortex element and boundary elements method*, in Ref. 87, 225–240.

[202] Wu, J.C. 1976, *Numerical boundary conditions for viscous flow problems*, AIAA J. **14**, 1042–1049.

[203] Ying, L.A. 1991, *Convergence of vortex methods for initial boundary value problems*, Adv. Math. **20**, 86–102.

[204] Zabusky, N.J., Hughes, M.H., and Roberts, K.V. 1979, *Contour dynamics for the Euler equations in two-dimensions*, J. Comput. Phys. **30**, 96–106.

[205] Zhu, J. 1989, *An adaptive vortex method for two-dimensional viscous and incompressible flows*, Ph.D. dissertation, New York University.

Index